Approximation, Complex Analysis, and Potential Theory

NATO Science Series

A Series presenting the results of scientific meetings supported under the NATO Science Programme.

The Series is published by IOS Press, Amsterdam, and Kluwer Academic Publishers in conjunction with the NATO Scientific Affairs Division

Sub-Series

I. **Life and Behavioural Sciences**	IOS Press
II. **Mathematics, Physics and Chemistry**	Kluwer Academic Publishers
III. **Computer and Systems Science**	IOS Press
IV. **Earth and Environmental Sciences**	Kluwer Academic Publishers

The NATO Science Series continues the series of books published formerly as the NATO ASI Series.

The NATO Science Programme offers support for collaboration in civil science between scientists of countries of the Euro-Atlantic Partnership Council. The types of scientific meeting generally supported are "Advanced Study Institutes" and "Advanced Research Workshops", and the NATO Science Series collects together the results of these meetings. The meetings are co-organized bij scientists from NATO countries and scientists from NATO's Partner countries – countries of the CIS and Central and Eastern Europe.

Advanced Study Institutes are high-level tutorial courses offering in-depth study of latest advances in a field.
Advanced Research Workshops are expert meetings aimed at critical assessment of a field, and identification of directions for future action.

As a consequence of the restructuring of the NATO Science Programme in 1999, the NATO Science Series was re-organized to the four sub-series noted above. Please consult the following web sites for information on previous volumes published in the Series.

http://www.nato.int/science
http://www.wkap.nl
http://www.iospress.nl
http://www.wtv-books.de/nato-pco.htm

Series II: Mathematics, Physics and Chemistry – Vol. 37

Approximation, Complex Analysis, and Potential Theory

edited by

N. Arakelian
Institute of Mathematics,
National Academy of Sciences of Armenia,
Yerevan, Armenia

and

P.M. Gauthier
Département de Mathématiques et de Statistique,
Université de Montréal,
Montréal, Québec, Canada

Technical Editor:

G. Sabidussi
Département de Mathématiques et de Statistique,
Université de Montréal,
Montréal, Québec, Canada

Kluwer Academic Publishers

Dordrecht / Boston / London

Published in cooperation with NATO Scientific Affairs Division

Proceedings of the NATO Advanced Study Institute on
Modern Methods in Scientific Computing and Applications
Montréal, Québec, Canada
3 to14 July 2000

A C.I.P. Catalogue record for this book is available from the Library of Congress.

ISBN 1-4020-0028-6

Published by Kluwer Academic Publishers,
P.O. Box 17, 3300 AA Dordrecht, The Netherlands.

Sold and distributed in North, Central and South America
by Kluwer Academic Publishers,
101 Philip Drive, Norwell, MA 02061, U.S.A.

In all other countries, sold and distributed
by Kluwer Academic Publishers,
P.O. Box 322, 3300 AH Dordrecht, The Netherlands.

Printed on acid-free paper

Table of Contents

Preface

In the summer of 1967, I moved to Montreal for my first (and only) job with my first (and only) wife and our first child (of six) just on time for the beginning of my first international conference, the yearly Séminaire de mathématiques supérieures, which that year was devoted to complex analysis. It turned out to be one of the best conferences I would ever attend. Since then, perhaps in the hope of recapturing the magic of 1967, I have organized two such SMS at the Université de Montréal, the first in 1993 and the second in 2000. For the latter, I had the great honour of having Academician Norair Arakelian as co-organizer. These are the proceedings.

At the 1967 SMS, Wolfgang Fuchs gave a series of lectures in which he promulgated the striking new results of Arakelian on complex approximation along with its applications to value distribution theory, previously 'available' only in Russian. The lectures of Fuchs changed the direction of my own research to approximation, which would remain my chief interest till the present day.

The courses given at SMS/NATO ASI-2000 were intended for advanced graduate students and mathematicians at an early stage of their carreer. However, some more 'mature' mathematicians also attended and found the session quite rewarding. In particular, we were blessed with the active participation of Waficka Al-Katifi and Walter Hayman, both of whom had participated in SMS-1967. The active participation of such senior mathematicians added immensely to the richness of the discussions and their avowed appreciation of the session strongly suggests that these proceedings should be of interest not only to budding analysts but to experts as well. Indeed, the lectures, while providing necessary background and references, also led to the most recent research in the topics covered.

Naturally, approximation was a major thrust of these courses. My own lectures served as an introduction to holomorphic approximation on Riemann surfaces. In Arakelian's lectures we saw how the theory of holomorphic approximation can be applied to solving problems in the value distribution theory of Rolf Nevanlinna. The lectures by David Armitage and Stephen Gardiner expounded the theory of harmonic approximation and delighted us with remarkable applications.

The study of subharmonic functions yields important information concerning both holomorphic functions and harmonic functions. Of course, harmonic functions are subharmonic and a most interesting example of a subharmonic function is $\log|f|$ where f is holomorphic. In fact, the lectures of David Drasin (who also participated in the SMS-1967) show that to a very great extent we can approximate subharmonic functions by functions of type $\log|f|$ where f is holomorphic. While subharmonic functions tell us interesting things about holomorphic ones, information may flow in the opposite direction as well. Thus, complex analysis and potential theory thrive in symbiosis.

The lectures of Thomas Ransford described an approach to potential theory via duality, the key idea being the notion of Jensen measure, an abstraction of the sub-mean-value property of subharmonic functions.

Besides subharmonic approximation, another way to generalize both holomorphic and harmonic approximation, is to consider various approximation problems associated with elliptic differential operators. This was the point of view adopted in the lectures by Thomas Bagby. In this context, the natural function spaces to consider are Sobolev spaces. A given function f may of course belong to various function spaces and the need may arise to approximate f simultaneously in more than one of these spaces. Arne Stray's lectures addressed this issue.

Approximation theory can be approached from two viewpoints: smoothing-type approximation and extension-type approximation. The first type of approximation seeks to approximate a given function by a nicer function defined on (more or less) the same domain whereas extension-type approximation seeks to approximate a given function by a function (of the same sort but) defined on a larger domain. All of the approximations in these lectures are of the extension-type with the exception of those discussed by Drasin, which are of smoothing type.

Since Louis de Branges confirmed the Bieberbach conjecture in 1985, the outstanding unanswered question in complex analysis has perhaps been that of determining the precise value of the Bloch constant. In 1990, after more than a half-century without progress, Mario Bonk made a major breakthrough by developing a new distortion theorem which allowed him to improve the known estimate on the Bloch constant. At SMS-2000 Bonk lectured on other (related) matters regarding hyperbolic spaces. Unfortunately, his lectures are not included in these Proceedings. However, the lectures of Chen Huai-hui gave an up-to-the-minute survey on the Bloch constant, including the best estimate at this time, due to Chen and myself.

Another series of lectures which are unfortunately not included in these Proceedings are those of Alano Ancona on the boundary Harnack principle. Subsequently and independently, Paul Koosis also gave a series of lectures in Montréal on the boundary Harnack principle, and produced a set of notes thereof. Those interested in this topic may wish to contact Paul Koosis for a copy of these notes.

I wish to express my sincere thanks to all the lecturers and participants for having helped to make this session such a success and most particularly to my co-organizer, Norair Arakelian. Special thanks are due to Ghislaine David, coordinator of the SMS, who contributed immeasurably to the preparation, mise-en-scène, and "aftermath". Also, my thanks go to Gert Sabidussi and Arik Loinaz for their excellent work in editing the present volume. The Organizing Committee is most grateful to NATO for financially supporting this session as an Advanced Study Institute. It is a pleasure to proclaim my love and appreciation to my wife, Sandy, for the hospitality she has extended to visitors throughout my carreer and in particular during this ASI. Finally, I extend my congratulations and appreciation to my friend, Aubert Daigneault, director of the SMS for many years, this having been his last. Bravo et merci, Aubert!

This preface began with my quest to recapture the magic of the SMS-1967. For me personally, the magic of SMS-1967 was in large part the work of Arakelian via the lectures of Fuchs. At SMS-1967, Arakelian was present in spirit and Fuchs in the flesh. At SMS-2000 it went the other way. At the excellent suggestion of my co-organizer, Academician Norair Arakelian, we dedicate the present Proceedings to the memory of Wolfgang Fuchs.

Paul M. Gauthier

Key to group picture

1 Egorov
2 Danielyan
3 Skhiri
4 Hanley
5 Caçao
6 Gögüs
7
8 Vodop'yanov
9 Boe
10 Hayman, Walter
11 Kotus
12 Hayman, Waficka
13 Gustafsson
14 Beznea
15 Yavrian
16 Poggi-Corradini
17 Yang
18 Chen, Yin
19 Jiang

20 Kosek
21 Bialas-Ciez
22 Levenberg
23 Rashkovskii
24 Kiepiela
25 Marcula
26 Castañeda
27 Kaptanoglu
28 Brzezina
29 Lavicka
30 Bensouda
31 Stoll
32 Schmieder
33 Szynal
34 Gal
35 Ryan
36 Pritsker
37 Becker
38 Luh

39 Schillings
40 Daigneault
41 Sabidussi
42 David
43 Michalska
44 Hirnyk
45 Radyna
46 Boivin
47 Lessard
48 Ransford
49 Stray
50 Armitage
51 Ancona
52 Gardiner
53 Bagby
54 Drasin
55 Arakelian
56 Gauthier
57 Bonk

Participants

Victor ANANDAM
Department of Mathematics
King Saud University
P.O. Box 2455
11451 Riyadh
Saudi Arabia
vanandam@ksu.edu.sa

Alano ANCONA
Mathématiques — Bâtiment 425
Université de Paris-Sud
91405 Orsay Cédex
France
ancona@math.u-psud.fr

Cristoph BECKER
FB IV Mathematik
Universität Trier
54286 Trier
Germany
beck4501@uni-trier.de

Charaf BENSOUDA
Faculté des sciences
Université Ibn Tofail
B.P. 133
Kenitra
Morocco

Lucian BEZNEA
Inst. Mathematics "Simion Stoilow"
Romanian Academy
P.O. Box 1-764
70700 Bucharest
Romania
beznea@stoilow.imar.ro

Leokadia BIALAS-CIEZ
Institute of Mathematics
Jagellonian University
ul. Reymonta 4
30-059 Krakow
Poland
bialas@im.uj.edu.pl

Bjarte BOE
Department of Mathematics
University of Bergen
Johannes Brunsgate 12
5008 Bergen
Norway
bjarte.boe@mi.uib.no

Mario BONK
Department of Mathematics
University of Michigan
525 East University Ave.
Ann Arbor, MI 48109-1109
USA
m.bonk@math.lsa.umich.edu

Miroslav BRZEZINA
Department of Applied Mathematics
Technical University of Liberec
Halkova 6
461 17 Liberec
Czech Republic
miroslav.brzezina@vslib.cz

Maria Isabel CAÇAO
Departamanto de Matematica
Universidade de Aveiro
Campus Santiago
3810-193 Aveiro
Portugal
isabelc@mat.ua.pt

Yin CHEN
Dép. de mathématiques et de statistique
Université Laval
Québec (QC), G1K 7P4
Canada
chen@mat.ulaval.ca

Arthur DANIELYAN
Department of Mathematics
University of South Florida
Tampa, FL 33620-5700
USA
adaniely@math.usf.edu

Juan Jesus DONAIRE
Departament de Matemàtiques
Universitat Autonoma de Barcelona
08193 Bellaterra-Barcelona
Spain
donaire@mat.uab.es

Alexandre EGOROV
Sobolev Institute of Mathematics
Pr. Akad. Koptyug 4
630090 Novosibirsk
Russia
yegorov@math.nsc.ru

Juan Carlos FARINA
Dpto. Analisis Matematico
Universitad de La Laguna
C/Astrofisico Fco. Sanchez
38271 La Laguna, Tenerife
Spain
jcfarina@ull.es

Richard FOURNIER
Dép. de mathématiques et de statistique
Université de Montréal
CP 6128 — Succ. Centre-ville
Montréal (QC), H3C 3J7
Canada
fournier@dms.umontreal.ca

Sorin GAL
Department of Mathematics
University of Oradea
Str. Armatei Romane 5
3700 Oradea
Romania
galso@math.uoradea.ro

Nihat Gökhan GÖGÜS
Department of Mathematics
Middle East Technical University
06531 Ankara
Turkey
ggogus@arf.math.metu.edu.tr

Anders GUSTAFSSON
Department of Mathematics
Umea University
90187 Umea
Sweden
anders.gustafsson@math.umu.se

Margaret M. HANLEY
Department of Mathematics
University College Dublin
4 Dublin
Ireland
mary.hanley@ucd.ie

Waficka HAYMAN
c/o Prof. Walter Hayman
Department of Mathematics
Imperial College
SW7 2B2 London
United Kingdom

Walter K. HAYMAN
Department of Mathematics
Imperial College
SW7 2B2 London
United Kingdom

Markiyan HIRNYK
Dept. of Mathematics and Statistics
Lviv Academy of Commerce
10, Tuhan-Baranovskyi St.
79008 Lviv
Ukraine
hirnyk@lac.lviv.ua

Baoguo JIANG
Dept. Mathematics — Middlesex College
University of Western Ontario
London (ON), N6A 5B7
Canada
bjiang@julian.uwo.ca

H. Turgay KAPTANOGLU
Department of Mathematics
Middle East Technical University
06531 Ankara
Turkey
kaptan@arf.math.metu.edu.tr

Katarzyna KIEPIELA
Department of Mathematics
Technical University Radom
J. Malczewskiego St. 29
26-600 Radom
Poland

Marta KOSEK
Instytut Matematyki
Uniwersytet Jagiellonski
ul. Reymonta 4
30-059 Krakow
Poland
marta.kosek@im.uj.edu.pl

Janina KOTUS
Department of Mathematics
Technical University of Warsaw
Plac Politechniki 1
00-661 Warsaw
Poland
janinak@snowman.impan.gov.pl

Roman LAVICKA
Mathematical Institute
Charles University
Sokolovska 83
186 75 Praha 8
Czech Republic
lavicka@karlin.mff.cuni.cz

Norman LEVENBERG
Department of Mathematics
Rawles Hall
Indiana University
Bloomington, IN 47405
USA
levenberg@scitec.auckland.ac.nz

Jose LLORENTE
Departament de Matemàtiques
Universitat Autonoma de Barcelona
08193 Bellaterra-Barcelona
Spain
gonzalez@math.uab.es

Wolfgang LUH
FB IV Mathematik
Universität Trier
54286 Trier
Germany
luh@uni-trier.de

Iwona MARCULA
Department of Applied Mathematics
M. Curie-Sklodowska University
Sq. Marie Curie-Sklodowska 5
20-031 Lublin
Poland
imarcula@ramzes.umcs.lublin.pl

Malgorzata MICHALSKA
Institute of Mathematics
M. Curie-Sklodowska University
Sq. Marie Curie-Sklodowska 1
20-031 Lublin
Poland
daisy@golem.umcs.lublin.pl

Konstantin OSIPENKO
Department of Mathematics
MATI — Russian State Tech. University
Orshanskaya, 3
121552 Moskow
Russia
konst@osipenko.mccme.ru

Pietro POGGI-CORRADINI
Department of Mathematics
Kansas State University
Manhattan, KS 66506
USA
pietro@math.ksu.edu

Mohamad R. POURYAYEVALI
Department of Mathematics
University of Isfahan
P.O. Box 81745-163
Isfahan
Iran

Igor PRITSKER
Department of Mathematics
Oklahoma State University
Stillwater, OK 74078-1058
USA
igor@math.okstate.edu

Aliaksandr RADYNA
Faculty of Mechanics & Mathematics
Belarusian State University
F. Skaryna Ave. 4
220050 Minsk
Belarus
radynoa@mmf.bsu.unibel.by

Alexander RASHKOVSKII
Institute for Low Temperature Physics
47 Lenin Pr.
61164 Kharkov
Ukraine
yarchyk@imath.kiev.ua

Dominic ROCHON
Dép. de mathématiques et de statistique
Université de Montréal
CP 6128 — Succ. Centre-ville
Montréal (QC), H3C 3J7
Canada
rochon@dms.umontreal.ca

Pamela RYAN
Department of Mathematics
Univ. of South Carolina, Columbia
Columbia, SC 29208-0001
USA
pryan000@math.sc.edu

Bettina SCHILLINGS
FB IV Mathematik
Universität Trier
54286 Trier
Germany
schi4501@uni-trier.de

Gerald SCHMIEDER
FB 6 Mathematik
Universität Oldenburg
PF 2503
26111 Oldenburg
Germany
schmieder@mathematik.uni-oldenburg.de

Haïkel SKHIRI
Dép. de mathématiques et de statistique
Université Laval
Québec (QC), G1K 7P4
Canada
skhiri@mat.ulaval.ca

Manfred STOLL
Department of Mathematics
University of South Carolina
Columbia, SC 29208
USA
stoll@math.sc.edu

Jan SZYNAL
Department of Applied Mathematics
M. Curie-Sklodowska University
Sq. Marie Curie-Sklodowska 5
20-031 Lublin
Poland
ysszynal@golem.unics.lublin.pl

Serguei VODOP'YANOV
Sobolev Institute of Mathematics
Akademika Koptyuga Ave. 4
630090 Novosibirsk
Russia
vodopis@math.nsc.ru

Chung-Chun YANG
Department of Mathematics
Hong Kong University of Sci. & Tech.
Clear Water Bay
Kowloon
Hong Kong, China
mayang@uxmail.ust.hk

Anush YAVRIAN
Department of Mathematics
Yerevan State University
Al. Manoogian Str. 1
375049 Yerevan
Armenia

Eduardo S. ZERON
Depto. Matematicas
CINVESTAV
Apdo. Postal 14-740
07000 México D.F.
México
eszeron@math.cinvestav.mx

Contributors

Norair ARAKELIAN
Institute of Mathematics
Armenian National Academy of Sciences
Marshall Bagramian Ave. 24b
375 019 Yerevan
Armenia
arakel@instmath.sci.am

David ARMITAGE
Department of Pure Mathematics
Queen's University of Belfalst
Belfast, BT7 1NN
Northern Ireland
d.armitage@queens-belfast.ac.uk

Thomas BAGBY
Department of Mathematics
Rawles Hall
Indiana University
Bloomington, IN 47405
USA
bagby@indiana.edu

André BOIVIN
Department of Mathematics, MC
University of Western Ontario
London (ON), N6A 5B7
Canada
boivin@uwo.ca

Nelson CASTAÑEDA
Department of Mathematical Sciences
Central Connecticut State University
P.O. Box 4010
New Britain, CT 06050-4010
USA
ncastaneda@earthlink.net

Huaihui CHEN
Department of Mathematics
Nanjing Normal University
Nanjing, Jiangsu 210024
China
hhchen@publicl.ptt.js.cn

David DRASIN
Department of Mathematics
Purdue University
West Lafayette, IN 47907-1395
USA
drasin@math.purdue.edu

Stephen J. GARDINER
Department of Mathematics
University College Dublin
4 Dublin
Ireland
stephen.gardiner@ucd.ie

Paul M. GAUTHIER
Dép. de mathématiques et de statistique
Université de Montréal
CP 6128 — Succ. Centre-ville
Montréal (QC), H3C 3J7
Canada
gauthier@dms.umontreal.ca

Thomas J. RANSFORD
Dép. de mathématiques et de statistique
Université Laval
Québec (QC), G1K 7P4
Canada
ransford@math.ulaval.ca

Arne STRAY
Department of Mathematics
University of Bergen
Johannes Brunsgate 12
5008 Bergen
Norway
stray@mi.uib.no

Approximation and value distribution

Norair ARAKELIAN

Institute of Mathematics
National Academy of Sciences of Armenia
375019 Yerevan
Armenia

Notes taken by
Anoush YAVRIAN

To the memory of Prof. Wolfgang Fuchs

Abstract

The aim of these lectures is to demonstrate the efficiency of methods and results of the theory of uniform and tangential approximation by entire functions in exploring the problems of the value distribution theory of R. Nevanlinna. It concerns the problems of constructing entire functions of finite order, having: (1) a given sequence of "small" entire functions as its deficient functions, with optimal estimates from below for the defects; (2) a given sequence of complex numbers as its (multiplicity) index values.

To examine the second problem, we present a new, purely analytic approach. Finally, we suggest an analytic method of construction of entire functions of finite order with joint deficient functions and index values.

1 Introduction

The methods and results of *complex approximation* theory in the plane represent a powerful instrument for investigating different problems in *analytic function* theory. Since this subject is too extensive, we will restrict ourselves in these lectures (as in [A6]) to considering only a few applications of the more specific theory of *uniform and tangential entire approximations*. The foundations of this last theory have been laid by Carleman [Ca], Roth [Ro], Keldysh [K] (see also [M1]) and other authors; on the further development of the theory see [A1, A2, A4, AG]. We will be restricting the area of the intended applications to the framework of the classical and extensively treated theory of *entire functions* of finite order. For these functions, we will discuss some problems of the *value distribution* theory due to R. Nevanlinna and developed by his successors (for the fundamentals of the theory see [N1, N2, Ha, GO]; references on further developments can be found in [G, GLO]).

For the class of meromorphic functions in the complex plane, the notions of *deficient* and *(multiplicity) index values* are fundamental. The solution of different types of inverse problems of Nevanlinna's theory by W. Hayman and W. Fuchs (see [Ha] and Drasin [D])

1

N. Arakelian and P.M. Gauthier (eds.), Approximation, Complex Analysis, and Potential Theory, 1–28.
© 2001 *Kluwer Academic Publishers. Printed in the Netherlands.*

has demonstrated the extreme sharpness its main statements (the two main theorems, the deficiency relation, etc.) in general. In this connection it is natural to investigate these notions for the case of entire (and meromorphic) functions of finite order, to see whether Nevanlinna's general theory remains still valid for them. Briefly—the answer is yes, and these lectures will provide some additional arguments to confirm this conclusion. Since the Nevanlinna theory itself is rather sophisticated, the discussion of the problem of sharpness of its different statements demands the usage of precise and powerful means.

The lectures may conveniently be divided into three parts.

1. The first part presents *preliminaries*—a brief extract from value distribution theory and tangential approximation theory. We present also (see Theorem 2.10 in subsection 2.3) a specific result on tangential approximation by entire functions on *"archipelagos"*—an appropriate infinite set of *"islands"*, accumulating at infinity—with optimal estimates of the growth of the approximating functions (see Lemma 3 in [AH], a refined version of Lemma 4 in [A3, A5]).

2. In the second part we test the generalized version of R. Nevanlinna's *deficiency relation*, as developed by Chuang Chi-Tai [Ch], Osgood [O], and Steinmetz [S] for so-called *"small"* functions instead of constants. Using the above-mentioned approximation theorem, we present new results on the existence of entire functions of finite order, whose set of *deficient functions* contains a given sequence of "small" functions, with optimal estimates from below for their *defects*. These results generalize not only the results of Arakelian [A3, A5], but also many subsequent results (see [D-W, G, E, Ru]). The other application to be examined in this part concerns the problem of the existence of *"universal"* entire functions of given order (see [B]).

3. In the third part we present a new construction of entire functions of finite order, having a given sequence of complex numbers as its (*multiplicity*) *index* values. In contrast to existing *geometric* (or quasi-conformal) and rather complicated constructions of such functions (a tradition originating with R. Nevanlinna), our constructions are purely analytic. They make it possible to obtain some new results in this direction. Moreover, we are combining the deficient and index values into one construction, again using approximation methods for this purpose.

The author dedicates these lectures to the memory of the late Professor Wolfgang Fuchs, who has made an inestimable contribution to the promotion of the ideas and methods of approximation theory and value distribution theory. We would especially like to mention the outstanding role played in this respect by his Montreal lectures of 1967 [F].

2   Preliminaries

2.1   Value distribution: basic notions and facts

The foundations of the value distribution theory of R. Nevanlinna are based on the notions of *deficient* and *index* values. In what follows, we will use these notions and some related notations (see [N1, N2, Ha, GO]).

Let f be a meromorphic function in $\mathbb{C}$. For $r \geq 0$ and $a \in \overline{\mathbb{C}}$, let $n(r, a)$ be the number of a-points (poles for $a = \infty$) of f, or roots of the equation $f(z) = a$ in $|z| \leq r$, counted with multiplicities. Also, let $n_1(r, a)$ be the number of *multiple* a-points of f in the same disc (i.e.

each a-point of multiplicity s is counted $s - 1$ times). Put

$$N(r, a) = \int_0^r \frac{n(t, a)}{t}\, dt, \qquad N_1(r, a) = \int_0^r \frac{n_1(t, a)}{t}\, dt, \qquad (2.1)$$

assuming $f(0) \neq a$; in case $f(0) = a$ and $r > 1$ the integration in (2.1) can be taken over $[1, r]$ (with some constants $c_a \geq 0$ added to the right hand sides in (2.1)). The *Nevanlinna characteristic* for $r \geq 0$ is the function

$$T(r) = m(r, \infty) + N(r, \infty), \qquad (2.2)$$

where, setting $x^+ = \max\{x, 0\}$,

$$m(r, \infty) = (2\pi)^{-1} \int_{-\pi}^{\pi} \log^+ \left| f(re^{it}) \right| dt$$

is the Nevanlinna *proximate function*. For $a \neq \infty$ it is defined by the formula

$$m(r, a) = (2\pi)^{-1} \int_{-\pi}^{\pi} \log^+ \left| f(re^{it}) - a \right|^{-1} dt.$$

If it is necessary to indicate the function f for which the above notations are being considered, it is customary to add f in the argument; for instance, $T(r)$ is replaced by $T(r, f)$.

The first fundamental theorem of the Nevanlinna theory states that

$$T(r) = m(r, a) + N(r, a) + O(1) \quad \text{as} \quad r \to \infty. \qquad (2.3)$$

The characteristic $T(r)$ reduces to $m(r, \infty)$ if f is an entire function. Then by (2.3), for any $a \in \mathbb{C}$ there exists a constant $L_a = L_a(f)$, such that for $r \geq 0$

$$N(r, a) - L_a < T(r, f) \leq \log^+ M(r, f), \qquad (2.4)$$

where

$$M(r, f) = \max_{|z|=r} |f(z)|.$$

For $a \in \overline{\mathbb{C}}$ the quantities

$$\delta(a, f) = \liminf_{r \to \infty} \frac{m(r, a)}{T(r, f)}, \qquad (2.5)$$

$$\theta(a, f) = \liminf_{r \to \infty} \frac{N_1(r, a)}{T(r, f)} \qquad (2.6)$$

represent the *(Nevanlinna) defect* and the *multiplicity* (or *ramification*) *index* of f at a, respectively. By (2.3) it follows that

$$\delta(a, f) = 1 - \limsup_{r \to \infty} \frac{N_1(r, a)}{T(r, f)}.$$

Obviously, $\delta(a, f) \geq 0$, $\theta(a, f) \geq 0$ and $\delta(a, f) + \theta(a, f) \leq 1$.

A number $a \in \mathbb{C}$ is a *deficient value* of f, if $\delta(a, f) > 0$, and a is said to be an *index value* of f, if $\theta(a, f) > 0$. The set $\mathbf{D}_f$ of deficient values and the set $\mathbf{I}_f$ of index values of a meromorphic function f are both at most countable. This follows from Nevanlinna's *deficiency relation*—the main corollary of the Second Fundamental Theorem (see [N2])—stating that

$$\sum_{a \in \overline{\mathbb{C}}} [\delta(a, f) + \theta(a, f)] \leq 2, \tag{2.7}$$

which in the case of an *entire* function f may be rewritten as

$$\sum_{a \in \mathbb{C}} [\delta(a, f) + \theta(a, f)] \leq 1. \tag{2.8}$$

2.1 Remark Let $\mu > 0$ and $s \in \mathbb{N}$. Then the function $f_1(z) = f(\mu z^s)$ satisfies

$$\delta(a, f_1) = \delta(a, f), \qquad \theta(a, f_1) = \theta(a, f), \qquad a \in \overline{\mathbb{C}}. \tag{2.9}$$

If f has order ρ (and normal type), then f_1 has order $s\rho$ (and normal type).

This statement follows from the equalities

$$N(r, a, f_1) = sN(\mu r^s, a, f),$$
$$N_1(r, a, f_1) = sN_1(\mu r^s, a, f),$$
$$T(r, f_1) = sT(\mu r^s, f).$$

The solution of Nevanlinna's inverse problem (see [Ha, D]) has demonstrated the sharpness of the relations (2.7), (2.8) in general, i.e. allowing meromorphic and entire functions of *infinite* order. It is of definite theoretical interest to investigate these notions for the case of entire (and meromorphic) functions of *finite* order, when the relations (2.7), (2.8) are not the only necessary conditions to be satisfied.

Now let f be a *transcendental* entire function and denote by $\mathcal{E}_f$ the set of relatively "small" entire functions a, i.e. the functions satisfying

$$T(r, a) = o(T(r, f)) \quad \text{as} \quad r \to \infty. \tag{2.10}$$

2.2 Example Obviously, $a \in \mathcal{E}_f$, if a is a *polynomial*.

The *defect* of a *function* $a \in \mathcal{E}_f$ (with respect to f) can be defined as in (2.5) by putting there $m(r, a) := m(r, 0, f - a)$. A function $a \in \mathcal{E}_f$ is said to be a *deficient function* of f, if $\delta(a, f) > 0$. We will retain the notation $\mathbf{D}_f$ also for the set of *deficient functions* a of f. The generalized (narrow) *deficiency relation* (see [Ch]; for the case of meromorphic functions see [O, S]) states that

$$\sum_{a \in \mathcal{E}_f} \delta(a, f) \leq 1. \tag{2.11}$$

Now suppose that f has *order* $\rho = \rho_f \in (0, \infty)$ and denote by σ_f, resp. τ_f, the *type* and the *lower type* of f with respect the order ρ. Using the relations of the characteristic T of an entire function with its maximum modulus function M (see [N2, Ch. 8, section 1, item 175]), one can see that the condition

$$\log^+ M(r, a) = o(r^\rho) \quad \text{as} \quad r \to \infty \tag{2.12}$$

implies (2.10) if $\tau_f > 0$; conversely, (2.10) implies (2.12) if $\sigma_f < \infty$. Thus the condition $a \in \mathcal{E}_f$ is equivalent to (2.12) provided $\rho_f \in (0, \infty)$ and $0 < \tau_f \leq \sigma_f < \infty$.

For the special case where f is an entire function of finite order and the functions a are polynomials, the relation (2.11) follows easily from E. Ullrich's known improvement of the deficiency relation (see [U] and also [Ha, Theorem 4.6]), stating that

$$\sum_{a \in \mathbb{C}} \delta(a, f) \leq \delta(0, f'). \tag{2.13}$$

For the proof it suffices to group together the polynomials a that have the same derivative, and apply (2.13) to each such group. For a finite sum of deficiencies of polynomials, repeated application of (2.13) in this manner yields the upper bound $\delta(0, f^{(m)}) \leq 1$ with some $m \in \mathbb{N}$.

Note in addition, as A. Edrei and W. Fuchs have proved (see [Ha, Theorem 4.13]), that an entire f has no finite *deficient values*, if $\rho_f \in [0, 1/2]$. Applying this to $f - a$, we obtain that the function f has also no *deficient function* a.

2.2   Some best approximation results

We will use the following notations. The interior, the closure and the boundary of a set $E \subset \mathbb{C}$ will be denoted by E°, $\bar{E}$ and ∂E, the distance of any two sets $E_1, E_2 \subset \mathbb{C}$ by $d(E_1, E_2)$. For an *open* set $\Omega \subset \mathbb{C}$, $H(\Omega)$ stands for the class of holomorphic functions in Ω. For a *closed* set $E \subset \mathbb{C}$ we denote $A(E) = C(E) \cap H(E^\circ)$, and by $H(E)$ the union of all classes $H(\Omega)$ with open sets $\Omega \supset E$. For a *compact* set $E \subset \mathbb{C}$, denote by $P(E)$ the uniform closure of the *polynomials* on E. Furthermore, put:

$$D(a, r) = \{z \in \mathbb{C} : |z - a| < r\} \quad \text{for} \quad a \in \mathbb{C}, \ r > 0;$$
$$D_r = D(a, r) \quad \text{for} \quad a = 0, \ \gamma_r := \partial D_r;$$
$$\Delta_\alpha = \{z \in \mathbb{C} : |\arg z| \leq \alpha/2\} \quad \text{for} \quad 0 < \alpha < 2\pi;$$
$$\Pi = \Delta_\pi = \{z \in \mathbb{C} : \operatorname{Re}(z) \geq 0\};$$
$$|\gamma| = \text{length } \gamma \text{ of a rectifiable curve } \gamma \subset \mathbb{C}.$$

Recall first some results on the possibility of uniform approximation by *polynomials* and *entire functions*. For a compact set $K \subset \mathbb{C}$, Mergelyan's approximation theorem states that $P(K) = A(K)$ if and only if $\mathbb{C} \setminus K$ is connected. It is customary to call such a set a *Mergelyan set*. A closed set $E \subset \mathbb{C}$ is said to be a set of *uniform entire approximation*, if the uniform closure of the entire functions on E coincides with $A(E)$. A necessary and sufficient condition for a closed set E to be a set of uniform entire approximation is that $\overline{\mathbb{C}} \setminus E$ must be connected and locally connected at the point ∞ (see [A1, A2]). For $f \in C(E)$ with $0 \in E$ we put

$$M(r, f) = \max_{z \in \overline{D}_r \cap E} |f(z)| \quad \text{for} \quad r > 0.$$

A closed set $E \subset \mathbb{C}$ is said to be a set of *tangential (entire) approximation*, if there exists a positive function $\sigma \in C(E)$ (a *speed* of tangential approximation) such that $\sigma(z) \to 0$ as $z \to \infty$, $z \in E$, and for any $f \in A(E)$ and $\varepsilon > 0$, there exists an entire function g satisfying

$$|f(z) - g(z)| < \varepsilon\sigma(z), \quad z \in E. \tag{2.14}$$

It follows from M. V. Keldysh's results (see [K, M1]) that each set E of uniform entire approximation is also a set of (stronger) tangential approximation, with any fixed speed of the form $\sigma(z) = \exp(-|z|^\eta)$, $\eta > 1/2$. In addition, if $E \subset \Delta_\alpha$ and $\rho \in (0, \pi/\alpha)$, then one can choose $\sigma(z) = \exp(-|z|^\rho)$. More information concerning the possible speed of tangential approximation can be found in [AG].

The construction of entire functions f having infinitely many deficient functions (see Theorem 3.1 below) is based on some tangential approximation results combined with the estimation of the growth of the approximating functions (cf. [A3, A5, F]). The following theorem is due to M. V. Keldysh (see [K, Theorem 4]; for the proof see [M1, Ch. 2, Theorem 3.5]).

2.3 Theorem *Let $F \in H(\Delta_\beta)$, $\rho \in [0, \pi\beta^{-1}]$ and $\varepsilon > 0$, $\delta > 0$ be arbitrary numbers. Then there exists an entire function G, satisfying*

$$|F(z) - G(z)| < \varepsilon\exp(-|z|^\rho) \quad \text{for} \quad z \in \Delta_{\beta-\delta},$$

and such that for $r \geq 0$, the growth of G is limited by the inequality

$$\log^+ M(r, G) < r^\eta \left\{ K + k \max_{t \in [1, kr]} \frac{t^\rho + \log^+ M(t, F)}{t^\rho} \right\},$$

where $\eta = \pi(2\pi - \beta)^{-1}$ and $K = K_{\varepsilon,\delta} > 0$, $k = k_\delta > 1$, are some constants.

An immediate consequence of Theorem 2.3 is the following:

2.4 Lemma *Let $\beta_\rho = \pi \min\{\rho^{-1}, 2 - \rho^{-1}\}$ for $\rho > 1/2$, and $0 < \alpha < \beta \leq \beta_\rho$. Then for $F \in H(\Delta_\beta)$, satisfying*

$$\log^+ M(r, F) = O(r^\rho) \quad \text{as} \quad r \to \infty,$$

and $\varepsilon > 0$, there exists an entire function G such that

$$|F(z) - G(z)| < \varepsilon\exp(-|z|^\rho) \quad \text{for} \quad z \in \Delta_\alpha, \tag{2.15}$$

and

$$\log^+ M(r, G) = O(r^\rho), \quad \text{as} \quad r \to \infty. \tag{2.16}$$

The next lemma (see [A5, Lemma 3]) is a version of M. V. Keldysh's fundamental lemma on rational approximation of the Cauchy kernel (see [M2]).

2.5 Lemma *Let Ω be a simply connected domain in $\mathbb{C}$, $\Omega \neq \mathbb{C}$, and let Γ be a rectifiable curve in Ω such that $\zeta, \zeta_0 \in \Gamma$. Then for $\varepsilon \in (0, 1]$ there exists a rational function Q_ζ with the only pole at ζ_0, satisfying*

$$\left| Q_\zeta(z) - \frac{1}{\zeta - z} \right| < \varepsilon \quad \text{for} \quad z \in \bar{\mathbb{C}} \setminus \Omega, \tag{2.17}$$

and

$$\log^+ |Q_\zeta(z)| < 40\{1 + \log^+[\varepsilon\, d(\zeta_0)]^{-1}\} \exp\{4I\} \qquad (2.18)$$

for $z \in \Omega \setminus D(\zeta_0, d_0)$, *where*

$$I = I_\Gamma = \int_\Gamma \frac{|dt|}{d(t)}, \qquad d(t) = d(t, \partial\Omega) \qquad for \quad t \in \Omega.$$

2.6 Corollary *Under the conditions of Lemma 2.5, let* $\{\Gamma_i\}_0^k$ *be a partition of* Γ *such that* $\zeta_0 \in \Gamma_0$, $\zeta \in \Gamma_k$, *and* $d(t) \geq d_i > 0$ *for* $t \in \Gamma_i$. *Then for* $z \in \Omega \setminus D(\zeta_0, d_0)$ *the estimate* (2.18) *may be replaced by*

$$\log^+ |Q_\zeta(z)| < 40 \left[1 + \log^+(\varepsilon d_0)^{-1}\right] \exp\left\{4 \sum_{i=0}^{k} d_i^{-1} |\Gamma_i|\right\}. \qquad (2.18')$$

2.7 Problem It would be interesting to obtain a *harmonic* analog of Lemma 2.5 in $\mathbb{R}^n$ with possible applications in harmonic function theory.

Actually, Lemma 2.5 is a result on best rational approximation of the Cauchy kernel and even a little more. The proof given in [A5] implies the following statement.

2.8 Proposition *Under the conditions of Lemma 2.5, for* $n \in \mathbb{N}$ *there exists a rational function* $\mathcal{R} = \mathcal{R}_n$ *of degree* n, *with the only pole at* ζ_0, *such that*

$$\left|\mathcal{R}_n(z) - \frac{1}{\zeta - z}\right| < \frac{2^7}{d(\zeta_0)} \exp\{6I - ne^{-2I}\}$$

for $z \in \bar{\mathbb{C}} \setminus \Omega$, *and*

$$|\mathcal{R}_n(z)| < \frac{4}{d(\zeta_0)} 2^{5n} \quad for \quad z \in \Omega \setminus D(\zeta_0, d_0).$$

For any two points $z, w \in \Omega$, the *quasi-hyperbolic* distance in Ω is defined by

$$\rho(z, w) := \inf_\gamma \int_\gamma \frac{|dt|}{d(t)},$$

where the infimum is taken over all rectifiable Jordan arcs $\gamma \subset \Omega$, joining the points z and w (see [GP]). It is obvious that the quantity $I = I_\Gamma$ in Lemma 2.5 and Proposition 2.8 can be replaced by $\rho(\zeta_0, \zeta)$. For another application of the notion of quasi-hyperbolic distance see [AS, section 3].

2.3 Tangential approximation on archipelagos

The approximation results presented above will allow us to formulate and prove a specific theorem on tangential approximation by entire functions on an appropriate infinite set of "islands" (tending to infinity), with optimal estimates for their growth, especially suitable for applications in the value distribution theory of entire functions. In this connection we will introduce some temporary notions.

An *island* E is a connected Mergelyan set in $\mathbb{C}$ such that E is a compact set in $\mathbb{C}$ and both E and $\mathbb{C}\backslash E$ are connected. An *archipelago* is a union of disjoint islands $\{E_n\}_1^\infty$, $E = \bigcup_{n=1}^\infty E_n$, satisfying the condition that each disk in $\mathbb{C}$ meets only a finite number of islands E_n. It is easy to see that any archipelago E is a set of *uniform* entire approximation and even more, E is a *Carleman set*, i.e. a set of tangential approximation with *arbitrary* speed σ in (2.14) (see [AG]).

For a number of applications of complex approximation theory (including those in value distribution theory) it is of interest not only to be assured of the *existence* of an entire function g satisfying (2.14) on an archipelago E, but also to have sufficient information on the growth of g (i.e. on the growth the function $M(r,g)$ as $r \to \infty$). It is important that the growth be as slow as possible.

In principle one can obtain some estimates of the function $M(r,g)$ in terms of the functions f, σ and the set E, in the spirit of Theorem 2.3, supposing f to be holomorphic on an archipelago larger than E. But in general, these estimates would be rough. In particular, it is hard to make sure that the function g be an entire function of *given* finite order. For these purposes it seems necessary to introduce some more canonical archipelagos instead of arbitrary ones, taking also into account that for possible applications they must have certain additional properties. The construction given below is a modification of the constructions suggested in [A3, A5] (see also [AH]).

1. For $2\rho > 1$ and $\beta = \beta_\rho$, as defined in Lemma 2.4, choose an $\alpha \in (0,\beta)$ and put $\alpha_k = \alpha \exp(-c_1 k)$ for $k \in \mathbb{N}$, where $c_1 > 8$ is a constant subject to further definition (it may depend only on ρ). Later on we will denote by c_m, $m \geq 2$, other similar constants. Furthermore, put $\Lambda_{k,\varepsilon} = \Lambda'_{k,\varepsilon} \cup \Lambda''_{k,\varepsilon}$ for $\varepsilon \in (0,1/2]$ and $k \in \mathbb{N}$, where

$$\Lambda'_{k,\varepsilon} = \{z : (2+\varepsilon)\alpha_k \leq \arg z \leq 2(1+\varepsilon)\alpha_k\},$$
$$\Lambda''_{k,\varepsilon} = \{z : 2(1-\varepsilon)\alpha_k \leq \arg z \leq (2-\varepsilon)\alpha_k\}.$$

Put also

$$\Lambda_{0,\varepsilon} = \{z : 2(1+\varepsilon)\alpha_1 \leq \arg z \leq \pi\},$$

and note that

$$\Lambda_{k,\varepsilon} \cap \Lambda_{j,\varepsilon} = \emptyset \quad \text{for} \quad \varepsilon \in (0,1/2), \; k \neq j,$$
$$\bigcup_{k=1}^\infty \Lambda_{k,\varepsilon} \subset \Delta_{6\alpha_1} \subset \Delta_\alpha \quad \text{for} \quad \varepsilon \in (0,1/2].$$

Next, given $n,k \in \mathbb{N}$ and $\varepsilon, \delta \in (0,1/2)$, put

$$S_{n,k}^\delta = \{z : 2^{(n-\delta)\alpha_k} \leq |z| \leq 2^{(n+1+\delta)\alpha_k}\},$$

and consider the "canonical" islands

$$E_{k,n}(\varepsilon,\delta) = \begin{cases} \Lambda'_{k,\varepsilon} \cap S_{n,k}^\delta, & \text{if } n \text{ is odd}, \\ \Lambda''_{k,\varepsilon} \cap S_{n,k}^\delta, & \text{if } n \text{ is even}. \end{cases}$$

Now for a subsequence $\{n_k\}_1^\infty$ of $\mathbb{N}$ we introduce a family of archipelagos

$$E(\varepsilon,\delta) = \bigcup_{k=1}^\infty \bigcup_{m=0}^\infty E_{k,n_k+m}(\varepsilon,\delta).$$

2.9 Remark The following property of the archipelagos $E(\varepsilon, \delta)$ is essential: for $k \in \mathbb{N}$ and $r \geq r_k := 2^{n_k \alpha_k}$, the intersection of $E(\varepsilon, \delta)$ with the circle γ_r contains an arc $\gamma_{k,r} \subset \Lambda_{k,\varepsilon}$ with $|\gamma_{k,r}| = (\varepsilon \alpha_k) r$. In addition, the islands $E_{k,n}(\varepsilon, \delta)$ forming $E(\varepsilon, \delta)$ are pairwise disjoint, and they are getting larger, if ε and δ are increasing.

Now fix $E_{k,n}^j = E_{k,n}(\varepsilon_j, \delta_j)$, $E_j = E(\varepsilon_j, \delta_j)$ with $\varepsilon_j = \delta_j = (j+1)/8$ and $j = 0$ and 1, so that $E_{k,n}^0 \subset E_{k,n}^1$, $E_0 \subset E_1^o$. It remains to impose some necessary restrictions on the sequence $\{n_k\}$. For this and for the subsequent discussion, it will be useful to introduce a function $\psi \in A(\Pi)$ by the formula

$$\psi(z) := \exp\{4iz^\rho\} \quad \text{for} \quad z \in \Pi. \tag{2.19}$$

Note that for $z \in \Lambda_{k,\varepsilon_1}$, $k \in \mathbb{N}$, it is easy to check the estimates

$$-2^{-1}|z|^\rho < -12\rho\alpha_k|z|^\rho < \log|\psi(z)| < -\alpha_k|z|^\rho. \tag{2.20}$$

We will subject the sequence $\{n_k\}_1^\infty$ in the definition of the sets E_0 and E_1 to the conditions

$$|\psi(\zeta)| < |\zeta|^{-2} \quad \text{for} \quad \zeta \in E_1, \tag{2.21}$$

$$\int_{\partial E_1} |\zeta|^{-2} |d\zeta| < 1. \tag{2.22}$$

The following theorem on tangential approximation on "canonical" archipelagos is adapted to several applications in value distribution theory.

2.10 Theorem *Let E_0 and E_1 be the archipelagos defined above for $\rho > 1/2$. For a function $\varphi \in A(E_1)$, satisfying the condition*

$$|\varphi(\zeta)||\psi(\zeta)| \leq 1 \quad for \quad \zeta \in E_1, \tag{2.23}$$

and for $\varepsilon > 0$, there exists an entire function f such that

$$\log^+ M(r, f) = O(r^\rho) \quad as \quad r \to \infty, \tag{2.24}$$

and

$$|f(z) - \varphi(z)| < \varepsilon|\psi(z)| \quad for \quad z \in E_0. \tag{2.25}$$

2.11 Remark Consider a sequence $A = \{a_k\}_1^\infty$ of entire functions satisfying condition (2.12) for given $\rho > 1/2$. Define a function $\varphi \in H(E_1)$ by putting

$$\varphi(\zeta) := a_k(\zeta) \quad \text{for} \quad \zeta \in \Lambda_{k,\varepsilon_1}, k \in \mathbb{N}. \tag{2.26}$$

Then φ will satisfy (2.23) if the sequence $\{n_k\}_1^\infty$ increases sufficiently rapidly (by (2.12) we can make sure that (2.23) holds in $\Lambda_{k,\varepsilon_1}$ for each fixed $k \in \mathbb{N}$, provided $|\zeta| \geq r_k$).

2. For the proof of Theorem 2.10 some additional preparation is necessary. Let $\zeta \in \partial E_1$, i.e. $\zeta \in \partial E_{k,n}^1$ for some $k, n \in \mathbb{N}$, and put $\zeta_0 = -2^{(n+1+\delta_1)\alpha_k}$. Consider the smaller arc of the circle $\gamma_{|\zeta|}$ joining ζ with $-|\zeta|$, and denote by γ_ζ its union with the interval $[\zeta_0, -|\zeta|]$. Since γ_ζ may intersect E_0, we modify γ_ζ to avoid this, in the following manner. If

$$\gamma_\zeta \cap E_1 \cap \Lambda_{i,\varepsilon_1} = l_i \neq \emptyset \quad \text{for some} \quad i \in [1, k],$$

then $l_i \in E^1_{i,m}$ for some $m \in \mathbb{N}$; if l_i is not a single point, we replace l_i by the arc L_i of $\partial E^1_{i,m}$, joining the endpoints of l_i and lying outside the disk $D_{|\zeta|}$. As a result we obtain the curve

$$\Gamma_\zeta := \left(\gamma_\zeta \setminus \bigcup_{i=1}^k l_i\right) \cup \bigcup_{i=1}^k L_i \subset (\mathbb{C} \setminus E_0) \cup \partial E_1,$$

joining the points ζ and ζ_0.

Consider a partition $\{\Gamma_{\zeta,i}\}_0^k$ of Γ_ζ, and put $\Gamma_{\zeta,i} = \Gamma_\zeta \cap \Lambda_{i,1/2}$. One can check that by our construction there exist constants $c_2 > 2$ and $c_3 \in (0,1)$ such that for $i \in [0,k]$

$$|\Gamma_{\zeta,i}| < c_2 \alpha_i |\zeta| \quad \text{and} \quad d(\Gamma_{\zeta,i}, E_0) = d_i > c_3 \alpha_i |\zeta| \tag{2.27}$$

with $\alpha_0 = \alpha_1$. It then follows from (2.27) that

$$\exp\left\{4 \sum_{i=0}^k d_i^{-1} |\Gamma_i|\right\} < \exp\left\{4c_2 c_3^{-1}\right\} := \frac{\alpha}{\alpha_k}, \tag{2.28}$$

if we fix the constant c_1 in the definition of the numbers α_k by putting $c_1 = 4c_2 c_3^{-1}$. Furthermore, we apply Lemma 2.5 and Corollary 2.6 with $\varepsilon = |\psi(\zeta)|^3$ to the simply connected domains

$$\Omega_\zeta = \bigcup_{i=0}^k \{z \in \mathbb{C} : d(z, \Gamma_{\zeta,i}) < d_i\}$$

and the partition of Γ_ζ mentioned above. We arrive at a function $Q(\zeta, z)$, defined on $\partial E_1 \times \Pi$, rational in z for each $\zeta \in \partial E_1$, and with poles in $\mathbb{C} \setminus \Pi$. Moreover, according to (2.17), (2.18'), (2.19)–(2.21), (2.27) and (2.28), $Q(\zeta, z)$ satisfies the inequalities

$$\left|Q(\zeta, z) - \frac{1}{\zeta - z}\right| < \left|\frac{\psi(\zeta)}{\zeta}\right|^2 \quad \text{for} \quad (\zeta, z) \in \partial E_1 \times (\mathbb{C} \setminus \Omega_\zeta), \tag{2.29}$$

and

$$|Q(\zeta, z)| < \exp\left[c_4 |\zeta|^\rho\right] \quad \text{for} \quad (\zeta, z) \in \partial E_1 \times \Pi. \tag{2.30}$$

2.12 Remark The inequalities (2.29) and (2.30) allow us to assume that the function $Q(\zeta, z)$ is piecewise constant in ζ, uniformly with respect to z. In addition, since $E_0 \subset \mathbb{C} \setminus \Omega_\zeta$ and $\Omega_\zeta \subset \mathbb{C} \setminus D_{\zeta|/2}$, the inequality (2.29) holds, in particular, if $\zeta \in \partial E_1$ and $z \in E_0 \cup (\Pi \cap D_{|\zeta|/2})$.

Proof of Theorem 2.10 By Lemma 2.4 and (2.20) it suffices to construct a function $f \in A(\Pi)$ satisfying the conditions (2.24) and (2.25) of Theorem 2.10. Put $L_r = \partial(E_1 \cap D_r)$ and $\Gamma_r = \partial(E_1 \setminus D_r)$ for $r \geq 0$, supposing these curves to be oriented positively with respect to those points of E_1^o that are separated by them. For $z \in \Pi \setminus \partial E_1$ consider the integrals

$$I_r(z) = \frac{1}{2\pi i} \int_{\Gamma_r} \frac{\varphi}{\psi}(\zeta) \left[Q(\zeta, z) - \frac{1}{\zeta - z}\right] d\zeta. \tag{2.31}$$

Obviously, if an integral I_r converges absolutely and uniformly on some subset of $\Pi \setminus \partial E_1$, then all integrals $I_{r'}$ with $r' < r$ have the same property. The absolute and uniform convergence of I_0 on each set $\bar{D}_{r'} \cap (\Pi \setminus \partial E_1)$, $r' > 0$, follows from (2.29) together with Remark 2.12 and (2.21)–(2.23), since for $r \geq 2r'$ the integrand in (2.31) does not exceed $|\zeta|^{-2}$. Thus any integral I_r is a holomorphic function on $(\Pi \setminus \partial E_1) \cup D_r$.

Analogously, using (2.21)–(2.23) and (2.29), we obtain from (2.31) that for $r > 0$,

$$|I_0(z)| < 2^{-1} \quad \text{for} \quad z \in E_0, \qquad |I_r(z)| < 2^{-1} \quad \text{for} \quad z \in \Pi \cap D_{r/2}. \tag{2.32}$$

Now introduce a holomorphic function

$$W(z) = I_0(z) + w(z) \quad \text{for} \quad z \in \Pi \setminus \partial E_1, \tag{2.33}$$

where $w(z) = (\varphi/\psi)(z)$, if $z \in E_1^o$ and $w(z) = 0$, if $z \in \Pi \setminus E_1$. By Cauchy's formula and theorem, we have for $r > 0$

$$w(z) = \frac{1}{2\pi i} \int_{L_r} \frac{w(\zeta)}{\zeta - z} \, d\zeta \quad \text{for} \quad z \in (\Pi \cap D_r) \setminus \partial E_1. \tag{2.34}$$

From (2.31), (2.33) and (2.34) we obtain for $z \in \Pi \setminus \partial E_1$ and $r > |z|$

$$W(z) = I_r(z) + \frac{1}{2\pi i} \int_{L_r} \frac{\varphi}{\psi}(\zeta) Q(\zeta, z) \, d\zeta = I_r(z) + J_r(z),$$

showing that actually $W \in H(\Pi)$. To estimate W on Π, take $r = 2|z|$. According to (2.19)–(2.23), (2.30) and (2.32), we obtain

$$|I_r(z)| + |J_r(z)| < 2^{-1} + \exp\{c_4|2z|^\rho\} \max_{\zeta \in L_{2|z|}} |\zeta\psi(\zeta)|^2,$$

which implies

$$|W(z)| < \exp\{c_5|z|^\rho\} \quad \text{for} \quad z \in \Pi. \tag{2.35}$$

We define the desired function $F \in H(\Pi)$, by putting $F = \psi W$. Now (2.24) follows from (2.35) and (2.19), and the inequality (2.25) from (2.32)–(2.34). □

3 Deficient functions

3.1 Statement of main results

The aim of this section is to present results on deficient functions and the "universality" property of entire functions of finite order (see [AH]). The following theorem throws considerable light on the question of possible sets of deficient functions and the possible sizes of their defects for the case mentioned earlier. Recall that $\mathbf{D}_f$ stands for the set of all deficient functions of a function f.

3.1 Theorem *For $\rho \in (1/2, \infty)$ and a sequence $\mathbf{A} = \{a_k\}_1^\infty$ of entire functions satisfying (2.12), there exists an entire function f of order ρ and normal type, with $\tau_f > 0$, such that $\mathbf{A} \subset \mathbf{D}_f$. Moreover, for some constant $c > 0$,*

$$\delta(a_k, f) > \exp(-ck) \quad \text{for} \quad k = 1, 2, \ldots. \tag{3.1}$$

For the case $a'_k = 0$ (i.e. a_k are *constants*) this result, with the exception of the estimates (3.1), has been obtained by Arakelian [A3] (see also [F]; a more detailed presentation can be found in [A5, GO]). The result has shown that the well-known conjecture of Nevanlinna (on the finiteness of the number of deficient values for entire functions of finite order) was false in general. The construction of the function f was based on methods and results of complex approximation theory, namely on tangential approximations of constants a_k on an appropriate archipelago, by an entire function f having optimal growth in $\mathbb{C}$.

In [A3] the following estimates were obtained instead of (3.1):

$$\delta(a_k, f) > \exp(-c\alpha_k^{-1}), \quad k \in \mathbb{N}, \tag{3.2}$$

where $\alpha_k > 0$ are arbitrary parameters with finite sum, so that (3.2) is actually rather close to (3.1). The estimates (3.1) for the case $a'_k = 0$ have been obtained by Eremenko [E], using a different approach. Here, for the proof of Theorem 3.1, we again use the approach developed in [A3], with appropriate modifications.

Theorem 3.1 induces some remarkable corollaries; the next result is its formal generalization.

3.2 Theorem *For $\rho \in (1/2, \infty)$ and a matrix $\{c_{n,k}\}_{n,k=1}^{\infty} = \{\mathbf{C}_n\}_1^{\infty}$ of entire functions $c_{n,k}$ satisfying (2.12), there exists an entire function f of order ρ and normal type, with $\tau_f > 0$, such that $\mathbf{C}_n \subset \mathbf{D}_{f^{(n-1)}}$ for $n \in \mathbb{N}$ (where $f^{(0)} = f$). Moreover, there is a constant $c > 0$ such that*

$$\delta(c_{n,k}, f^{(n-1)}) > \exp(-cnk) \quad \text{for} \quad n, k \in \mathbb{N}. \tag{3.3}$$

Note that for matrices with *constant* elements $c_{n,k}$, the possibility of constructing a function f satisfying the conditions of Theorem 3.2 (except (3.3)) by the methods developed in [A3], was proposed in [Hi], and simultaneously realized by a group of authors [D-W].

The next result does not follow directly from Theorem 3.1, but rather from the concrete construction of the function f in that theorem. It concerns the question, not yet sufficiently studied, of the construction of "universal" entire functions having given growth in $\mathbb{C}$ (see [B]; related topics see [Lu]).

3.3 Definition An entire function f is said to be *universal* with respect to a sequence $\{z_m\}_1^{\infty} \subset \mathbb{C}$, $z_m \to \infty$ as $m \to \infty$, if for any compact $K \subset \mathbb{C}$ with connected complement and any function $\varphi \in A(K)$, there is a subsequence of $\{f_m\}_1^{\infty}$ (where $f_m(z) = f(z + z_m)$ for $z \in \mathbb{C}$) which converges to φ uniformly on K.

3.4 Theorem *For given $\rho > 1/2$, there is a universal entire function f of order ρ and normal type.*

It follows from the proof of Theorem 3.4, presented below, that actually there exists an entire function f of order $\rho > 1/2$ and normal type, satisfying the statements of Theorems 3.4 and 3.1 (and consequently of Theorem 3.2) simultaneously. Moreover, all the derivatives $f^{(n)}$, $n \in \mathbb{N}$, of the function f are again universal entire functions of order ρ and normal type (with respect to the same fixed sequence $\{z_m\}_1^{\infty}$).

3.5 Problem Is the statement of Theorem 3.4 still valid for $\rho \leq 1/2$? There are some arguments supporting this assumption.

3.6 Proposition *Theorem 3.1 (as well as Theorems 3.2 and 3.4) can be extended to the case of entire functions f of order $\rho > 1/2$, having a "proximate" growth ν. This means that*

$$\log M(r, f) = O(\nu(r)) \quad as \quad r \to \infty$$

for some function $\nu \in C^1(\mathbb{R}_+)$, $\nu > 0$, satisfying

$$\lim_{r \to \infty} \frac{r\nu'(r)}{\nu(r)} = \rho.$$

Actually this condition means that the function $\rho(r) = (\log r)^{-1} \log \nu(r)$, $r \geq 1$, is a *proximate order* (apply L'Hopital's rule to $\rho(r)$ to see that $\rho(r) \to \rho(= \rho_\nu)$ as $r \to \infty$). In the above-mentioned theorems one may therefore also consider entire functions f having *minimal* or *maximal* ρ-type (but always having normal ν-type!). Most difficulties in the proof arise in the case $\rho_\nu = 1$. Consider in this connection the following:

3.7 Problem Does there exist a simple construction of an entire function ω satisfying

$$\log M(r, \omega) = O(\nu(r)) \quad as \quad r \to \infty,$$

with $\rho_\nu = 1$, such that for some $\varepsilon > 0$ and some constant $c > 0$,

$$-c\nu(|z|) < \log |\omega(z)| < -\nu(|z|) \quad for \quad z \in \Delta_\varepsilon?$$

For $\rho_\nu > 1/2$, $\rho_\nu \neq 1$, the result follows from the known asymptotic of the Weierstrass canonical products with zeros on a ray (see [GO, Ch. 1] and [GLO, Ch. 1, section 2]).

Although it is not directly related to our general subject we mention here the following problem which is in the same spirit as the preceding one (see [A7, subsection 1.4 and cf. with Lemma 1]).

3.8 Problem For an arbitrary proximate growth ν with $\rho_\nu > 0$ construct an entire function $\omega = \omega_\nu$ satisfying, for some constant $c > 0$, the inequalities

$$-c\nu(|z|) < \log |\omega(z)| < -\nu(|z|), \quad if \quad |\operatorname{Im} z| \leq 1,$$

$$\log M(r, \omega) < cr \int_1^r \frac{\nu(t)}{t^2} dt + c \quad for \quad r \geq 1.$$

3.2 Proof of the main theorems

Proof of Theorem 3.1 We apply Theorem 2.10 with $\varepsilon = 1$ to the function $\varphi \in H(\Pi)$ defined by formula (2.26). We arrive at an entire function f satisfying

$$|f(z) - \varphi(z)| < |\psi(z)| \quad for \quad z \in E_0, \tag{3.4}$$

and having a growth at most of the order ρ and finite type, so that

$$T(r, f) < \sigma r^\rho \quad for \quad r > 1. \tag{3.5}$$

The function f is the required function. In fact, by the definition of E_0 and the function φ (see also Remark 2.11 and (2.26)), we have that for each $k \in \mathbb{N}$ and $r > r_k$ there exists an arc $\gamma_{k,r}$ such that $|\gamma_{k,r}| = (4^{-1}\alpha_k)r$ and $\varphi(z) = a_k(z)$ for $z \in \gamma_{k,r}$. It then follows from (3.4), (2.19), (2.20) and (2.26), that for $k \in \mathbb{N}$ and $r \geq r_k$,

$$\left|(f - a_k)(re^{it})\right| < \exp(-\alpha_k r^\rho) \quad \text{for} \quad t \in e_{k,r},$$

where $e_{k,r}$ has Lebesgue measure $\alpha_k/4$. From this (using the notational conventions of Nevanlinna theory) it follows that

$$m(r, a_k, f) > (2\pi)^{-1} \int_{e_{k,r}} \log^+ \left|(f - a_k)(re^{it})\right|^{-1} dt > c_6 \alpha_k^2 r^\rho. \tag{3.6}$$

From (3.6) and (3.5) we finally obtain that for $k \in \mathbb{N}$ and some constant $c > 0$,

$$\delta(a_k, f) = \liminf_{r \to \infty} \frac{m(r, a_k, f)}{T(r, f)} > (c_6/\sigma)\alpha_k^2 > \exp(-ck).$$

In addition, (3.5), (3.6) and (2.12) for $a = a_k$ imply $\rho_f = \rho$ and $\tau_f > 0$. $\qquad\square$

Proof of Theorem 3.2 Let $A = \{a_{n,k}\}_{n,k=1}^\infty$ be a matrix of entire functions satisfying (2.12). By (2.13),

$$\delta(a_{n,k}, f) \leq \delta(a_{n,k}^{(n-1)}, f^{(n-1)}) \quad \text{for} \quad n, k \in \mathbb{N}.$$

We obtain the result from Theorem 3.1 by imposing on the elements $a_{n,k}$ of $\mathbf{A}$ the conditions $a_{n,k}^{(n-1)} = c_{n,k}$. Then the functions $a_{n,k}$ will also satisfy condition (2.12), and (3.3) follows from (3.1), since one can represent $a_{n,k}$ as a sequence with index $\leq nk$. $\qquad\square$

Proof of Theorem 3.4 Consider the set $\mathcal{P}$ of all polynomials with complex rational coefficients, $\mathcal{P} = \{P_m\}_1^\infty$. By Mergelyan's theorem (see subsection 2.2), the algebra $P(K)$ of all polynomials (and consequently $\mathcal{P}$) is uniformly dense in $A(K)$ for each compact set $K \subset \mathbb{C}$ with connected complement. Thus for the proof it is enough to construct an entire function f of order $\rho > 1/2$ and normal type, such that for $m \in \mathbb{N}$,

$$|f(z) - P_m(z - z_m)| < \varepsilon_m \quad \text{for} \quad z \in \overline{D}(z_m, r_m), \tag{3.7}$$

with some sequences $\varepsilon_m \to 0$, $z_m \to \infty$ and $r_m \to +\infty$.

For the proof we will again apply Theorem 2.10. As in the proof of Theorem 3.1 we define the function φ on E_1 by formula (2.26), except on the islands $E_{1,n}^1$, $n \geq n_1$, lying in the sector $\Lambda_1(\varepsilon_1)$. On these islands we redefine φ in the following manner. Let $\overline{D}(\zeta_n, \tau_n)$ be the closed disk, inscribed in $E_{1,n}^1$, $n \geq n_1$. Choose a sequence $\{q_m\}_1^\infty$, $q_1 = n_1$, $q_m \uparrow +\infty$ as $m \uparrow +\infty$, and define

$$\varphi(z) = P_m(z - \zeta_n) \quad \text{for} \quad z \in E_{1,n}^1, \; n \in [q_m, q_{m+1}). \tag{3.8}$$

We assume $P_1 \equiv 0$ and let P_m be a polynomial of degree k_m with the sum of the moduli of its coefficients equal to s_m. Then

$$|\varphi(z)| \leq s_m |z - \zeta_n|^{k_m} \leq s_m (3|z|)^{k_m} \quad \text{for} \quad z \in E_{1,n}^1, \; n \in [q_m, q_{m+1}),$$

since $|z|/|\zeta_n| \le 2^{2\alpha_1} < 2$ for $z \in E^1_{1,n}$. Imposing on q_m the condition

$$s_m(5r)^{k_m} < \exp\{\alpha_1 r^\rho\} \quad \text{for} \quad r \ge 2^{(q_m-1/2)\alpha_1},$$

we obtain by (2.20) that φ satisfies condition (2.23) also on $E_1 \cap \Lambda_1(\varepsilon_1)$.

Applying Theorem 2.10 with $\varepsilon = 1$, we arrive at an entire function f of order ρ and of normal type, satisfying on E_0 the inequalities (2.24) and (2.25); actually, f also satisfies the conditions of Theorem 3.1. From (2.25), taking into account (3.8) and (2.20), we obtain the desired inequalities (3.7), by putting

$$z_m = \zeta_{q_m}, \qquad r_m = 2^{-\alpha_1}\tau_{q_m}, \qquad \varepsilon_m = \exp\{-4^{-1}\alpha_1 r_m^\rho\},$$

and noting that $\overline{D}(z_m, r_m) \subset E_0$ for $m \in \mathbb{N}$. $\qquad\qquad\qquad\qquad\qquad\qquad\qquad\qquad\square$

4 Index values

At the present time, the distribution of deficient values is much more investigated than the *index values* (see the monographs [N1, N2, Ha, GO] and the surveys [G, GLO]). Some authors have expressed the opinion that the investigation of the first notion is more important than the second. This seems to be doubtful, since both notions have a definite, and important, *geometric* interpretation in the structure of the Riemann surface of a meromorphic function. Perhaps the main difference is that formally the definition of the index values seems to be more closely related to the geometric interpretation, whereas the definition of the deficient values has a more *quantitative* character. As a result (following the tradition originating with R. Nevanlinna's monographs [N1, N2]), mainly *analytic* methods have been used for constructing functions with *deficient* values (see [GO, Ch. 4, sections 4–5]), and purely *geometric (quasi-conformal)* methods for the construction of functions with *index* values (see [GO, Ch. 7, section 8]), based on the preliminary construction of their Riemann surfaces with presupposed properties.

The aim of this section is to present some purely analytic and simple means for constructing entire functions of given finite positive order that have infinite sets of index values (see subsection 4.1), and of functions of zero order with an arbitrary finite set of index values (see subsection 4.3). This approach makes it possible to obtain some new results. In subsection 4.2 we combine deficient and index values into a single construction, again using approximation methods for this purposes.

4.1 Indices of functions of positive order

In this subsection we present a purely analytic proof of the following existence theorem on multiplicity indices of entire functions of finite order. Also, we obtain some estimates from below for the indices. Recall that $\mathbf{I}_f$ denotes the set of index values of a function f.

4.1 Theorem *Given a number $\rho > 0$ and a sequence $\mathbf{A} = \{a_k\}_1^\infty \subset \mathbb{C}$, there exists an entire function $f = f_\rho$ of order ρ and normal type such that $\mathbf{A} \subset \mathbf{I}_f$.*

Proof 1. Define $s = s_\rho$ and $\lambda = \lambda_\rho$ by putting

$$s := \min\{p \in \mathbb{N} : p \ge 4\rho\}, \qquad \lambda := s\rho^{-1}. \tag{4.1}$$

It suffices to prove the theorem for $4\rho \leq 1$, when $s = 1$ and $\lambda = \rho^{-1} \geq 4$. Otherwise we define $f_\rho(z) = f_{\lambda^{-1}}(z^s)$ and apply Remark 2.1 with (2.9).

Now consider the product

$$B_\rho(z) := \prod_{q=1}^{\infty} (1 - 2^{-\lambda q} z)^{2^q} \tag{4.2}$$

which defines an entire function. Note that

$$\log M(r, B_\rho) = \int_0^{\infty} \frac{n(rt)}{t(t+1)} \, dt \quad \text{for} \quad r \geq 0, \tag{4.3}$$

where $n(t) = 0$ for $t \in [0, 2^\lambda)$ and

$$n(t) = 2^{q+1} - 2 \quad \text{for} \quad t \in [2^{\lambda q}, 2^{\lambda(q+1)}), \ q \in \mathbb{N}.$$

This implies

$$(t^\rho - 2)^+ \leq n(t) < 2t^\rho \quad \text{for} \quad t \geq 0,$$

and then it follows from (4.3) that for some constant $c_\rho > 0$,

$$\log^+ M(r, B_\rho) < c_\rho r^\rho \quad \text{for} \quad r \geq 0. \tag{4.4}$$

By (4.4), B_ρ is a function of order at most ρ and normal type, but actually B_ρ has exactly the order ρ and a normal type $\geq \lambda$, since for $a = 0$ and $r \geq 2^\lambda$, (2.4) implies

$$\log^+ M(r, B_\rho) > \lambda r^\rho - 2 \log r - L_0.$$

Now introduce the product

$$\mathbb{B}(z) = \prod_{p=1}^{\infty} B_\rho \left(\frac{z}{z_{p,p}} \right) = \prod_{q=2}^{\infty} \prod_{p=1}^{q-1} \{b_{p,q}(z)\}^{2^{q-p}}, \tag{4.5}$$

with $b_{p,q}(z) = 1 - (z_{p,q})^{-1} z$, where $z_{p,q} = \mu_p 2^{\lambda q}$ and $\mu_p = i \exp(2^{1-p}\pi i)$. Obviously, the product (4.5) converges local-uniformly in $\mathbb{C}$, and defines an entire function $\mathbb{B}$ with zeros at points $z_{p,q} = \mu_p 2^{\lambda q}$ of degree 2^{q-p} correspondingly. By (4.4) and (4.5),

$$\log^+ M(r, \mathbb{B}) < c_\rho r^\rho \quad \text{for} \quad r \geq 0. \tag{4.6}$$

Note also that by (2.4), $\mathbb{B}$ has order ρ and normal type, since it is the case for the function $N(r, 0, \mathbb{B})$.

2. For the further construction we need estimates of $|\mathbb{B}|$ from below outside some neighborhood Ω of its zeros. Given $\delta > 0$, put

$$\mathbb{D}_{p,q}(\delta) := \{z \in \mathbb{C} : |b_{p,q}(z)| < \delta\},$$

and define Ω (with $\delta_p = 2^{-p}$) by

$$\Omega = \bigcup_{q=2}^{\infty} \bigcup_{p=1}^{q-1} \mathbb{D}_{p,q}(3\delta_{p+1}). \tag{4.7}$$

Since $|\mu_1 - \mu_2| = \sqrt{2}$ and

$$|\mu_{p-1} - \mu_p| < 2|\mu_p - \mu_{p+1}| \quad \text{for} \quad p \geq 2,$$

it follows that

$$|\mu_p - \mu_{p+1}| > 3(\delta_{p+1} + \delta_{p+2}) \quad \text{for} \quad p \in \mathbb{N}.$$

This implies

$$|\mu_p - \mu_k| \geq 3(\delta_{p+1} + \delta_{k+1}) \quad \text{if} \quad p \neq k,$$

and using that

$$3\delta_{p+1} < (2^\lambda - 1)(2^\lambda + 1)^{-1} \quad \text{if} \quad p \geq 1, \ \lambda \geq 4,$$

we obtain

$$\overline{\mathbb{D}}_{p,q}(3\delta_{p+1}) \cap \overline{\mathbb{D}}_{k,n}(3\delta_{k+1}) = \emptyset, \quad \text{if} \quad (k,n) \neq (p,q).$$

In particular, it follows that

$$\partial\Omega = \bigcup_{q=2}^{\infty} \bigcup_{p=1}^{q-1} \partial\mathbb{D}_{p,q}(3\delta_{p+1}). \tag{4.8}$$

Now fix a pair $(k,n) \in \mathbb{N}^2$ with $n > k$, and let $z \in \partial\mathbb{D}_{k,n}(\delta_k) = \gamma_{k,n}$, i.e. $|b_{p,q}(z)| = \delta_k$. From the definition of the functions $b_{p,q}$ we derive

$$|b_{p,q}(z)| \geq \begin{cases} 2^{\lambda(n-q)-1} - 1, & \text{if } q < n, \\ \delta_p, & \text{if } q = n, \\ 1 - 2^{1-\lambda(q-n)}, & \text{if } q > n. \end{cases}$$

For $z \in \gamma_{k,n}$ this implies

$$\prod_{p=1}^{q-1} |b_{p,q}(z)|^{2^{q-p}} \geq \begin{cases} (2^{\lambda(n-q)-1} - 1)^{2^q-2}, & \text{if } q < n, \\ 2^{-2\cdot 2^n}, & \text{if } q = n, \\ (1 - 2^{1-\lambda(q-n)})^{2^q-2}, & \text{if } q > n. \end{cases} \tag{4.9}$$

Now note that for $n > 2$,

$$\prod_{j=1}^{n-2} (2^{\lambda j - 1} - 1)^{2^{n-j}-2} > (2^{2\lambda-1}b_\lambda)^{2^n} 2^{-\lambda(n^2+n+2)},$$

where

$$b_\lambda = \prod_{q=1}^{\infty} (1 - 2^{1-\lambda q})^{2^q+1}.$$

Using this, we obtain from (4.5) and (4.9) that for $n > \max\{k, 2\}$,

$$|\mathbb{B}(z)| \geq (2^{2\lambda-3}b_\lambda)^{2^n} 2^{-\lambda(n^2+n+2)} \quad \text{for} \quad z \in \gamma_{k,n}.$$

From here, it follows the existence of a constant $\beta = \beta_\lambda > 0$, so that for $k, n \in \mathbb{N}$ with $n > k$,

$$|\mathbb{B}(z)| > \beta(2^{2\lambda-4}b_\lambda)^{2^n} \quad \text{for} \quad z \in \gamma_{k,n}. \tag{4.10}$$

Note that $b_\lambda \to 1$ as $\lambda \to \infty$, and (4.10) implies for some $\lambda_0 > 1$ and $k, n \in \mathbb{N}$, $n > k$, that

$$|\mathbb{B}(z)| > \beta 2^{2^{n+1}} \quad \text{for} \quad z \in \gamma_{k,n}, \quad \lambda \geq \lambda_0. \tag{4.11}$$

For λ_0 we can in fact take the concrete value $\lambda_0 = 4$. To see this, consider a sequence $\{x_q\}_{q=1}^\infty \subset (0,1)$ and note that the inequality

$$\prod_{q=1}^m (1 - x_q) > 1 - \sum_{q=1}^m x_q$$

is satisfied for any $m \in \mathbb{N}$ and hence also for $m = \infty$; of course, it is of interest only if the right-hand side is positive, in which case it is easily derivable using induction on m. Applying this inequality to the product b_λ, we obtain for $\lambda \geq 4$ the desired estimate:

$$b_\lambda > 1 - 2(2^\lambda - 1)^{-1} - 4(2^\lambda - 2)^{-1} > 2^{-1} > 2^{6-2\lambda}.$$

3. Now let $\mathbf{A} = \{a_k\}_1^\infty \subset \mathbb{C}$ be the sequence in Theorem 4.1. Note that one can always repeat each a_k as many times in $\mathbf{A}$ as is necessary to make sure that the condition

$$|\mathbf{A}| := \sup_{n \geq 2} \left\{ 2^{-2^n} \sum_{k=1}^{n-1} |a_k| \right\} < +\infty. \tag{4.12}$$

is satisfied.

Let $g_{k,n}$ be the *principal part* of the Laurent expansion of the meromorphic function $1/\mathbb{B}$ about a point $z_{k,n}$ with $k, n \in \mathbb{N}$, $n > k$, representing a rational function of degree 2^{n-k} with the only pole at the point $z_{k,n}$. Denote by $p_{k,n}$ the partial sum of degree $2^n - 1$ of the Taylor series expansion of $g_{k,n}$ about the origin. We introduce a Mittag-Leffler type series

$$g(z) = \sum_{n=2}^\infty \sum_{k=1}^{n-1} a_k \{ g_{k,n}(z) - p_{k,n}(z) \}. \tag{4.13}$$

Our first aim is to ascertain the absolute and local uniform convergence of the series (4.13) in $\mathbb{C}$ under condition (4.12). Note first the formula

$$g_{k,n}(z) = -\frac{1}{2\pi i} \int_{\gamma_{k,n}} \frac{d\zeta}{\mathbb{B}(\zeta)(\zeta - z)},$$

valid for $z \in \mathbb{C} \setminus \overline{\mathbb{D}}_{k,n}(\delta_k)$, where the circle $\gamma_{k,n}$ is oriented positively with respect to its center $z_{k,n}$. From this one can derive for $z \in \mathbb{C} \setminus \overline{\mathbb{D}}_{k,n}(\delta_k)$ the following useful formula:

$$g_{k,n}(z) - p_{k,n}(z) = -\frac{1}{2\pi i} \int_{\gamma_{k,n}} (z/\zeta)^{2^n} \frac{d\zeta}{\mathbb{B}(\zeta)(\zeta - z)}. \tag{4.14}$$

To estimate this integral for $z \in \mathbb{C} \setminus \mathbb{D}_{k,n}(3\delta_{k+1})$ we apply (4.11). Noting that

$$|\zeta - z| \geq \delta_{k+1}|z_{k,n}|, \qquad |\zeta| \geq 2^{-1}|z_{k,n}|, \qquad \text{if} \quad \zeta \in \gamma_{k,n},$$

we obtain

$$|g_{k,n}(z) - p_{k,n}(z)| \leq 2\beta^{-1} \left(2^{-\lambda n}|z|\right)^{2^n}.$$

Using then (4.12) and putting $c = 2\beta^{-1}\|\mathbf{A}\|$, for $z \in \mathbb{C} \setminus \cup_{k=1}^{n-1}\mathbb{D}_{k,n}(3\delta_{k+1})$ we arrive at

$$\sum_{k=1}^{n-1} |a_k\{g_{k,n}(z) - p_{k,n}(z)\}| \leq c\left(2^{-\lambda n}|z|\right)^{2^n}. \tag{4.15}$$

Now let $0 < r \leq 2^{\lambda n - 2}$, for $n \geq m$. Then by (4.15) the series (4.13) converges uniformly and absolutely for $|z| \leq r$. Thus g is a meromorphic function in $\mathbb{C}$ with poles at the points $z_{k,n}$ (for $k, n \in \mathbb{N}$ and $n > k$) of degree 2^{n-k}, if $a_k \neq 0$ (if $a_k = 0$ for some $k \in \mathbb{N}$, the corresponding points $z_{k,n}$ would be regular points of g). In addition, we see that outside the open set Ω (see (4.7)) the series (4.13) has a *majorant* power series

$$G(z) = \sum_{n=2}^{\infty} c_{2^n} z^{2^n}, \qquad c_{2^n} = c2^{-\lambda n 2^n},$$

representing an entire function in $\mathbb{C}$. Using known formulas for the order and the type of an entire function in terms of the coefficients of its power series representation (see [Le]), we obtain that the function G has in $\mathbb{C}$ the order $\rho = \lambda^{-1}$ and the type $\sigma = e^{-1}\lambda$. Thus outside the set Ω the growth of the meromorphic function g does not exceed the order ρ and a normal type. The same statement is true also for the *entire* function $f := g\mathbb{B}$, i.e.

$$\log^+ |f(\zeta)| < \sigma_1|\zeta|^\rho + c_1 \quad \text{for} \quad \zeta \in \mathbb{C} \setminus \Omega, \tag{4.16}$$

with some positive constants σ_1 and c_1.

Now let $z \in \Omega$, so that $z \in \mathbb{D}(z_{k,n}, 3\delta_{k+1})$ for some fixed k and $n > k$. Since (4.16) holds also on $\partial\Omega$, then by (4.8) it holds for $\zeta \in \partial\mathbb{D}_{k,n}(3\delta_{k+1})$, in which case $|\zeta| < 3|z|$. The maximum modulus principle yields

$$\log^+ |f(z)| < (3^\rho\sigma_1)|z|^\rho + c_1 \quad \text{for} \quad z \in \Omega.$$

From this and (4.16) we finally obtain (with $\sigma = 3^\rho\sigma_1$)

$$T(r, f) \leq \log^+ M(r, f) < \sigma r^\rho + c_1 \quad \text{for} \quad r \geq 0. \tag{4.17}$$

Let us check now that the function $f = \mathbb{B}g$,

$$f(z) = \sum_{n=2}^{\infty} \sum_{k=1}^{n-1} a_k\mathbb{B}(z)\{g_{k,n}(z) - p_{k,n}(z)\}, \tag{4.18}$$

is the desired entire function. Fix any $k \in \mathbb{N}$, and let $n > k$. From (4.18) and the definition of the principal part $g_{k,n}$ near $z_{k,n}$, we have the representations:

$$f(z) = a_k\mathbb{B}(z)g_{k,n}(z) + f_{k,n}(z),$$
$$1/\mathbb{B}(z) = g_{k,n}(z) + h_{k,n}(z),$$

with $f_{k,n}$ and $h_{k,n}$ holomorphic at $z_{k,n}$, so that

$$f(z) = a_k + \mathbb{B}(z)\{f_{k,n}(z) - h_{k,n}(z)\} \tag{4.19}$$

near a point $z_{k,n}$. We see from (4.19) that the point $z_{k,n}$ is an a_k-point of f with multiplicity 2^{n-k}. For the number of *multiple* a_k-points of f this implies the estimate

$$n_1(r, a_k) \geq n_1(2^{-\lambda k} r, 0, B_\lambda) \quad \text{for} \quad r \geq r_k,$$

with $r_k := 2^{\lambda k + \lambda}$. Supposing now $r \geq r_k$ and $r \in [2^{\lambda j}, 2^{\lambda j + \lambda})$, $j \in \mathbb{N}$, we obtain

$$n_1(r, a_k) \geq 2^{j+1-k} - 2 - j > 2^{-k} r^\rho - 2 - \log r,$$

which for $r \geq r_k$ and a constant $c_k > 0$ implies (see (2.1))

$$N_1(r, a_k) \geq \rho^{-1} 2^{-k} r^\rho - \log^2 r - c_k. \tag{4.20}$$

Now apply (2.4) and put $k = 1$ in (4.20). We obtain that f is exactly of the order ρ and a lower type $\geq (2\rho)^{-1}$. Finally, by the definition (2.6)' of the *multiplicity index* $\theta(b_k, f)$ and (4.17), (4.20), it follows for $c = (\rho\sigma)^{-1}$ and $k \in \mathbb{N}$ that

$$\theta(a_k, f) \geq \liminf_{r \to \infty} \frac{N_1(r, a_k)}{\log^+ M(r, f)} \geq c 2^{-k} > 0. \tag{4.21}$$

This concludes the proof of Theorem 4.1.   □

4.2 Remark Assuming that the sequence $\mathbf{A}$ satisfies condition (4.12), one can add in Theorem 4.1 that estimates of the form (4.21) hold. Otherwise let (4.12) be satisfied after repeating each term a_k of $\mathbf{A}$ some m_k times, $k \in \mathbb{N}$. Then, putting $M_k = m_1 + m_2 + \cdots + m_k$, (4.21) implies the estimate

$$\theta(a_k, f) > c 2^{-M_k} \quad \text{for} \quad k \in \mathbb{N}.$$

4.3 Remark The function $f = f_\rho$ constructed above to satisfy the conditions of Theorem 4.1 has been defined, in terms of the parameters (4.1), by formula (4.18) for $\rho \leq 1/4$, and by $f_\rho(z) = f_{\lambda^{-1}}(z^s)$, $z \in \mathbb{C}$, if $\rho > 1/4$. Let us analogously extend the function $\mathbb{B} = \mathbb{B}_\rho$ (defined for $\rho \leq 1/4$ by (4.5)), for the case $\rho > 1/4$ by the formula $\mathbb{B}_\rho(z) = \mathbb{B}_{\lambda^{-1}}(z^s)$, $z \in \mathbb{C}$. Then it follows from the actual construction of f_ρ that for any $\rho > 0$, $\mu > 0$ and an arbitrary entire function G of order at most ρ and normal type, the formula

$$\mathbb{F}(z) := f_\rho(\mu z) - G(z) \mathbb{B}_\rho(\mu z), \quad z \in \mathbb{C}, \tag{4.22}$$

defines an entire function satisfying the conditions of Theorem 4.1, i.e. $\mathbb{F}$ has order ρ, normal type, and $\mathbf{A} \subset \mathbf{I}_f$.

4.2 Joint deficient and index values

In this subsection we construct entire functions of finite order with infinite sets of both deficient and index values.

4.4 Theorem *Let $\rho \in (1/2, \infty)$, $\mathbf{A} = \{a_k\}_1^\infty$ be a sequence of entire functions satisfying condition (2.12), and let $\mathbf{B} = \{b_k\}_1^\infty \subset \mathbb{C}$. Then there exists an entire function $\mathbb{F}$ of order ρ and normal type, with lower type $\tau_{\mathbb{F}} > 0$, such that $\mathbf{A} \subset \mathbf{D}_{\mathbb{F}}$ and $\mathbf{B} \subset \mathbf{I}_{\mathbb{F}}$. Moreover, the defects satisfy the estimates (3.1); in addition, the indices satisfy (4.21) provided (4.12) holds for $\mathbf{B}$.*

Actually, Theorem 4.4 is a synthesis of Theorems 3.1 and 4.1. As in Theorem 3.1, we need here the condition that $\rho > 1/2$.

Proof For the proof we will use Remark 4.3 and will define the desired function $\mathbb{F}$ by formula (4.22), where $f = f_\rho$ satisfies the condition $\mathbf{B} \subset \mathbf{I}_f$ and thus $\mathbf{B} \subset \mathbf{I}_{\mathbb{F}}$. In order to choose an entire function G in (4.22) so as to ensure that the function $\mathbb{F}$ satisfies the condition $\mathbf{A} \subset \mathbf{D}_{\mathbb{F}}$, we need some estimates for $|\mathbb{B}_\rho|$ from below. Depending on ρ we distinguish two cases.

For the case $\rho \leq 1/4$, we represent the product (4.2) (with $\lambda = 1/\rho$) by the formula

$$\log B_\rho(z) = -z \int_0^\infty \frac{n(t)}{t(t-z)} \, dt, \quad z \in \mathbb{C} \setminus [0, \infty). \tag{4.23}$$

Now note that for $\delta \in (0, \pi/2)$ and $t \in [0, \infty)$

$$|t - z| \geq (\delta/\pi)(t + |z|), \quad z \in \mathbb{C} \setminus \Delta_{2\delta}^o.$$

Then, putting $\delta = \pi/4$, we derive from (4.23) for $z \in \mathbb{C} \setminus \Delta_{\pi/2}^o$ the estimate

$$|\log B_\rho(z)| \leq 4|z| \int_0^\infty \frac{n(t)}{t(t + |z|)} \, dt = \log M(|z|, B_\rho).$$

In particular this implies (with the constant $c_\rho > 0$ from (4.4)) that

$$\log |B_\rho(z)| \geq -4c_\rho |z|^{1/\lambda} \quad \text{for} \quad z \in \mathbb{C} \setminus \Delta_{\pi/2}^o.$$

For the product $\mathbb{B}_\rho$, defined in (4.5), it then follows that

$$\log |\mathbb{B}_\rho(z)| \geq -4c_\rho |z|^{1/\lambda} \quad \text{for} \quad z \in \Delta_{\pi/2}. \tag{4.24}$$

For the case $\rho > 1/4$ (including the case $\rho > 1/2$ of our theorem with $s = s_\rho \geq 2$), we obtain from (4.24) and some constant $k_\rho > 0$ the inequality

$$\log |\mathbb{B}_\rho(z)| \geq -k_\rho |z|^\rho \quad \text{for} \quad z \in \Delta_{\pi/(2s)}. \tag{4.25}$$

Now it follows from (4.6) and (4.25) for sufficiently small $\mu > 0$, that the entire function $z \to B_\rho(\mu z)$ has order ρ and normal type, and that for $z \in \Delta_{\pi/(2s)}$ it satisfies

$$-\frac{1}{2}|z|^\rho \leq \log |\mathbb{B}_\rho(\mu z)| < \frac{1}{2}|z|^\rho. \tag{4.26}$$

Now let β_ρ be as in Lemma 2.4 and $\alpha \in (0, \beta)$, where $\beta \leq \min\{\beta_\rho, \pi/(2s)\}$. Then the archipelago E_1, constructed in subsection 2.3, point 1, and figuring in Theorem 2.10, will satisfy the condition $E_1 \subset \Delta_\alpha \subset \Delta_\beta \subset \Delta_{\pi/(2s)}$.

Consider the function $\varphi \in H(E_1)$ defined by formula (2.26) and satisfying condition (2.23). Apply Theorem 2.10 to φ with $\varepsilon = 1/2$ (in analogy with the proof of Theorem 3.1), and let f be the entire function from that theorem, satisfying (2.24) and (2.25). Write the function $\mathbb{F}$, defined by formula (4.22), as

$$\mathbb{F}(z) = [F(z) - G(z)]\mathbb{B}_\rho(\mu z) + f(z),$$

where

$$F(z) = \frac{f_\rho(\mu z) - f(z)}{\mathbb{B}_\rho(\mu z)}.$$

Obviously, $F \in H(\Delta_\beta)$. By (2.24), (4.26) and Theorem 4.1, F satisfies the conditions of Lemma 2.4. Applying this lemma to F with $\varepsilon = 1/2$, we arrive at an entire function G satisfying (2.15) and (2.16). In addition, we have by (2.15) and (4.26) that

$$|\mathbb{F}(z) - f(z)| < \exp(-|z|^\rho)|\mathbb{B}_\rho(\mu z)| < \frac{1}{2}\exp\{-2^{-1}|z|^\rho\}$$

holds for $z \in \Delta_\alpha$. From this and (2.15), (2.20), (2.25) it follows that in (3.4) and (3.5) one can replace the function f by $\mathbb{F}$. The repetition of the final part of the proof of Theorem 3.1 (in subsection 3.2) then completes the proof. □

4.3 Indices of functions of order zero

In this subsection we present an analytic construction of entire functions of order zero, having a given finite set of complex numbers as its *index* values. The method allows to construct entire functions f whose growth is as slow as we can reasonably wish. Moreover, all these functions satisfy the condition $T(r, f) = O(\log^2 r)$ as $r \to \infty$.

4.5 Definition We say a function $\varphi : I = [1, \infty) \to \mathbb{R}$ is of *class* $\Lambda(\varphi \in \Lambda_I)$, if $\varphi \in C^1(I)$, $\varphi(1) = 0$, $\varphi(t) \uparrow \infty$ as $t \uparrow \infty$, and the function $t \to t\varphi'(t)$ is non-increasing on I.

4.6 Example The following functions $t \to \varphi(t)$ belong to Λ_I for $t \in I$: $\varphi(t) = \log^\alpha t$ for $\alpha \in (0, 1]$, or $\varphi(t) = \log^\alpha[\log(et)]$ for $\alpha > 0$. One can continue this process, putting $\varphi(t) = \log^\alpha[\log_p(e_p t)]$ for $\alpha > 0$ and $p \in \mathbb{N}$, where $\log_1 = \log$, $\log_p = \log(\log_{p-1})$, $e_1 = e$ and $e_p = e_{p-1}^e$.

We will use the following property of the functions $\varphi \in \Lambda(I)$:

$$\varphi(t_1 t_2) \leq \varphi(t_1) + \varphi(t_2) \quad \text{for} \quad t_1, t_2 \in I. \tag{4.27}$$

This follows from the inequality $\varphi'(t) \leq t_1^{-1}\varphi'(t_1^{-1}t)$ for $1 \leq t_1 \leq t$, by integrating it over the interval $[t_1, t_1 t_2]$.

Put $\varphi(t) = 0$ for $t \in [0, 1)$, and associate with $\varphi \in \Lambda$ the function

$$N_\varphi(r) = \int_0^r \frac{\varphi(t)}{t} dt \quad \text{for} \quad r \geq 0. \tag{4.28}$$

Note that the condition $\varphi \in \Lambda_I$ implies $\varphi(r) \leq c\log^+ r$ for $r \geq 0$ with $c = \varphi'(1) > 0$. Applying L'Hopital's rule, we see that

$$\log r = o(N_\varphi(r)), \qquad \varphi(r) = o(N_\varphi(r)), \qquad \text{as} \quad r \to \infty.$$

On the other hand, (4.28) implies for $r \geq 1$

$$N_\varphi(r) \leq \varphi(r)\log r \leq c\log^2 r,$$

so that the growth of $N_\varphi(r)$ will not substantially exceed the growth of $\log r$ (the functions $\varphi \in \Lambda_I$ in Example 4.6 have slow growth in this sense). Note also that N_φ is a *slowly varying function* (see [GLO, Ch 1, section 1]), i.e.,

$$\lim_{r \to \infty} \frac{N_\varphi(sr)}{N_\varphi(r)} = 1 \quad (s > 1), \tag{4.29}$$

since $N_\varphi(sr) - N_\varphi(r) \leq \varphi(sr)\log s$ for any $s > 1$.

4.7 Theorem *Let $\varphi \in \Lambda_I$ and let $\mathbf{A} = \{a_k\}_1^m$ be a finite set of complex numbers. Then there exists an entire function f and positive constants c', c'' and r' such that:*

(i) $$c'N_\varphi(r) \leq \log^+ M(r,f) \leq c''N_\varphi(r) \quad \text{for} \quad r \geq r'; \tag{4.30}$$

(ii) $$\mathbf{A} \subset \mathbf{I}_f; \tag{}$$

(iii) $$\theta(a_k, f) \geq \frac{1}{2m} \quad \text{for} \quad 1 \leq k \leq m. \tag{4.31}$$

Proof Define an increasing sequence $\{r_n\}_1^\infty \subset I$ by the condition $\varphi(r_n) = n$. Then

$$\frac{1}{r_n} \leq \exp\left\{-\frac{n}{c}\right\} \quad \text{for} \quad n \in \mathbb{N} \tag{4.32}$$

with $c = \varphi'(1)$. By (4.32) the product

$$B(z) = \prod_{n=1}^\infty \left(1 - \frac{z}{r_n}\right) \tag{4.33}$$

converges local-uniformly in $\mathbb{C}$ and defines an entire function with zeros r_n, $n \in \mathbb{N}$. Since $n(t, 0, B) = [\varphi(t)]$ (the integer part of $\varphi(t)$), we obtain for $r \geq 0$ a relation of the function N_φ (see (4.28)) with the counting function $N(r, 0, B)$:

$$N_\varphi(r) - \log^+ r < N(r, 0, B) \leq N_\varphi(r). \tag{4.34}$$

Furthermore, from (4.33) it follows by (4.27) that

$$\log^+ M(r, B) = r\int_0^\infty \frac{[\varphi(t)]}{t(t+r)}\, dt < N_\varphi(r) + \varphi(r) + c. \tag{4.35}$$

Combining (4.34) and (2.4) yields an estimate in the opposite direction:

$$\log^+ M(r, b) \geq T(r, B) > N_\varphi(r) - \log^+ r - L_0, \tag{4.36}$$

and by (4.35) and (4.36) we arrive at

$$\lim_{r \to \infty} \frac{T(r,B)}{N_\varphi(r)} = \lim_{r \to \infty} \frac{\log^+ M(r,B)}{N_\varphi(r)} = 1. \tag{4.37}$$

Now put $\mu_p = \exp\{2pm^{-1}\pi i\}$ for $p \le m$, $m \ge 2$, and introduce the product

$$\mathbb{B}(z) = \prod_{p=1}^{m} B^2\left(\frac{z}{\mu_p}\right) = \prod_{q=1}^{\infty}\left[1 - \left(\frac{z}{r_q}\right)^m\right]^2, \tag{4.38}$$

having zeros of degree 2 at the points $z_{p,q} = \mu_p r_q$ for $p \le m$ and $q \ge 1$. Obviously by (4.38) and (2.4),

$$2mN(r,0,B) - L_0 \le T(r,\mathbb{B}) \le \log^+ M(r,\mathbb{B}) \le 2m\log^+ M(r,B),$$

which together with (4.34) and (4.37) implies

$$\lim_{r \to \infty} \frac{T(r,\mathbb{B})}{N_\varphi(r)} = \lim_{r \to \infty} \frac{\log^+ M(r,\mathbb{B})}{N_\varphi(r)} = 2m. \tag{4.39}$$

In particular, for any $c_1 > 2m$ there exists $c_2 > 0$ such that

$$\log^+ M(r,\mathbb{B}) < c_1 N_\varphi(r) + c_2 \quad \text{for} \quad r \ge r_1. \tag{4.40}$$

Now fix $0 < \delta < \min\{m^{-1}, 1 - r_1^{-1}\}$, and for $p \le m$, $q \ge 1$ consider the discs

$$\mathbb{D}_{p,q}(\delta) := \{z \in \mathbb{C} : |1 - z/z_{p,q}| < \delta\}.$$

Note first that

$$\overline{\mathbb{D}}_{p,q}(2\delta) \cap \overline{\mathbb{D}}_{k,n}(2\delta) = \emptyset, \quad \text{if} \quad (p,q) \ne (k,n).$$

Estimate now $|\mathbb{B}(z)|$ from below for $z \in \gamma_{k,n} := \partial\mathbb{D}_{k,n}(\delta)$ (i.e. $|1 - (z_{k,n})^{-1}z| = \delta$) with some fixed $k \le m$ and $n \ge 1$. One can see

$$|\mathbb{B}(z)| \ge B_n^{2m} \quad \text{with} \quad B_n = \prod_{q=1}^{\infty} d_q, \tag{4.41}$$

where $d_q = d_q(n)$ is defined as follows: $d_n = \delta$ and for $\tau = 1 - \delta$,

$$d_q = \begin{cases} \dfrac{\tau r_n}{r_q} - 1 & \text{for } q < n, \\[2mm] 1 - \dfrac{r_n}{\tau r_q} & \text{for } q > n, \end{cases}$$

Puting $t_1 = r_p$, $t_2 = r_q$ in (4.27) we note that

$$r_p r_q \le r_{p+q} \quad \text{for all} \quad p, q \in \mathbb{N}, \tag{4.42}$$

whence

$$d_q \ge \frac{\tau r_n}{r_q}\left(1 - \frac{1}{\tau r_{n-q}}\right) \quad \text{for} \quad q < n, \tag{4.43}$$

$$d_q \ge 1 - \frac{1}{\tau r_{n-q}} \quad \text{for} \quad q > n. \tag{4.44}$$

It follows from (4.43) and (4.44) that

$$B_n \geq B^2(1/\tau)\frac{(\tau r_n)^{n-1}}{r_1 r_2 \cdots r_{n-1}}. \tag{4.45}$$

By (4.42), $r_k r_{n-k} \leq r_n$ for $k < n$, so that $r_1 r_2 \cdots r_{n-1} \leq r_n^{(n-1)/2}$. This, together with (4.45) and (4.32), yields

$$B_n^{2m} > B^{4m}(1/\tau)(\tau^2 r_n)^{m(n-1)} > c_3 2^n \quad \text{for} \quad n \in \mathbb{N},$$

with a $c_3 > 0$ depending only on m and δ. By (4.41) we arrive at

$$\inf_{k \leq m} \min_{z \in \gamma_{k,n}} |\mathbb{B}(z)| > c_3 2^n \quad \text{for} \quad n \in \mathbb{N}. \tag{4.46}$$

Define the desired entire function f with prescribed growth of order zero by putting $f = \mathbb{B}g$, where g is the meromorphic function defined by the formula (cf. (4.13)):

$$g(z) = \sum_{n=1}^{\infty} \sum_{k=1}^{m} a_k g_{k,n}(z). \tag{4.47}$$

Here $g_{k,n}$ is the principal part the Laurent expansion of the function $1/\mathbb{B}$ near the point $z_{k,n}$; this time $g_{k,n}$ is a rational function with the only pole (of degree 2) at the point $z_{k,n}$.

We will now show that condition (4.46) guarantees the local-uniform convergence of the series (4.47) in $\mathbb{C}$, so that g represents in fact a meromorphic function with possible poles (of degree 2) at the points $z_{k,n}$ if $a_k \neq 0$ for the corresponding k. For this we will use the formula

$$g_{k,n}(z) = -\frac{1}{2\pi i} \int_{\gamma_{k,n}} \frac{d\zeta}{\mathbb{B}(\zeta)(\zeta - z)}, \quad z \in \mathbb{C} \setminus \overline{\mathbb{D}}_{k,n}(\delta). \tag{4.48}$$

Then for $\zeta \in \gamma_{k,n}$ and $z \in \mathbb{C} \setminus \mathbb{D}_{k,n}(2\delta)$

$$|\zeta - z| \geq \delta r_n, \qquad |\zeta| \geq (1 - \delta)r_n > 1,$$

so that, using the estimate (4.46) in (4.48), we obtain

$$|g_{k,n}(z)| \leq c_4 2^{-n} \quad \text{for} \quad z \in \mathbb{C} \setminus \mathbb{D}_{k,n}(2\delta).$$

This estimate demonstrates the local-uniform convergence of the series (4.47) in $\mathbb{C}$.

Putting $M = |a_1| + \cdots + |a_m|$, we get

$$|g(z)| < Mc_4, \quad \text{if} \quad z \notin \Omega := \bigcup_{k=1}^{m} \bigcup_{n=1}^{\infty} \mathbb{D}_{k,n}(2\delta). \tag{4.49}$$

Together with (4.40) this gives

$$\log^+ |f(\zeta)| < c_1 N_\varphi(|\zeta|) + c_5 \quad \text{for} \quad \zeta \in \mathbb{C} \setminus \Omega. \tag{4.50}$$

Now let $z \in \Omega$, i.e., $z \in \mathbb{D}_{k,n}(2\delta)$ for some $k \leq m$ and $n \geq 1$. Since (4.50) holds also for $\zeta \in \partial\mathbb{D}_{k,n}(2\delta)$ with $|\zeta| \leq s|z|$ for some $s = s_\delta > 1$, the maximum modulus principle yields

$$\log^+ |f(z)| < c_1 N_\varphi(s|z|) + c_5 \quad \text{for} \quad z \in \Omega.$$

Application of property (4.29) to this inequality and to (4.50) shows that given any $\sigma > 2m$ there exists a constant $c_6 > 0$ such that

$$\log^+ |f(z)| < \sigma N_\varphi(|z|) + c_6 \quad \text{for} \quad z \in \mathbb{C}.$$

From this we conclude that

$$T(r, f) \leq \log^+ M(r, f) < \sigma N_\varphi(r) + c_6. \quad \text{for} \quad r \geq 0 \qquad (4.51)$$

Now fix an integer $k \leq m$ and consider all corresponding points $\{z_{k,n}\}_{n=1}^\infty$ lying on the same ray from the origin. By the definition of the functions $g_{k,n}$ and g (see (4.47)) and the relation $f = \mathbb{B}g$, we have, near each point $z_{k,n}$, the representations

$$f(z) = a_k \mathbb{B}(z) g_{k,n}(z) + \mathbb{B}(z) f_{k,n}(z),$$
$$\frac{1}{\mathbb{B}(z)} = g_{k,n}(z) + h_{k,n}(z),$$

where $f_{k,n}$ and $h_{k,n}$ are holomorphic functions at $z_{k,n}$. This implies that near $z_{k,n}$,

$$f(z) = a_k + \mathbb{B}(z)[f_{k,n}(z) - a_k h_{k,n}(z)],$$

so that $z_{k,n}$ is an a_k-point of f of multiplicity 2. For the number of *multiple* a_k-points of f this gives the estimate $n_1(r, a_k) \geq [\varphi(r)]$. By (4.34),

$$N_1(r, a_k) > N_\varphi(r) - \log^+ r, \quad r \geq 0. \qquad (4.52)$$

Note also that for $r \geq 0$, (2.4) implies

$$\log^+ M(r, f) \geq T(r, f) \geq N_1(r, a_k) - L_{a_k}. \qquad (4.53)$$

Now (4.30) follows from (4.51)–(4.53). Finally, the definition (2.6) of the multiplicity index and (4.51), (4.52) together with (4.29) imply $\theta(a_k, f) \geq 1/\sigma$. Since $\sigma > 2m$ is an arbitrary constant, we arrive at $\theta(a_k, f) \geq 1/(2m)$. □

4.8 Remark For any fixed integer $l > 2$ the estimate for $\theta(a_k, f)$ in Theorem 4.7 can be replaced by

$$\theta(a_k, f) \geq \frac{1}{m}\left(1 - \frac{1}{l}\right) \quad \text{for} \quad 1 \leq k \leq m,$$

which is stronger than (4.31). For this, simply replace B^2 in definition (4.38) by B^l. Note also that the conditions $\theta(a_k, f) \geq 1/m$ for $k = 1, 2, \ldots, m$ will imply $\theta(a_k, f) = 1/m$ for all k, and the set of index values of f will coincide with $\mathbf{A}$.

References

[A1]   N. U. Arakelian, On uniform and asymptotic approximation by entire functions on unbounded closed sets, *Soviet Math. Dokl.* **157** (1) (1964), 9–11.

[A2]   N. U. Arakelian, On uniform approximation by entire functions on closed sets, *Izv. Akad. Nauk SSSR Ser. Mat.* **28** (1964), 1187–1206.

[A3]   N. U. Arakelian, Entire functions of finite order with an infinite set of deficient values, *Soviet Math. Dokl.* **170** (5) (1966), 999–1002.

[A4]   N. U. Arakelian, Uniform and tangential approximations by analytic functions, *Izv. Akad. Nauk Armjan. SSR Ser. Mat.* **3** (4–5) (1968), 273–285. English transl.: *Amer. Math. Soc. Transl. Ser. 2* **122** (1984), 85–97.

[A5]   N. U. Arakelian, Entire and analytic functions of limited growth with an infinite number of deficient values, *Izv. Acad. Nauk Armjan. SSR Ser. Mat.* **5** (6) (1970), 486–506.

[A6]   N. U. Arakelian, Approximation complexe et propriétés des fonctions analytiques, *Actes, Congrès International des Mathématiciens (Nice 1970), t. 2*, Gauthier-Villars, Paris, 1971, 595–600.

[A7]   N. U. Arakelian, On uniform and asymptotic approximation on the real axis by entire functions of restricted growth, *Mat. Sb.* **113** (**155**) (9) (1980), 3–40. English transl.: *Math. USSR-Sb.* **41** (1) (1982), 1–32.

[AG]   N. U. Arakelian and P. M. Gauthier, On tangential approximation by holomorphic functions, *Izv. Akad. Nauk Armyan. SSR Ser. Mat.* **17** (6) (1982), 421–441.

[AH]   N. U. Arakelian and A. M. Hakobian, Entire functions of finite order with infinite sets of deficient functions, *Proc. of Israel Math. Conf.* (to appear).

[AS]   N. U. Arakelian and H. Shahgholian, Propagation of smallness for harmonic and analytic functions in arbitrary domains, *Bull. London Math. Soc.* **31** (1999), 671–678.

[B]   G. D. Birkhoff, Démonstration d'un théorème élémentaire sur les fonctions entières, *C. R. Acad. Sci. Paris* **189** (1929), 473–475.

[Ca]   T. Carleman, Sur un théorème de Weierstrass, *Ark. Mat. Astron. Fysik* **20** (4) (1927), 1–5.

[Ch]   Chuang Chi-Tai, Une généralisation d'une intégrale de Nevanlinna, *Sci. Sinica* **13** (6) (1964), 887–895.

[D]   D. Drasin, The inverse problem of the Nevanlinna theory, *Acta Math.* **138** (1977), 83–151.

[D-W]   D. Drasin, Zhang Guanghou, Yang Lo, and A. Weitsman, Deficient functions and their derivatives, *Proc. Amer. Math. Soc.* **82** (4) (1981), 607–612.

[E]   A. Eremenko, A counterexample to the Arakelian conjecture, *Bull. Amer. Math. Soc.* **27** (1) (1992), 159–164.

[F]   W. H. J. Fuchs, *Théorie de l'approximation des fonctions d'une variable complexe*, Sém. Math. Sup. **26**, Les Presses de l'Université de Montréal, 1968.

[G]   A. A. Goldberg, Meromorphic functions, in: *Mathematical Analysis 10*, Akad. Nauk SSSR VINITI, Moscow, 1973, 5–97. English transl.: *J. Soviet Math.* **4** (1975), 157–216.

[GLO] A. A. Goldberg, B. Ya. Levin, and I. V. Ostrovskii, Entire and meromorphic functions,
       in: *Complex Analysis. One Variable*. Results of Sci. and Technol. Contemp. Problems
       of Math. **85**, VINITI, Moscow, 1991, 5–186. English transl. in: *Complex Analysis I*,
       Encyclopaedia Math. Sci., Springer, Berlin, 1995, 1–193.

[GO]   A. A. Goldberg and I. V. Ostrovskii, *Value Distribution of Meromorphic Functions*,
       Nauka, Moscow, 1970.

[GP]   F. W. Gehring and B. P. Palka, Quasiconformally homogeneous domains, *J. Anal.
       Math.* **30** (1976), 172–199.

[Ha]   W. K. Hayman, *Meromorphic Functions*, Clarendon Press, Oxford, 1964.

[Hi]   M. A. Hirnik (M. A. Girnik), On deficiencies of derivatives of an entire function,
       Ukrain. Mat. Zh. **33** (4) (1981), 510–513. English transl.: *Ukrainian Math. J.* **33** (4)
       (1982), 390–392.

[K]   M. V. Keldysh, On approximation of holomorphic functions by entire functions, *Dokl.
       Akad. Nauk SSSR* **47** (4) (1945), 243–245.

[Le]   B. Ya. Levin, *Distribution of Zeros of Entire Functions*, Transl. Math. Monogr. **5**,
       Amer. Math. Soc., Providence, RI, 1964.

[Lu]   W. Luh, Holomorphic monsters, *J. Approx. Theory* **53** (2) (1988), 128–144.

[M1]   S. N. Mergelyan, Uniform approximation of functions of a complex variable, *Uspekhi
       Mat. Nauk* **7**, No. 2(48) (1952), 31–122. English transl.: *Amer. Math. Soc. Transl.* **3**
       (1962).

[M2]   S. N. Mergelyan, On completeness of systems of analytic functions, *Uspekhi Mat.
       Nauk* **8**, No. 4(56) (1953), 3–63. English transl.: *Amer. Math. Soc. Transl. Ser. 2* **19**
       (1962), 109.

[N1]   R. Nevanlinna, *Le théorème de Picard-Borel et la théorie des fonctions méromorphes*,
       Gauthier-Villars, Paris, 1929.

[N2]   R. Nevanlinna, *Eindeutige analytische Funktionen*, Springer, Berlin, 1936.

[O]   S. F. Osgood, Sometimes effective Thue-Siegel-Roth-Nevanlinna bounds, or better, *J.
       Number Theory* **21** (3) (1964), 347–389.

[Ro]   A. Roth, Approximationseigenschaften und Strahlengrenzwerte meromorpher und
       ganzer Funktionen, *Comment. Math. Helv.* **11** (1938), 77–125.

[Ru]   A. Russakovskii, Approximation by entire functions and Arakelian-type examples for
       moving targets, *Illinois J. Math.* **40** (3) (1996), 439–452.

[S]   N. Steinmetz, Eine Verallgemeinerung des zweiten Nevanlinnaschen Hauptsatzes, *J.
       Reine Angew. Math.* **368** (1986), 134–141.

[U]   E. Ullrich, Über die Ableitung einer meromorphen Funktion, *Sitzungsber. Preuss.
       Akad. Wiss., Math.-Phys. Kl.* (1929), 592–608.

Uniform and tangential harmonic approximation

David H. ARMITAGE

Department of Pure Mathematics
Queen's University of Belfast
Belfast BT7 1NN
Northern Ireland

Abstract

We give a survey of Runge-type harmonic approximation theorems and the techniques used to prove them. The emphasis is on generalizations of Walsh's classical theorem concerning uniform harmonic approximation on compact sets to the case where approximation takes place on certain non-compact sets: both uniform and tangential approximation are treated. We also give some applications of the theory to the construction of harmonic functions exhibiting various kinds of unexpected behaviour. The course is partly intended to provide preparatory material for S. J. Gardiner's course "Harmonic approximation and applications", published in this volume.

1 Introduction

Let E be a non-empty subset of the Euclidean space $\mathbb{R}^n$, where $n \geq 2$. We say that a real-valued function u is *harmonic near* E if u is harmonic on some open set containing E. The earliest reasonably general theorem on harmonic approximation is that of Walsh [41] from 1929 which states that if u is harmonic near K, where K is compact and $\mathbb{R}^n \setminus K$ is connected, then u can be approximated uniformly on K by functions that are harmonic on the whole of $\mathbb{R}^n$ (and hence by harmonic polynomials). The techniques used to prove Walsh's theorem yield a slight generalization: if K is a compact subset of an open set Ω in $\mathbb{R}^n$ and every bounded component of $\mathbb{R}^n \setminus K$ intersects $\mathbb{R}^n \setminus \Omega$, then every function u that is harmonic near K can be approximated uniformly on K by functions harmonic on Ω. A landmark in the theory of harmonic approximation is a further generalization in which the compact set K is replaced by a set E that is relatively closed in Ω but not necessarily compact. Provided that the pair (Ω, E) satisfies certain mild topological hypotheses, functions harmonic near E can be approximated uniformly on E by functions harmonic on Ω. This development dates from the early 1980s and is due to Gauthier, Goldstein and Ow ([26] for $n = 2$ and [27] for $n \geq 3$; see also the lecture notes by Gauthier and Hengartner [28]). About a decade later, Armitage and Goldstein [7, 10] showed that in the theorem of Gauthier-Goldstein-Ow the quality of the approximation can be improved: that is to say, tangential (better than uniform) approximation can be achieved.

In these notes we discuss the techniques used to prove the results outlined above. We also give an account of some of the many applications of harmonic approximation theorems.

N. Arakelian and P.M. Gauthier (eds.), Approximation, Complex Analysis, and Potential Theory, 29–71.

Our discussion is restricted to Runge-type approximation theorems, in which the function
u to be approximated is always harmonic near the set E on which the approximation is to
be valid; we do not treat Arakelian-type approximation, in which u is merely assumed to
be continuous on E and harmonic on E° (although Arakelian-type theorems are given in
our main sources [27, 7, 10]). The proofs we give of Runge-type harmonic approximation
theorems are elementary in the sense that they depend on multivariate calculus but involve
almost no potential theory (if we deem the beginning of potential theory to be the study
of superharmonic functions). The important role played by potential theory in harmonic
approximation is apparent in S. J. Gardiner's article [25] in this volume. Some refinements of
the results presented in these notes can be found in Gardiner's monograph [24], which gives
a comprehensive account of harmonic approximation in the Euclidean space setting up to
1995.

Many applications of harmonic approximation theorems have been found and continue
to be found. In these notes we include some such applications, with "strange behaviour" as
a common theme. We construct, for example, harmonic functions on the whole of $\mathbb{R}^n$ that
decay rapidly to zero on every algebraic curve going to infinity and have zero integral on
every $(n-1)$-dimensional hyperplane. The extreme simplicity of these constructions is quite
striking.

The scheme of these notes is as follows. In Section 2 we give some general theory of
harmonic functions, culminating in harmonic Laurent expansions, and in Section 3 we use
such expansions to prove pole-pushing lemmas. Section 4 deals with Walsh-type theorems
concerning harmonic approximation on compact sets. In Sections 5, 6 and 7 we show that if
u is harmonic near a relatively closed subset E of an open set Ω, then u can be approximated
on E by functions v that are harmonic on Ω except for isolated singularities. Different types
of approximation are treated in these three sections. In Section 5 we restrict ourselves to
the case where $n \geq 3$ and show that the fusion result in [27], which was used to achieve
uniform approximation, can be slightly strengthened to produce approximations that are, in
a certain weak sense, tangential near infinity (that is to say, assuming E to be unbounded,
we can arrange for the error $|(u - v)(x)|$ to tend to zero as x tends to infinity along E). In
Section 6 we show how, with further modifications to the technique of [27], introduced in
[7], it is possible to achieve a much stronger type of tangential approximation near infinity.
Section 7 is based on [10] and shows how tangential approximation near all, or nearly all,
points of $\partial\Omega \cap \partial E$ simultaneously can be achieved. In Section 8 we show that, under certain
topological hypotheses, the approximating functions introduced in the earlier sections can
be taken to be free of singularities; for this we use pole-pushing. Some applications of the
theorems in Section 8 are given in Sections 9, 10 and 11.

2   Preliminaries

In order to make these notes reasonably self-contained, we give in this section the main results
from the general theory of harmonic functions that we shall need. Most of this material will
be well known to readers who are familiar with the theory of harmonic functions as presented
in [33, 32, 13, 6] for example. While we try to indicate clearly how results and concepts
are related to one another, we do not usually give detailed proofs, referring instead to [6,
Chapters 1, 2] and occasionally to other literature. We do, however, give details about basic

results that have special importance for harmonic approximation, such as those concerning the convergence of harmonic polynomial expansions in a ball and harmonic Laurent expansions in an annular domain.

2.1 Basic notation

Points of $\mathbb{R}^n$ are denoted by $x, y, \ldots$, and the Cartesian coordinates of x by $(x_1, \ldots, x_n)$. We use the usual inner product and norm on $\mathbb{R}^n$:

$$\langle x, y \rangle = x_1 y_1 + \cdots + x_n y_n, \qquad ||x|| = \sqrt{\langle x, x \rangle}.$$

The open ball of centre x and radius r is denoted by $B(x, r)$ and its boundary by $S(x, r)$. We write λ for n-dimensional Lebesgue measure and σ for $(n-1)$-dimensional surface measure. Some constants are defined as follows:

$$\lambda_n = \lambda(B(0, 1)), \qquad \sigma_n = \sigma(S(0, 1)), \qquad c_n = (\sigma_n \max\{1, n-2\})^{-1}.$$

Differentiation in the direction of the exterior normal to an open set with a smooth one-sided boundary is denoted by $\partial/\partial n_e$. We use standard multi-index notation: if $\alpha = (\alpha_1, \ldots, \alpha_n)$ is an ordered n-tuple of non-negative integers, then

$$|\alpha| = \alpha_1 + \cdots + \alpha_n, \qquad D^\alpha = \partial^{|\alpha|}/\partial x_1^{\alpha_1} \cdots \partial x_n^{\alpha_n}.$$

As usual, $C(E)$ denotes the vector space of all real-valued functions that are continuous on a subset E of $\mathbb{R}^n$, and $C^m(E)$, where $m \in \mathbb{N}$, denotes the space of functions f such that $D^\alpha f \in C(E)$ whenever $|\alpha| \leq m$; also $C^\infty(E) = \bigcap_m C^m(E)$. The Laplacian and gradient operators are denoted by Δ and ∇:

$$\Delta = \partial^2/\partial x_1^2 + \cdots + \partial^2/\partial x_n^2, \qquad \nabla = (\partial/\partial x_1, \ldots, \partial/\partial x_n).$$

To indicate that a differential operator acts on the jth argument ($j = 1, 2$) of a function of two vector variables, we attach a subscript j to the operator. The letter C, sometimes subscripted, will be used to denote a positive constant, not necessarily the same on any two occurrences; to indicate that C depends on $a, b, \ldots$, we write $C = C(a, b, \ldots)$. Throughout these notes Ω will be a non-empty open subset of $\mathbb{R}^n$. Several equivalent definitions of harmonicity are possible (see [6, Theorems 1.2.2, 1.3.3]): for our purposes it is convenient to define a function u to be *harmonic* on Ω if $u \in C^2(\Omega)$ and $\Delta u \equiv 0$. If E is any non-empty subset of $\mathbb{R}^n$, then $\mathcal{H}(E)$ denotes the vector space of functions u that are harmonic near E (that is, on some open set containing E). In particular, $u \in \mathcal{H}(\Omega)$ if and only if u is harmonic on Ω.

2.2 Green's formula

A well-known theorem in multivariate calculus states that for a large class of bounded open sets W in $\mathbb{R}^n$ with smooth boundary ∂W the following formula holds whenever $f \in C^1(\overline{W})$ and $g \in C^2(\overline{W})$:

$$\int_{\partial W} f \frac{\partial g}{\partial n_e} \, d\sigma = \int_W (\langle \nabla f, \nabla g \rangle + f \Delta g) \, d\lambda. \tag{2.1}$$

We refer to (2.1) as *Green's formula* and say that W is admissible if (2.1) holds for every pair of such functions f, g. Sufficient conditions for admissibility are given in [32, p. 22]. We see immediately that if W is admissible and $f, g \in C^2(\overline{W})$, then

$$\int_{\partial W} \left(f \frac{\partial g}{\partial n_e} - g \frac{\partial f}{\partial n_e} \right) d\sigma = \int_W (f \Delta g - g \Delta f) \, d\lambda. \qquad (2.2)$$

We define a function U on $\mathbb{R}^n \times \mathbb{R}^n$ by

$$U(x, y) = \begin{cases} -\log \|x - y\| & (n = 2) \\ \|x - y\|^{2-n} & (n \geq 3) \end{cases} \quad (x \neq y), \qquad U(x, x) = +\infty.$$

A calculation shows that for fixed y the function $U(\cdot, y) = U(y, \cdot)$ is harmonic on $\mathbb{R}^n \setminus \{y\}$. Moreover, any harmonic function on $\mathbb{R}^n \setminus \{y\}$ depending only on $\| \cdot -y\|$ has the form $a + bU(\cdot, y)$ for some constants a, b (see [6, p. 2]). We call U the *Newtonian* (or, in the case $n = 2$, *logarithmic*) *kernel* of $\mathbb{R}^n$. With y fixed, the function $U(\cdot, y)$ is sometimes called the *fundamental harmonic function with pole y*.

The following elementary result is important in the proofs of fusion results, which play an essential role in harmonic approximation on non-compact sets (see Sections 5–7).

2.1 Theorem (Green's representation) *Let W be an admissible open set. If $f \in C^1(\overline{W}) \cap C^2(W)$ and V is a function given by*

$$V(x, y) = U(x, y) + h(x, y),$$

where $h(x, \cdot) \in \mathcal{H}(\overline{W})$ for each fixed x in W, then

$$f(x) = c_n \int_{\partial W} \left(V(x, y) \frac{\partial f}{d n_e}(y) - f(y) \frac{\partial}{\partial n_e} V(x, y) \right) d\sigma(y)$$
$$- c_n \int_W V(x, y) \Delta f(y) \, d\lambda(y) \quad (x \in W).$$

We sketch a proof. Let x be fixed in W and let r be such that $\overline{B(x, r)} \subset W$. Applying (2.2) on the admissible set $W \setminus \overline{B(x, r)}$ and noting the harmonicity of $V(x, \cdot)$ there, we obtain

$$\int_{\partial W} + \int_{S(x,r)} \left(V(x, y) \frac{\partial f}{\partial n_e}(y) - f(y) \frac{\partial}{\partial n_e} V(x, y) \right) d\sigma(y)$$
$$- \int_{W \setminus \overline{B(x,r)}} V(x, y) \Delta f(y) \, d\lambda(y) = 0.$$

Easy estimates show that the integral over $S(x, r)$ in this equation tends to $-f(x)/c_n$ as $r \to 0+$, and the result follows.

2.3 Mean values and the minimum principle

If $f \in C(S(x, r))$, then we denote the mean value of f on $S(x, r)$ by $\mathcal{M}(f; x, r)$: thus

$$\mathcal{M}(f; x, r) = \frac{1}{\sigma_n r^{n-1}} \int_{S(x,r)} f \, d\sigma.$$

Similarly, if $f \in C(\overline{B(x,r)})$, then we write $\mathcal{A}(f;x,r)$ for the mean value of f on $B(x,r)$:

$$\mathcal{A}(f;x,r) = \frac{1}{\lambda_n r^n} \int_{B(x,r)} f \, d\lambda.$$

These mean values are related by the equation

$$\mathcal{A}(f;x,r) = \frac{n}{r^n} \int_0^r t^{n-1} \mathcal{M}(f;x,t) \, dt. \tag{2.3}$$

Let u be a function in $\mathcal{H}(\overline{B(x,r)})$. Applying Green's formula to the functions 1 and u on $B(x,t)$, we obtain

$$\int_{S(x,t)} \frac{\partial u}{\partial n_e} \, d\sigma = 0 \quad (0 < t \leq r),$$

which implies that $(d/dt)\mathcal{M}(u;x,\cdot) = 0$ on $(0,r]$ and hence that $\mathcal{M}(u;x,\cdot)$ is constant there. By continuity the value of the constant is $u(x)$. Hence, by (2.3)

$$u(x) = \mathcal{M}(u;x,r) = \mathcal{A}(u;x,r). \tag{2.4}$$

In fact this mean value property, together with continuity, characterizes harmonicity: $u \in \mathcal{H}(\Omega)$ if and only if $u \in C(\Omega)$ and $u(x) = \mathcal{M}(u;x,r)$ whenever $\overline{B(x,r)} \subset \Omega$; the same is true with $\mathcal{A}$ in place of $\mathcal{M}$; for details, see [6, Theorem 1.2.2].

We write $\partial^\infty \Omega$ for the boundary of Ω in the one-point compactification $\mathbb{R}^n \cup \{\infty\}$ of $\mathbb{R}^n$. Thus $\infty \in \partial^\infty \Omega$ if and only if Ω is unbounded. It follows easily from (2.4) that a harmonic function u cannot attain a local minimum at a point x unless u is constant on some neighbourhood of x. This is a local form of the minimum principle, from which the following global form is easily deduced (see [6, p. 5]).

2.2 Theorem (Minimum principle) *If $u \in \mathcal{H}(\Omega)$ and*

$$\liminf_{x \to y} u(x) \geq M \quad (y \in \partial^\infty \Omega),$$

where $M \in \mathbb{R}$, then $u \geq M$ on Ω.

An obvious corollary is the *maximum principle*: if $u \in \mathcal{H}(\Omega)$ and $\limsup_{x \to y} u(x) \leq M$ for all y in $\partial^\infty \Omega$, then $u \leq M$ on Ω. In particular, if $u \in \mathcal{H}(\Omega)$ and u vanishes continuously on $\partial^\infty \Omega$, then $u \equiv 0$.

2.4   The Poisson integral

The *Poisson kernel* $\mathcal{K}$ of $B(x_0,r)$ is defined by

$$\mathcal{K}(x,y) = \frac{1}{\sigma_n r} \frac{r^2 - \|x - x_0\|^2}{\|x - y\|^n} \quad (x \in B(x_0,r); \; y \in S(x_0,r)).$$

A calculation shows that $\mathcal{K}(\cdot,y) \in \mathcal{H}(B(x_0,r))$ for each fixed y in $S(x_0,r)$. Also, for each multi-index α, the partial derivative $D_1^\alpha \mathcal{K}$ is continuous on $B(x_0,r) \times S(x_0,r)$. It follows that if f is a σ-integrable function on $S(x_0,r)$, then the *Poisson integral* $\mathcal{I}_f$ of f, defined by

$$\mathcal{I}_f(x) = \int_{S(x_0,r)} \mathcal{K}(x,y) f(y) \, d\sigma(y) \quad (x \in B(x_0,r)),$$

is harmonic on $B(x_0, r)$. It is not hard to show that if f is continuous at a point z in $S(x_0, r)$, then $\mathcal{I}_f(x) \to f(z)$ as $x \to z$; see [6, Theorem 1.3.3]. In particular, if $u \in C(\overline{B(x_0, r)}) \cap \mathcal{H}(B(x_0, r))$, then $u - \mathcal{I}_u$ is harmonic on $B(x_0, r)$ and vanishes continuously on $S(x_0, r)$. Hence, by the minimum principle, we obtain the following result:

2.3 Theorem (Poisson integral representation) *If $u \in C(\overline{B(x_0, r)}) \cap \mathcal{H}(B(x_0, r))$, then $u = \mathcal{I}_u$ on $B(x_0, r)$.*

Though sufficient for our purposes, Theorem 2.3 is rather weak; for more general results, see [6, pp. 9, 10].

The Poisson integral representation is the key to many important properties of harmonic functions. For each fixed y in $S(x_0, r)$ the multiple Taylor series of $\mathcal{K}(\cdot, y)$ centred at x_0 converges absolutely and locally uniformly in $B(x_0, Cr)$, where C is an absolute constant. Using convergence properties of this Taylor series together with Theorem 2.3, we deduce that if $u \in C(\overline{B(x_0, r)}) \cap \mathcal{H}(B(x_0, r))$, then the multiple Taylor series of u centred at x_0 converges to u in $B(x_0, Cr)$. Thus harmonic functions are real-analytic. The details of this argument with $C = \sqrt{2} - 1$ are not difficult (see [6, pp. 28, 29]). The best possible value of C is $1/\sqrt{2}$; this was proved independently by Hayman [31] and Kiselman [34], and Siciak [39] gave a further refinement.

It follows in particular that if u is harmonic on Ω, then so also is any partial derivative $D^\alpha u$. Moreover, by estimating $D_1^\alpha \mathcal{K}$ and using Theorem 2.3, we can estimate $D^\alpha u$ for any u in $\mathcal{H}(\overline{B(x_0, r)})$. Then, by a standard covering argument, we can deduce the following result (see [6, Corollary 1.4.3]).

2.4 Corollary *For each compact subset K of Ω and each multi-index α there exists a constant $C = C(\Omega, K, \alpha)$ with the following property: if $u \in \mathcal{H}(\Omega)$ and $|u| \leq M$ on Ω, then $|D^\alpha u| \leq CM$ on K.*

A particular case ($|\alpha| = 1$, $K = \{x_0\}$), with an explicit value for C is given in the next result (see [6, pp. 13, 14]).

2.5 Corollary *Let u be harmonic on $B(x_0, r)$.*

(i) *If $|u| \leq M$ on $B(x_0, r)$, then*

$$\|\nabla u(x_0)\| \leq \frac{2nM}{r}.$$

(ii) *If $u > 0$ on $B(x_0, r)$, then*

$$\|\nabla u(x_0)\| \leq \frac{n}{r} u(x_0).$$

By Corollary 2.4, if (u_m) is a locally uniformly convergent sequence of harmonic functions, then so also is $(D^\alpha u_m)$ for each multi-index α. This observation leads easily to the following result.

2.6 Corollary *If (u_m) is a sequence in $\mathcal{H}(\Omega)$ converging locally uniformly to a limit function u on Ω, then $u \in \mathcal{H}(\Omega)$ and $D^\alpha u_m \to D^\alpha u$ as $m \to \infty$ locally uniformly on Ω for each multi-index α.*

2.5   Harmonic polynomials

For each non-negative integer m let $\mathcal{P}_m$ denote the vector space of all real-valued homogeneous polynomials of degree m on $\mathbb{R}^n$, and let $\mathcal{H}_m$ denote the subspace consisting of all harmonic elements of $\mathcal{P}_m$. We define $d_m = \dim \mathcal{H}_m$. Clearly $d_0 = 1$ and $d_1 = m$. Also

$$d_m = \dim \mathcal{P}_m - \dim \mathcal{P}_{m-2} = \binom{m+n-1}{m} - \binom{m+n-3}{m-2} \quad (m \geq 2); \qquad (2.5)$$

see [6, Corollary 2.1.4]. Simplifying (2.5) and using Stirling's formula, we find that $d_m = O(m^{n-2})$ as $m \to \infty$. The spaces $\mathcal{H}_m$ are mutually orthogonal in the sense that

$$\mathcal{M}(HI; 0, r) = 0 \quad (H \in \mathcal{H}_j; \ I \in \mathcal{H}_k; \ j \neq k; \ r > 0). \qquad (2.6)$$

The proof of this uses the homogeneity of the polynomials and (2.2); see [6, Lemma 2.2.1].

Let x be a point of $\mathbb{R}^n \setminus \{0\}$ and let f be a function defined at least on some ball $B(0, r)$. We say that f is x-axial if $f(y)$ depends only on $\|y\|$ and $\langle x, y \rangle$. For such x we define $J_0(x, \cdot) = 1$ on $\mathbb{R}^n$,

$$J_m(x, y) = \begin{cases} \|y\|^m \cos(m\theta) & (n = 2; \ m \geq 1; \ y \in \mathbb{R}^2 \setminus \{0\}) \\ \|y\|^m P_m^{((n-2)/2)}(\cos \theta)/P_m^{((n-2)/2)}(1) & (n \geq 3; \ m \geq 1; \ y \in \mathbb{R}^n \setminus \{0\}), \end{cases} \qquad (2.7)$$

where $\theta = \cos^{-1}(\langle x, y \rangle / \|x\| \|y\|)$ and $P_m^{(\mu)}$ is the ultraspherical polynomial of degree m and order μ in the notation of Szegö [40]. (We note that, for the case $n = 3$, we have $P_m^{(1/2)} = P_m$, the Legendre polynomial; see [40, formula (4.7.2)].) We also define $J_m(x, 0) = 0$ when $m \geq 1$. It is clear that $J_m(x, \cdot)$ is an x-axial element of $\mathcal{P}_m$. A calculation, based when $n \geq 3$ on the explicit formula for $P_m^{(\mu)}$ given in [40, formula (4.7.31)], shows that $J_m(x, \cdot)$ is harmonic (see [6, p. 54]). Also $|P_m^{(\mu)}(t)| \leq P_m^{(\mu)}(1)$ when $\mu > 0$ and $-1 \leq t \leq 1$; see [40, formulae (4.7.3) and (7.33.1)]. Hence, for all dimensions n,

$$|J_m(x, y)| \leq \|y\|^m \quad (y \in \mathbb{R}^n). \qquad (2.8)$$

Up to scalar multiplication $J_m(x, \cdot)$ is the unique x-axial element of $\mathcal{H}_m$; see [6, Theorem 2.3.4]. We shall need the following more general uniqueness result. Elementary proofs, each depending upon the real-analyticity of harmonic functions, are given in [6, Lemma 2.4.1] and [5, Lemma 3].

2.7 Lemma If u is a y-axial harmonic function on $B(0, \|y\|)$, where $y \in \mathbb{R}^n \setminus \{0\}$, and $u(ty) = 0$ for each t in $(-1, 1)$, then $u \equiv 0$.

We shall say that a series $\sum f_n$ of functions on a set E is *Weierstrass convergent* on E if $\sum \sup_E |f_n| < +\infty$, and $\sum f_n$ will be called *locally Weierstrass convergent* on E if $\sum f_n$ is Weierstrass convergent on every compact subset of E. This (non-standard) terminology is suggested by the Weierstrass M-test.

2.8 Theorem (Harmonic polynomial expansion) If $u \in \mathcal{H}(B(x_0, r))$, then there exist polynomials H_m in $\mathcal{H}_m$, uniquely determined by u, such that

$$u(x) = \sum_{m=0}^{\infty} H_m(x - x_0) \quad (x \in B(x_0, r)),$$

and the series is locally Weierstrass convergent on $B(x_0, r)$.

In outlining the proof of this, we may suppose that $B(x_0, r) = B(0,1)$, since harmonicity is preserved by translations and dilations. We treat first the special case where $u = \mathcal{K}(\cdot, y)$ for some y in $S(0,1)$. If $-1 < t < 1$, then

$$\sigma_n \mathcal{K}(ty, y) = (1-t^2)(1-t)^{-n} = (1-t^2) \sum_{m=0}^{\infty} \binom{m+n-1}{m} t^m = \sum_{m=0}^{\infty} d_m t^m;$$

see (2.5). We write $v = \sum_{m=0}^{\infty} d_m J_m(y, \cdot)$. Since $d_m = O(m^{n-2})$ as $m \to \infty$, we see by (2.8) that this series is locally Weierstrass convergent on $B(0,1)$, and therefore $v \in \mathcal{H}(B(0,1))$. Thus $\sigma_n \mathcal{K}(\cdot, y) - v$ is a y-axial function in $\mathcal{H}(B(0,1))$ vanishing on the line-segment $\{ty : -1 < t < 1\}$. Hence, by Lemma 2.7, we have $\sigma_n \mathcal{K}(\cdot, y) = v$.

We now suppose that $u \in \mathcal{H}(\overline{B(0,1)})$. By the Poisson integral representation and the results of the preceding paragraph,

$$u(x) = \sum_{m=0}^{\infty} \frac{d_m}{\sigma_n} \int_{S(0,1)} J_m(y,x) u(y) \, d\sigma(y) \quad (x \in B(0,1)) \tag{2.9}$$
$$= \sum_{m=0}^{\infty} H_m(x), \quad \text{say.}$$

Differentiation under the integral sign in (2.9) shows that $\Delta H_m \equiv 0$ and $D^\alpha H_m \equiv 0$ when $|\alpha| \neq m$. Thus $H_m \in \mathcal{H}_m$. Also, $|H_m(x)| \le d_m \|x\|^m \sup |u|$, so $\sum H_m$ is locally Weierstrass convergent on $B(0,1)$. A simple dilation argument extends the result to the case where $u \in \mathcal{H}(B(0,1))$. The uniqueness assertion follows easily from a standard uniqueness result for single-variable power series.

2.6 The Kelvin transform

For each point x in $\mathbb{R}^n \setminus \{0\}$, we define $x^* = x\|x\|^{-2}$. Thus $\|x\|\|x^*\| = 1$ and $(x^*)^* = x$. We write $\tilde\Omega = \{x : x^* \in \Omega\}$ and, given a function $f \colon \Omega \to \mathbb{R}$, we define

$$f^*(x) = \|x\|^{2-n} f(x^*) \quad (x \in \tilde\Omega).$$

The mapping $f \mapsto f^*$ is called the *Kelvin transform*. A calculation (see [6, pp. 20, 21]) shows that the Kelvin transform preserves harmonicity: if $u \in \mathcal{H}(\Omega)$, then $u^* \in \mathcal{H}(\tilde\Omega)$.

We note that if $H \in \mathcal{H}_m$, then by homogeneity

$$H^*(x) = \|x\|^{2-n} H(x\|x\|^{-2}) = \|x\|^{2-n-2m} H(x) \quad (x \in \mathbb{R}^n \setminus \{0\}).$$

2.9 Theorem *If $u \in \mathcal{H}(\mathbb{R}^n \setminus \overline{B(x_0, r)})$ and*

$$u(x) = \begin{cases} o(\log \|x\|) & (n=2) \\ o(1) & (n \ge 3) \end{cases} \quad (x \to \infty),$$

then there exist polynomials I_m in $\mathcal{H}_m$, uniquely determined by u, such that

$$u(x) = \sum_{m=0}^{\infty} \|x - x_0\|^{2-n-2m} I_m(x - x_0), \tag{2.10}$$

and the series is Weierstrass convergent on $\mathbb{R}^n \setminus B(x_0, R)$ for each $R > r$.

We indicate a proof in the case where $B(x_0, r) = B(0, 1)$. The function u^* is harmonic on $B(0, 1) \setminus \{0\}$ and $u^*(x) = o(U(x, 0))$ as $x \to 0$. Using the minimum principle, we can show that u^* has a harmonic continuation to $B(0, 1)$; see [6, Theorem 1.3.7]. Hence, by Theorem 2.8, there exist polynomials I_m in $\mathcal{H}_m$ such that $u^* = \sum I_m$, so

$$u(x) = ||x||^{2-n} u^*(x^*) = \sum_{m=0}^{\infty} ||x||^{2-n} I_m(x^*) = \sum_{m=0}^{\infty} ||x||^{2-n-2m} I_m(x) \quad (||x|| > 1).$$

The results about convergence and uniqueness in Theorem 2.9 follow from those in Theorem 2.8.

2.10 Corollary (Schwarz Lemma) *Let a, M, r be positive numbers and suppose that $u \in \mathcal{H}(\mathbb{R}^n \setminus B(0, r))$. If $|u| \leq M$ on $S(0, r)$ and $u(x) = O(||x||^{-a})$ as $x \to \infty$, then*

$$|u(x)| \leq C(a, n)(r/||x||)^a M \quad (||x|| > r).$$

The proof we give is a slight simplification of that in [7, pp. 323, 324] (or see [24, Lemma 5.1]); for an alternative approach, see [14, Section 2]. We may assume that $r = 1$. By the maximum principle $|u(x)| \leq M$ when $||x|| > 1$, so it suffices to prove the result with $||x|| > 2$. By Theorem 2.9,

$$u(x) = \sum_{m=0}^{\infty} ||x||^{2-n-2m} I_m(x) \quad (||x|| > 1),$$

where $I_m \in \mathcal{H}_m$. The proof of Theorem 2.9 is easily modified to show that the series is Weierstrass convergent on $\mathbb{R}^n \setminus B(0, 1)$, since $u \in \mathcal{H}(\mathbb{R}^n \setminus B(0, 1))$. Hence by the orthogonality property (2.6) and the homogeneity of each I_m,

$$\mathcal{M}(u^2; 0, t) = \sum_{m=0}^{\infty} t^{4-2n-4m} \mathcal{M}(I_m^2; 0, t) = \sum_{m=0}^{\infty} t^{4-2n-2m} \mathcal{M}(I_m^2; 0, 1) \quad (t \geq 1).$$

Since $\mathcal{M}(u^2; 0, t) = O(t^{-2a})$ as $t \to +\infty$, it follows that $\mathcal{M}(I_m^2; 0, 1) = 0$ and hence $I_m \equiv 0$ when $m + n - 2 < a$, so

$$\mathcal{M}(u^2; 0, t) \leq t^{-2a} \sum_{m=0}^{\infty} \mathcal{M}(I_m^2; 0, 1) \leq M^2 t^{-2a}.$$

Hence, writing $||x|| = \rho$ and $A = \{y : \rho/2 < ||y|| < 3\rho/2\}$, we find by using the mean value property and the Cauchy-Schwarz inequality that when $\rho > 2$,

$$(u(x))^2 \leq \mathcal{A}(u^2; x, \rho/2)$$
$$\leq \frac{2^n}{\lambda_n \rho^n} \int_A u^2 \, d\lambda$$
$$= \frac{2^n n}{\rho^n} \int_{\rho/2}^{3\rho/2} t^{n-1} \mathcal{M}(u^2; 0, t) \, dt$$
$$\leq C(a, n) \rho^{-2a} M^2.$$

2.7   The harmonic Laurent expansion

We define

$$A(x_0; r_1, r_2) = \{x \in \mathbb{R}^n : r_1 < ||x - x_0|| < r_2\} \quad (0 \le r_1 < r_2 \le +\infty).$$

2.11 Lemma *If $u \in \mathcal{H}(A(x_0; r_1, r_2))$, then there are unique functions u_1 in $\mathcal{H}(A(x_0; r_1, +\infty))$ and u_2 in $\mathcal{H}(B(x_0, r_2))$ and a real number b such that $u_1(x) = O(||x||^{1-n})$ as $x \to \infty$ and*

$$u(x) = u_1(x) + u_2(x) + bU(x, x_0) \quad (x \in A(x_0; r_1, r_2)). \tag{2.11}$$

In indicating the proof of this lemma, we suppose that $x_0 = 0$ and treat the case where $0 < r_1 < r_2 < +\infty$ and $u \in \mathcal{H}(\overline{A(0; r_1, r_2)})$. The general result can be obtained by a limiting process and translation (see [6, Lemma 2.5.2] for details). By Theorem 2.1, we have $u = v_1 + v_2$ on $A(0; r_1, r_2)$, where

$$v_j(x) = c_n \int_{S(0,r_j)} \left(U(x, y) \frac{\partial u}{\partial n_e}(y) - u(y) \frac{\partial}{\partial n_e} U(x, y) \right) d\sigma(y),$$

with $\partial/\partial n_e$ denoting differentiation in the direction of the exterior normal to $A(0; r_1, r_2)$. The integrand is continuous on $(\mathbb{R}^n \setminus S(0, r_j)) \times S(0, r_j)$ and harmonic as a function of x in $\mathbb{R}^n \setminus S(0, r_j)$ for each fixed y in $S(0, r_j)$. It follows that $v_j \in \mathcal{H}(\mathbb{R}^n \setminus S(0, r_j))$.

In the case $n = 2$, it is easy to show that

$$v_1(x) = -b \log ||x|| + O(||x||^{-1}) \quad (x \to \infty), \qquad b = -\frac{1}{2\pi} \int_{S(0,r_1)} \frac{\partial u}{\partial n_e} d\sigma,$$

so (2.11) holds with $u_1 = v_1 - bU(\cdot, 0)$ and $u_2 = v_2$.

In the case $n \ge 3$, it is clear that $v_1(x) \to 0$ as $x \to \infty$, so by Theorem 2.9, the function v_1 has a series representation of the form (2.10) with $x_0 = 0$. The term corresponding to $m = 0$ in this series has the form $bU(x, 0)$, and $v(x) - bU(x, 0) = O(||x||^{1-n})$ as $x \to \infty$. Again (2.11) holds with $u_1 = v_1 - bU(\cdot, 0)$ and $u_2 = v_2$.

If u_1', u_2' are functions having all the properties of u_1, u_2, then $u_2 - u_2'$ has a harmonic continuation to $\mathbb{R}^n$ vanishing at ∞, so $u_2 = u_2'$ and the uniqueness asserted in the lemma follows.

Putting together Lemma 2.11 and Theorems 2.8, 2.9, we immediately obtain the following result.

2.12 Theorem (Harmonic Laurent expansion) *If $u \in \mathcal{H}(A(x_0; r_1, r_2))$, then there exist unique polynomials H_m, I_m in $\mathcal{H}_m$ such that*

$$u(x) = \sum_{m=0}^{\infty} H_m(x - x_0) + \sum_{m=1}^{\infty} ||x - x_0||^{2-n-2m} I_m(x - x_0) + bU(x, x_0), \tag{2.12}$$

where $b \in \mathbb{R}$. If $r_1 < \rho_1 < \rho_2 < r_2$, then the first series is Weierstrass convergent on $B(x_0, \rho_2)$ and the second on $A(x_0; \rho_1, +\infty)$.

If u is harmonic on $B(x_0,r) \setminus \{x_0\}$ but not on $B(x_0,r)$ for some r, then we say that u has an *isolated singularity* at x_0. Such a function u has a unique expansion of the form (2.12). We call the function u', given by

$$u'(x) = \sum_{m=1}^{\infty} \|x - x_0\|^{2-n-2m} I_m(x - x_0) + bU(x, x_0) \quad (x \in \mathbb{R}^n \setminus \{x_0\}),$$

the *principal part* of u at x_0. We note that $u' \in \mathcal{H}(\mathbb{R}^n \setminus \{x_0\})$ and $u - u'$ has a harmonic continuation to $B(x_0,r)$. If $u' \equiv 0$, then we say that x_0 is a *removable singularity* of u.

3 Pole-pushing

Let u be a harmonic or holomorphic function with an isolated singularity at a point x_0. The technique of pole-pushing allows us to approximate u, outside a suitable neighbourhood of x_0 by a similar function v having a singularity near, but not at, x_0. The technique can often be used recursively to yield an approximating function free of singularities. The idea is very old: in the holomorphic context it appears in Runge's famous paper [37] of 1885. It is also very simple, requiring nothing more than convergence properties of Laurent expansions. As well as being important for the proofs of general theorems on harmonic (and holomorphic) approximation (see Sections 4 and 8), pole-pushing can be applied directly to construct functions exhibiting unexpected behaviour (see Sections 9 and 10).

3.1 Lemma *If $0 \le r_1 < r_2 < +\infty$ and $u \in \mathcal{H}(A(y; r_1, +\infty))$, then for each pair of positive numbers a and ε there exists a function v in $\mathcal{H}(\mathbb{R}^n \setminus \{y\})$ such that*

$$|(u-v)(x)| < \varepsilon(1 + \|x\|)^{-a} \quad (\|x - y\| > r_2).$$

By Theorem 2.12 there exist a function h in $\mathcal{H}(\mathbb{R}^n)$, a real number b and polynomials I_m in $\mathcal{H}_m$ such that

$$u(x) = h(x) + bU(x,y) + \sum_{m=1}^{\infty} \|x - y\|^{2-n-2m} I_m(x - y) \quad (\|x - y\| > r_1),$$

and the series is Weierstrass convergent on $\overline{A(y; r_2, +\infty)}$. Let C_1 denote the constant in Corollary 2.10 and let C_2 be such that

$$C_1 C_2 \left(\frac{r_2(1 + \|x\|)}{\|x - y\|} \right)^2 < 1 \quad (\|x - y\| \ge r_2).$$

If k is sufficiently large, then the function v, defined by

$$v(x) = h(x) + bU(x,y) + \sum_{m=1}^{k} \|x - y\|^{2-n-2m} I_m(x - y),$$

which is harmonic on $\mathbb{R}^n \setminus \{y\}$, satisfies $|u - v| < C_2\varepsilon$ on $S(y, r_2)$ and $v(x) = O(\|x - y\|^{-a})$ as $x \to \infty$. By Corollary 2.10,

$$|(u-v)(x)| \le C_1 C_2 \varepsilon (r_2/\|x - y\|)^a < \varepsilon(1 + \|x\|)^{-a} \quad (\|x - y\| > r_2).$$

We write Ω^* for the Alexandroff (one-point) compactification of Ω and denote the ideal boundary point of Ω by $\mathcal{A}$. We recall that Ω^* is topologized by declaring a subset A of Ω^* to be open if either A is an open subset of Ω or $\Omega^* \setminus A$ is a compact subset of Ω. In the case where $\Omega = \mathbb{R}^n$, we continue to write ∞ for $\mathcal{A}$. If $y \in \Omega$ and $f \colon [0, +\infty) \to \Omega$ is a continuous function such that $f(0) = y$ and $f(t) \to \mathcal{A}$ as $t \to +\infty$ (that is to say, for every compact subset K of Ω there exists a number t_0 such that $f(t) \in \Omega \setminus K$ when $t > t_0$), then we call the image $f([0, +\infty))$ a *path connecting y to $\mathcal{A}$*. We say that an open subset T of Ω is a *tract connecting y to $\mathcal{A}$* if T contains some path connecting y to $\mathcal{A}$.

3.2 Lemma *If T is a tract connecting a point y in Ω to $\mathcal{A}$ and $u \in \mathcal{H}(\Omega \setminus \{y\})$, then for each pair of positive numbers a and ε there exists a function v in $\mathcal{H}(\Omega)$ such that*

$$|(u - v)(x)| < \varepsilon(1 + ||x||)^{-a} \quad (x \in \Omega \setminus T).$$

To prove this, let $u = u_1 + u_2$, where $u_1 \in \mathcal{H}(\Omega)$ and is u_2 is the principal part of u at y. Let $(B_k) = (B(y_k, r_k))$ be a sequence of open balls such that

$$y \in B_1, \qquad y_k \in B_{k+1}, \qquad B_k \subset T \quad (k \in \mathbb{N})$$

and $B_k \to \mathcal{A}$ (that is, each compact subset of Ω meets only finitely many B_k). We define $y_0 = y$ and $v_0 = u_2$. Applying Lemma 3.1 repeatedly, we find that for each non-negative integer k there exists a function v_k in $\mathcal{H}(\mathbb{R}^n \setminus \{y_k\})$ such that

$$|(v_{k+1} - v_k)(x)| < \varepsilon 2^{-k-1}(1 + ||x||)^{-a} \quad (x \in \mathbb{R}^n \setminus B_{k+1}).$$

It follows that (v_k) converges locally uniformly on Ω to a harmonic limit function v'. Defining $v = u_1 + v'$ on Ω, we see that $v \in \mathcal{H}(\Omega)$ and

$$|(u - v)(x)| \le \sum_{k=0}^{\infty} |(v_{k+1} - v_k)(x)| < \varepsilon(1 + ||x||)^{-a} \quad (x \in \Omega \setminus T).$$

The following example, an easy consequence of Lemma 3.2, shows that a harmonic function on $\mathbb{R}^n$ can be small on a very large set. It is the basis for constructions given in Sections 8 and 9.

3.3 Example Let T be a tract in $\mathbb{R}^n$ joining 0 to ∞. For each positive number a there exists a function v in $\mathcal{H}(\mathbb{R}^n)$ such that $v \ne 0$ and

$$|v(x)| < (1 + ||x||)^{-a} \quad (x \in \mathbb{R}^n \setminus T).$$

To show this, we choose a non-null element H of $\mathcal{H}_m$, where $m + n - 2 \ge a$, and define $u(x) = ||x||^{2-n-2m} H(x)$. Then $u \in \mathcal{H}(\mathbb{R}^n \setminus \{0\})$ and, multiplying by a suitable constant if necessary, we can arrange that $|u(x)| < 1/2(1 + ||x||)^{-a}$ for all x in $\mathbb{R}^n \setminus T$. Let x_0 be a point in $\mathbb{R}^n \setminus T$ such that $u(x_0) \ne 0$, and let $\varepsilon = \min\{1/2, |u(x_0)|\}$. By Lemma 3.2 there exists a function v in $\mathcal{H}(\mathbb{R}^n)$ such that $|(u - v)(x)| < \varepsilon(1 + ||x||)^{-a}$ for each x in $\mathbb{R}^n \setminus T$. Hence $|v(x)| < (1 + ||x||)^{-a}$ for each such x and $v(x_0) \ne 0$.

4   Uniform harmonic approximation on compact sets

If E is a non-empty subset of $\mathbb{R}^n$, we write $\mathcal{S}(E)$ for the space of functions that are harmonic on some open set containing E except possibly for isolated singularities.

4.1 Theorem *Let K be a compact subset of $\mathbb{R}^n$. If $u \in \mathcal{H}(K)$ and $\varepsilon > 0$, then there exists a function v in $\mathcal{H}(K) \cap \mathcal{S}(\mathbb{R}^n)$ such that $|u - v| < \varepsilon$ on K.*

To prove this, let W be an admissible open set such that $K \subset W$ and $u \in \mathcal{H}(\overline{W})$. By Green's representation,

$$u(x) = \int_{\partial W} F(x, y)\, d\sigma(y) \quad (x \in W), \tag{4.1}$$

where

$$F(x, y) = c_n \left(U(x, y) \frac{\partial u}{\partial n_e}(y) - u(y) \frac{\partial}{\partial n_e} U(x, y) \right) \quad (x \in \mathbb{R}^n \setminus \partial W, y \in \partial W).$$

Since F is continuous on the compact set $K \times \partial W$, we can approximate the integral in (4.1) uniformly on K by a Riemann sum. Thus there exist points $y_1, \ldots, y_m$ in ∂W and numbers $a_1, \ldots, a_m$ such that

$$\left| u(x) - \sum_{k=1}^{m} a_k F(x, y_k) \right| < \varepsilon \quad (x \in K).$$

Since $F(\cdot, y_k) \in \mathcal{H}(\mathbb{R}^n \setminus \{y_k\})$ and $y_k \in \mathbb{R}^n \setminus K$, we can take $v = \sum_{k=1}^{m} a_k F(\cdot, y_k)$.

4.2 Theorem *Let K be a compact subset of Ω such that every bounded component of $\mathbb{R}^n \setminus K$ meets $\mathbb{R}^n \setminus \Omega$. If $u \in \mathcal{H}(K)$ and $\varepsilon > 0$, then there exists a function v in $\mathcal{H}(\Omega)$ such that $|u - v| < \varepsilon$ on K.*

The proof of Theorem 4.1 shows that there are points $y_1, \ldots, y_m$ in $\mathbb{R}^n \setminus K$ and functions $u_1, \ldots, u_m$ such that $u_k \in \mathcal{H}(\mathbb{R}^n \setminus \{y_k\})$ and

$$\left| u(x) - \sum_{k=1}^{m} u_k(x) \right| < \frac{1}{2}\varepsilon \quad (x \in K).$$

By the topological hypothesis of Theorem 4.2, for each k there is a tract T_k in $\Omega \setminus K$ connecting y_k to $\mathcal{A}$. Hence by Lemma 3.2, there exists a function v_k in $\mathcal{H}(\Omega)$ such that $|u_k - v_k| < 2^{-k-1}\varepsilon$ on $\Omega \setminus T_k$. Defining $v = \sum_{k=1}^{m} v_k$, we see that $v \in \mathcal{H}(\Omega)$ and

$$|u - v| \leq \left| u - \sum_{k=1}^{m} u_k \right| + \sum_{k=1}^{m} |u_k - v_k| < \varepsilon \quad \text{on } K.$$

In the case where $\Omega = \mathbb{R}^n$, the topological hypothesis in Theorem 4.2 is satisfied if (and only if) $\mathbb{R}^n \setminus K$ is connected. If this condition holds, then a function u in $\mathcal{H}(K)$ can be uniformly approximated on K by functions in $\mathcal{H}(\mathbb{R}^n)$. Since, also, a function in $\mathcal{H}(\mathbb{R}^n)$ is uniformly approximated on K by the partial sums of its polynomial expansion, we obtain the following corollary of Theorem 4.2.

4.3 Corollary (Walsh's theorem [41]) *Let K be a compact subset of $\mathbb{R}^n$ such that $\mathbb{R}^n \setminus K$ is connected. If $u \in \mathcal{H}(K)$ and $\varepsilon > 0$, then there exists a harmonic polynomial H such that $|u - H| < \varepsilon$ on K.*

Walsh's theorem may be compared with Runge's classical theorem on holomorphic approximation ([37], or see [19, p. 201, Exercise 4]). Runge's theorem says that if K is a compact subset of $\mathbb{C}$, then the connectedness of $\mathbb{C} \setminus K$ is necessary and sufficient for every function holomorphic on a neighbourhood of K to be uniformly approximable on K by (holomorphic) polynomials. Walsh's theorem merely asserts that the connectedness of $\mathbb{R}^n \setminus K$ is sufficient for the conclusion of the theorem. The question arises as to whether the topological hypothesis in Walsh's theorem, or in Theorem 4.2, is necessary. It is not, as the following example shows.

4.4 Example Let $K = S(0, 1)$. If $u \in \mathcal{H}(K)$ (or even if $u \in C(K)$) and $\varepsilon > 0$, then there exists a harmonic polynomial H such that $|H - u| < \varepsilon$ on K.

To see this, we note that by the Stone-Weierstrass theorem, there is some polynomial P on $\mathbb{R}^n$ such that $|u - P| < \varepsilon$ on K. Also, there is a harmonic polynomial H such that $H = P$ on K (see [6, Corollary 2.1.3]).

On the other hand, the connectedness of $\mathbb{R}^n \setminus K$ in Walsh's theorem cannot be dispensed with entirely.

4.5 Example Let $K = \overline{A(0; 1, 2)}$. If $u = U(\cdot, 0)$ and ε is a sufficiently small positive number, then there is no function v in $\mathcal{H}(\mathbb{R}^n)$ such that $|u - v| < \varepsilon$ on K.

If there were such a function v, then we would have

$$|v(0) - \mathcal{M}(u; 0, t)| = |\mathcal{M}(v - u; 0, t)| < \varepsilon \quad (1 \leq t \leq 2),$$

which is impossible if ε is small.

The refinement of Theorem 4.1 corresponding to Runge's theorem involves the potential theoretical concept of thinness which is beyond the scope of these notes. We simply state the result, in which $\hat{K}$ denotes the *hull* of K, that is, the union of K with all the bounded components of $\mathbb{R}^n \setminus K$.

4.6 Theorem *Let K be a compact subset of Ω. The following are equivalent:*

(i) *for each u in $\mathcal{H}(K)$ and each positive number ε there exists v in $\mathcal{H}(\Omega)$ such that $|u - v| < \varepsilon$ on K;*

(i) *$\Omega \setminus \hat{K}$ and $\Omega \setminus K$ are thin at the same points of K.*

In the case $n = 2$, condition (ii) reduces to the simple topological condition that $\partial \hat{K} = \partial K$. For the concept of thinness, we refer to [6, Section 7.2]. Theorem 4.6 is a special case of a theorem of Gardiner [23] (see also [24, Theorem 1.10]). It can also be deduced from results of Bliedtner and Hansen [16], which were proved in an abstract setting.

5 Uniform approximation on non-compact sets by harmonic functions with isolated singularities

The main technique used for passing from approximation on compact sets to approximation on non-compact sets is fusion. It was introduced in the holomorphic context by Roth [36]. Inspired by Roth's work, Gauthier, Goldstein and Ow ([26] for $n = 2$; [27] for $n \geq 3$; see also the lecture notes [28]) proved fusion results for harmonic functions. In this section we give a fusion result and then use it to prove a theorem concerning approximation on sets that are not necessarily compact. The approximating functions will be harmonic, except possibly for isolated singularities. We will show in Section 8 how, under suitable topological conditions, pole-pushing can be used to produce approximants free of singularities. Here it is convenient to confine our attention to the case where $n \geq 3$. The following fusion theorem is a slight strengthening of a result given in the lecture notes of Gauthier and Hengartner [28, Lemma 2.2.9].

5.1 Theorem (Fusion—preliminary form) *Suppose that $n \geq 3$. Let K, E_1 be compact sets with $K \neq \emptyset$ and let E_2 be a closed set such that $E_1 \cap E_2 = \emptyset$. There is a constant C with the following property: if $u_1, u_2 \in S(\mathbb{R}^n) \cap \mathcal{H}(K)$ and $|u_1 - u_2| < \varepsilon$ on K, then there exists u in $S(\mathbb{R}^n) \cap \mathcal{H}(K)$ such that*

$$|(u - u_k)(x)| < C\varepsilon(1 + ||x||)^{2-n} \qquad (x \in K \cup E_k; \ k = 1, 2). \qquad (5.1)$$

Because of the factor $(1 + ||x||)^{2-n}$ in (5.1), this theorem is somewhat stronger than the original version [27] which had no such factor. Only minor modifications to the original proof are required in order to obtain this strengthening. This slightly improved fusion result enables us to give a correspondingly improved approximation theorem (Theorem 5.2 below). We note that Theorem 5.1 is false with $n = 2$. (Take, for example,

$$E_1 = \overline{B(0,1)}, \qquad E_2 = \overline{A(0;2,+\infty)}, \qquad K = \overline{A(0;1,2)}, \qquad u_1 = 0, \qquad u_2(x) = \log ||x||.)$$

However, as we shall see in Theorem 6.4, with the mild additional assumption that $K \cup E_1 \cup E_2 \neq \mathbb{R}^n$, a much stronger fusion result is valid for all dimensions. Inequalities such as (5.1) are to be understood to mean that $u - u_k$ has, at worst, removable singularities in $K \cup E_k$ and that the inequality holds when $u - u_k$ is continuously extended to such points.

In proving Theorem 5.1, we may suppose that $u_2 = 0$. For if the result is established in this case, then in the general case there will be a function v in $S(\Omega)$ such that

$$|(v - (u_1 - u_2))(x)| < C\varepsilon(1 + ||x||)^{2-n} \quad (x \in K \cup E_1)$$

and

$$|v(x)| < C\varepsilon(1 + ||x||)^{2-n} \quad (x \in K \cup E_2),$$

and (5.1) will be satisfied with $u = v + u_2$. Thus we need to show that there exists u in $S(\mathbb{R}^n)$ such that

$$|(u - u_1)(x)| < C\varepsilon(1 + ||x||)^{2-n} \quad (x \in K \cup E_1) \qquad (5.2)$$

and

$$|u(x)| < C\varepsilon(1 + ||x||)^{2-n} \quad (x \in K \cup E_2). \qquad (5.3)$$

Let W_1, W_2 be bounded admissible open sets such that

$$E_1 \subset W_1, \qquad \overline{W}_1 \subset W_2, \qquad \overline{W}_2 \subset \mathbb{R}^n \setminus E_2.$$

We may assume that u_1 has no singularities in $\partial W_1 \cup \partial W_2$. To see this, suppose that u_1 has such singularities $x_1, \ldots, x_m$ and let p_j be the principal part of u_1 at x_j. By Lemma 3.1, there exist points $y_1, \ldots, y_m$ in $\mathbb{R}^n \setminus (K \cup \partial W_1 \cup \partial W_2)$ and functions q_j in $\mathcal{H}(\mathbb{R}^n \setminus \{y_j\})$ such that

$$|(p_j - q_j)(x)| < \frac{\varepsilon}{m}(1 + \|x\|)^{2-n} \quad (x \in K \cup E_1;\ j = 1, \ldots, m).$$

Let $u' = u_1 - \sum_{j=1}^m (p_j - q_j)$. Then $u' \in \mathcal{S}(\mathbb{R}^n) \cap \mathcal{H}(K)$ and u' has no singularities in $\partial W_1 \cup \partial W_2$. Also,

$$|(u_1 - u')(x)| < \varepsilon(1 + \|x\|)^{2-n} \quad (x \in K \cup E_1).$$

If we can find a function u in $\mathcal{S}(\mathbb{R}^n) \cap \mathcal{H}(K)$ such that (5.2) and (5.3) hold with u' in place of u_1 and with some constant C_1, then the result will hold in full generality with $C = C_1 + 1$.

Let r be such that $K \subset B(0, r)$. Since $|u_1| < \varepsilon$ on K, there is an open set V such that $K \subset V \subseteq B(0, r)$ and $|u_1| < \varepsilon$ on $\overline{V}$. We can also arrange that $u_1 \in \mathcal{H}(\overline{V})$ and that $V \cup W_1$ is admissible. (See Figure 1.)

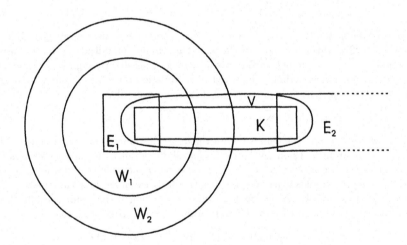

Figure 1

Let ϕ be a function in $C^\infty(\mathbb{R}^n)$ such that $0 \le \phi \le 1$ on $\mathbb{R}^n$ with $\phi = 1$ on W_1 and $\operatorname{supp} \phi \subset W_2$. We define

$$\phi = \begin{cases} \phi u_1 & \text{on } W_2 \\ 0 & \text{on } \mathbb{R}^n \setminus W_2. \end{cases}$$

Thus $\phi = u_1$ on $E_1, \phi = 0$ on E_2 and $|\phi - u_1| \le |u_1|$ on K. Since $|u_1| < \varepsilon$ on $\overline{V}$, we find that

$$|(\phi - u_1)(x)| \le C_2\varepsilon(1 + ||x||)^{2-n} \quad (x \in K \cup E_1) \tag{5.4}$$

$$|\phi(x)| \le C_2\varepsilon(1 + ||x||)^{2-n} \quad (x \in K \cup E_2), \tag{5.5}$$

where $C_2 = (1 + r)^{n-2}$.

Let S denote the set of singularities of u_1 in W_1 and let $S_\delta = \{x : \text{dist}(x, S) \le \delta\}$. We assume that δ is small enough to ensure that S_δ is a disjoint union of balls contained in W_1. Let $W = W_1 \cup V \cup (\mathbb{R}^n \setminus \overline{W}_2)$ and let $x \in W \setminus S$. If δ is such that $x \in W \setminus S_\delta$ and R is such that $\{x\} \cup \overline{W}_2 \subset B(0, R)$, then applying Theorem 2.1 on $(W \cap B(0, R)) \setminus S_\delta$, we find that

$$\psi(x) = w_1(x) + s_2(x) - c_n \int_V U(x, y)\Delta\psi(y)\, d\lambda(y), \tag{5.6}$$

where

$$w_1(x) = c_n \int_{\partial W} \left(\psi(y)\frac{\partial}{\partial n_e}U(x, y) - U(x, y)\frac{\partial\psi}{\partial n_e}(y) \right) d\sigma(y)$$

and

$$s_2(x) = c_n \int_{\partial S_\delta} \left(\psi(y)\frac{\partial}{\partial n_e}U(x, y) - U(x, y)\frac{\partial\psi}{\partial n_e}(y) \right) d\sigma(y);$$

here $\partial/\partial n_e$ denote differentiation in the direction of the exterior normal to $W \setminus S_\delta$. We define

$$w_3(x) = -2c_n \int_{\partial W} u_1(y)U(x, y)\frac{\partial\phi}{\partial n_e}(y)\, d\sigma(y). \tag{5.7}$$

By Green's formula,

$$\begin{aligned} w_3(x) = -2c_n \int_V &(u_1(y)\langle \nabla_2 U(x, y), \nabla\phi(y)\rangle \\ &+ U(x, y)\langle \nabla u_1(y), \nabla\phi(y)\rangle + u_1(y)U(x, y)\Delta\phi(y))\, d\lambda(y). \end{aligned} \tag{5.8}$$

The integral in (5.6) is equal to

$$\int_V U(x, y)\left(u_1(y)\Delta\phi(y) + 2\langle \nabla u_1(y), \nabla\phi(y)\rangle\right) d\lambda(y). \tag{5.9}$$

From (5.6), (5.8) and (5.9) we obtain

$$\psi(x) = w_1(x) + s_2(x) + w_3(x) + I(x), \tag{5.10}$$

where

$$I(x) = c_n \int_V u_1(y)\left(U(x, y)\Delta\phi(y) + 2\langle \nabla_2 U(x, y), \nabla\phi(y)\rangle\right) d\lambda(y).$$

There is a constant C_3 such that $c_n|\Delta\phi| < C_3$ and $2c_n||\nabla\phi|| < C_3$ on $\mathbb{R}^n$. Hence

$$|I(x)| < C_3\varepsilon \int_V (U(x, y) + ||\nabla_2 U(x, y)||)\, d\lambda(y)$$

$$= C_3\varepsilon \int_V (||x - y||^{2-n} + (n - 2)||x - y||^{1-n})\, d\lambda(y). \tag{5.11}$$

If $||x|| \leq 2r$, then $V \subset B(x, 3r)$ and the integral in (5.11) is less than some constant $C_4 = C_4(n, r)$. If $||x|| > 2r$, then $||x - y|| > C_5(r)(1 + ||x||)$ for all y in V. Hence, by (5.10) and (5.11), in both cases we have

$$|\psi(x) - (w_1(x) + s_2(x) + w_3(x))| < C_6 \varepsilon (1 + ||x||)^{2-n}. \tag{5.12}$$

An application of Green's formula to a set of the form $(S_{\delta_2} \setminus S_{\delta_1})^\circ$ shows that s_2 is independent of δ and therefore belongs to $\mathcal{H}(\mathbb{R}^n \setminus S)$. We now wish to approximate $w_1 + w_3$ by an element of $\mathcal{S}(\mathbb{R}^n)$. We have

$$(w_1 + w_3)(x) = \int_{\partial W} F(x, y) d\sigma(y), \tag{5.13}$$

where

$$F(x, y) = c_n \left(\psi(y) \frac{\partial}{\partial n_e} U(x, y) - U(x, y) \frac{\partial \psi}{\partial n_e}(y) - 2u_1(y)U(x, y)\frac{\partial \phi}{\partial n_e}(y) \right). \tag{5.14}$$

The function F is continuous on $(\mathbb{R}^n \setminus \partial W) \times \partial W$; and for each fixed y in ∂W, the function $F(\cdot, y)$ is harmonic on $\mathbb{R}^n \setminus W$. Hence $w_1 + w_3 \in \mathcal{H}(\mathbb{R}^n \setminus \partial W)$. Let ω be a bounded open set such that $\partial W \cap \operatorname{supp} \psi \subset \omega$ and $\omega \cap (K \cup E_1 \cup E_2) = \emptyset$; see Figure 2. On the compact set $\partial \omega$ the integral in (5.13) can be uniformly approximated by a Riemann sum s_1:

$$|(w_1 + w_3 - s_1)(x)| < \varepsilon \quad (x \in \partial \omega),$$

where $s_1 = \sum_{j=1}^m a_j F(\cdot, y_j)$ for some numbers $a_1, \ldots, a_m$ and points $y_1, \ldots, y_m$ in $\partial W \cap \operatorname{supp} \psi$. We note that $F(x, y) = O(||x||^{2-n})$ as $x \to \infty$, uniformly for y in $\partial W \cap \operatorname{supp} \psi$, so $(w_1 + w_3 - s_1)(x) = O(||x||^{2-n})$ as $x \to \infty$. Hence by the maximum principle, $|w_1 + w_3 - s_1| < \varepsilon$ on $\mathbb{R}^n \setminus \overline{\omega}$, and by Corollary 2.10,

$$|(w_1 + w_3 - s_1)(x)| < C_7 \varepsilon (1 + ||x||)^{2-n} \quad (x \in \mathbb{R}^n \setminus \overline{\omega}). \tag{5.15}$$

We note that $s_1 + s_2 \in \mathcal{S}(\mathbb{R}^n) \cap \mathcal{H}(K)$. Writing $u = s_1 + s_2$ and collecting together (5.4), (5.5), (5.12) and (5.15), we find that (5.2) and (5.3) hold.

The uniform version of the following theorem (that is, the version with the factor $(1 + ||x||)^{2-n}$ absent from (5.16)) is due to Gauthier, Goldstein and Ow [27].

5.2 Theorem *Suppose that* $n \geq 3$. *let* E *be a relatively closed proper subset of* Ω. *If* $u \in \mathcal{H}(E)$ *and* $\varepsilon > 0$, *then there exists* v *in* $\mathcal{S}(\Omega) \cap \mathcal{H}(E)$ *such that*

$$|(u - v)(x)| < \varepsilon (1 + ||x||)^{2-n} \quad (x \in E). \tag{5.16}$$

To prove this, we start with a sequence of bounded open sets (Ω_m) such that $\overline{\Omega}_m \subset \Omega_{m+1}$ $(m \in \mathbb{N})$, $\bigcup_{m=1}^\infty \Omega_m = \Omega$ and $\Omega_1 \cap E \neq \emptyset$. We define $F_m = \overline{\Omega}_m \cap E$. Let C_m be the constant of Theorem 5.1 corresponding to the assignments $E_1 = \overline{\Omega}_m$, $E_2 = \overline{E} \setminus \Omega_{m+1}$ and $K = F_{m+1}$. We may assume that $C_m > 1$. Let (δ_m) be a sequence such that

$$0 < \delta_m < \varepsilon 2^{-m-2} C_m^{-1} \quad (m \in \mathbb{N}). \tag{5.17}$$

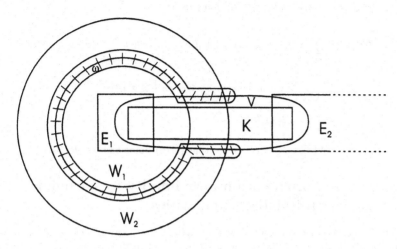

Figure 2

Since $u \in \mathcal{H}(F_{m+1})$, it follows from Theorem 4.1 that there exists w_m in $\mathcal{S}(\mathbb{R}^n)$ such that

$$|(u - w_m)(x)| < \delta_m(1 + ||x||)^{2-n} \quad (x \in F_{m+1}) \tag{5.18}$$

and hence

$$|(w_{m+1} - w_m)(x)| < 2\delta_m(1 + ||x||)^{2-n} \quad (x \in F_{m+1}). \tag{5.19}$$

By Theorem 5.1 and (5.19), there exists v_m in $\mathcal{S}(\mathbb{R}^n)$ such that

$$|(v_m - w_m)(x)| < 2\delta_m C_m(1 + ||x||)^{2-n} \quad (x \in \overline{\Omega}_m \cup F_{m+1}) \tag{5.20}$$

and

$$|(v_m - w_{m+1})(x)| < 2\delta_m C_m(1 + ||x||)^{2-n} \quad (x \in E); \tag{5.21}$$

for (5.21) note that $E \subseteq (\overline{E} \setminus \Omega_{m+1}) \cup F_{m+1}$. We define

$$v = w_1 + \sum_{m=1}^{\infty} (v_m - w_m). \tag{5.22}$$

To see that $v \in \mathcal{S}(\Omega)$, let k be a positive integer and note that by (5.20) and (5.17),

$$|(v_m - w_m)(x)| < 2^{-m-1}\varepsilon \quad (x \in \Omega_k; m \geq k),$$

so $\sum_{m=k}^{\infty}(v_m - w_m)$ is a uniformly convergent series of harmonic functions on Ω_k. Hence $v \in \mathcal{S}(\Omega_k)$ and so, since k is arbitrary, $v \in \mathcal{S}(\Omega)$.

If $x \in F_k$, then by (5.21), (5.18), (5.20) and (5.17),

$$|(u-v)(x)| \le \sum_{m=1}^{k-1} |(v_m - w_{m+1})(x)| + |(w_m - u)(x)| + \sum_{m=k}^{\infty} |(v_m - w_m)(x)|$$

$$\le \left(\sum_{m=1}^{k-1} 2\delta_m C_m + \delta_k + \sum_{m=k}^{\infty} 2\delta_m C_m \right) (1 + ||x||)^{2-n}$$

$$< \frac{5}{8}\varepsilon(1 + ||x||)^{2-n}.$$

Since k is arbitrary and $\bigcup_{k=1}^{\infty} F_k = E$, we see that (5.16) holds and $v \in \mathcal{H}(E)$.

6 Tangential approximation near infinity by harmonic functions with isolated singularities

The proof of Theorem 5.1 concerning fusion exploits the decay of the Newtonian kernel U: the bound $C\varepsilon(1 + ||x||)^{2-n}$ for the error $|(u - u_k)(x)|$ in (5.1) arises from the fact that $U(x,y) = O(||x||^{2-n})$ uniformly as $x \to \infty$ when y lies in a prescribed compact set and $n \ge 3$. This suggests that Theorem 5.1, and hence Theorem 5.2, could be strengthened if a more rapidly decaying kernel could be found to replace U. To give a simple illustration, we define

$$U_0(x,y) = U(x,y) - U(x,0).$$

Then $U_0(x,y) = O(||x||^{1-n})$ uniformly as $x \to \infty$ when y lies in a prescribed compact set, and this holds for any $n \ge 2$. Like U, the kernel U_0 is harmonic as a function of each argument separately, except for the obvious singularities. The proof of Theorem 5.1 can be reworked with U_0 in place of U throughout to show that its conclusion holds with $(1 + ||x||)^{1-n}$ in place of $(1 + ||x||)^{2-n}$, provided that $K \cup E_1 \cup E_2 \ne \mathbb{R}^n$. (With this proviso, we may suppose without loss of generality that $0 \notin K \cup E_1 \cup E_2$ and thus circumvent difficulties which would otherwise arise because of the singularity of $U_0(\cdot,y)$ at 0.) Then, with only minor modifications to its proof, Theorem 5.2 can be correspondingly improved and extended to cover the case $n = 2$. To obtain further improvements, we require more elaborately modified kernels. The kernel U_p, which is introduced below, is essentially the same as that used in the proof of the subharmonic version of the Weierstrass representation theorem given in Chapter 4 of Hayman and Kennedy's book [32]. The presentation we give here is based on [7] (or see [24, Chapter 5]).

Recalling the x-axial homogeneous harmonic polynomials $J_m(x, \cdot)$ introduced in Section 2, we start by defining

$$P_m(x,y) = ||x||^{2-n-m} J_m(x,y) \quad (x, y \in \mathbb{R}^n; \ x \ne 0; \ m \in \mathbb{N}).$$

We also define $P_0(x,y) = U(x,0)$; thus $P_0(x,y) = ||x||^{2-n} J_0(x,y)$ when $n \ge 3$ but not when $n = 2$. The properties of P_m that we shall need are summarized in the following lemma.

6.1 Lemma   (i) $P_m(x, \cdot) \in \mathcal{H}_m \quad (x \in \mathbb{R}^n \setminus \{0\}; \ m \ge 0)$;

(ii) $P_m(\cdot, y) \in \mathcal{H}(\mathbb{R}^n \setminus \{0\}) \quad (y \in \mathbb{R}^n; \ m \ge 0)$;

(iii) $|P_m(x,y)| \le ||x||^{2-n-m}||y||^m$ $(x, y \in \mathbb{R}^n;\ x \ne 0; m+n \ge 3)$.

Part (i) is immediate since $J_m(x, \cdot) \in \mathcal{H}_m$, and part (iii) is merely a restatement of (2.8). For (ii), we note that by the definition (2.7) of J_m and the homogeneity of $J_m(y, \cdot)$,

$$P_m(x, y) = ||y||^m ||x||^{2-n-2m} J_m(y, x) = ||y||^m ||x||^{2-n} J_m(y, x^*),$$

where $x^* = x||x||^{-2}$. Hence $P_m(\cdot, y) \in \mathcal{H}(\mathbb{R}^n \setminus \{0\})$, since the Kelvin transform preserves harmonicity.

6.2 Lemma

$$U(x, y) = \sum_{m=0}^{\infty} b_m P_m(x, y) \quad (||y|| < ||x||),$$

where $b_0 = 1$ and

$$b_m = -1/m \quad (m \ge 1;\ n = 2), \qquad b_m = \binom{m+n-3}{m} \quad (m \ge 1;\ n \ge 3).$$

To prove this, we fix x in $\mathbb{R}^n \setminus \{0\}$ and define

$$v(y) = \sum_{m=0}^{\infty} b_m P_m(x, y) \quad (||y|| < ||x||).$$

The series is locally Weierstrass convergent on $B(0, ||x||)$ in view of Lemma 6.1 (iii) and the fact that $b_m = O(m^{n-3})$ as $m \to \infty$. Hence v is harmonic on $B(0, ||x||)$. Since $U(x, \cdot)$ and v are x-axial, it will follow from Lemma 2.7 that $U(x, \cdot) = v$ if we show that $U(x, tx) = v(tx)$ when $-1 < t < 1$. For such t, we have

$$v(tx) = U(x, 0) + \sum_{m=1}^{\infty} b_m ||x||^{2-n} t^m$$
$$= \begin{cases} -\log((1-t)||x||) & (n = 2) \\ ||x - tx||^{2-n} & (n \ge 3) \end{cases}$$
$$= U(x, tx),$$

as required.

For each non-negative integer p, we define

$$Q_p(x, y) = \sum_{m=0}^{p} b_m P_m(x, y) \quad (x, y \in \mathbb{R}^n;\ x \ne 0)$$

and

$$U_p(x, y) = U(x, y) - Q_p(x, y).$$

6.3 Lemma   (i)

$$Q_p \in C^{\infty}((\mathbb{R}^n \setminus \{0\}) \times \mathbb{R}^n);$$
$$U_p \in C^{\infty}(\{(x, y) \in \mathbb{R}^n \times \mathbb{R}^n : x \notin \{0, y\}\}).$$

(ii) *For x fixed in $\mathbb{R}^n \setminus \{0\}$ and y fixed in $\mathbb{R}^n$,*

$$Q_p(x, \cdot) \in \mathcal{H}(\mathbb{R}^n), \qquad\qquad Q_p(\cdot, y) \in \mathcal{H}(\mathbb{R}^n \setminus \{0\}),$$
$$U_p(x, \cdot) \in \mathcal{H}(\mathbb{R}^n \setminus \{x\}), \qquad U_p(\cdot, y) \in \mathcal{H}(\mathbb{R}^n \setminus \{0, y\}).$$

(iii)

$$|U_p(x, y)| \le C\|x\|^{-n-p+1}\|y\|^{p+1} \quad (\|y\| < \|x\|/2);$$
$$\|\nabla_2 U_p(x, y)\| \le C\|x\|^{-n-p+1}\|y\|^p \quad (\|y\| < \|x\|/4),$$

where $C = C(n, p)$.

The properties in (i) and (ii) follow immediately from the corresponding properties of U and P_m. For the first inequality in (iii), we note that if $\|y\| < \|x\|/2$, then by Lemmas 6.2 and 6.1 (iii),

$$
\begin{aligned}
|U_p(x, y)| &\le \sum_{m=p+1}^{\infty} |b_m| |P_m(x, y)| \\
&\le \|x\|^{-n-p+1}\|y\|^{p+1} \sum_{m=p+1}^{\infty} |b_m| (\|y\|/\|x\|)^{m-p-1} \\
&\le C_1 \|x\|^{-n-p+1}\|y\|^{p+1},
\end{aligned}
$$

where $C_1 = 2^{p+1} \sum_{m=p+1}^{\infty} 2^{-m} |b_m|$. Since $U_p(x, \cdot) \in \mathcal{H}(\mathbb{R}^n \setminus \{x\})$, we can apply Corollary 2.5 (i) to obtain

$$\|\nabla_2 U_p(x, y)\| \le 8n\|y\|^{-1} \sup_{B(y, \|y\|/4)} |U_p(x, \cdot)| \le 8nC_1\|x\|^{-n-p-1}\|y\|^p$$

when $\|y\| < \|x\|/4$.

6.4 Theorem (Fusion—first modification) *Let a, ε be positive numbers, let K, E_1 be compact sets with $K \ne \emptyset$, and let E_2 be a closed set such that $E_1 \cap E_2 = \emptyset$ and $K \cup E_1 \cup E_2 \ne \mathbb{R}^n$. There is a constant C with the following property: if $u_1, u_2 \in \mathcal{S}(\mathbb{R}^n) \cap \mathcal{H}(K)$ and $|u_1 - u_2| < \varepsilon$ on K, then there exists u in $\mathcal{S}(\mathbb{R}^n) \cap \mathcal{H}(K)$ such that*

$$|(u - u_k)(x)| < C\varepsilon(1 + \|x\|)^{-a} \quad (x \in K \cup E_k; \ k = 1, 2).$$

The proof of Theorem 6.4 is modelled closely on that of Theorem 5.1. We indicate the changes that are needed. The hypothesis that $K \cup E_1 \cup E_2 \ne \mathbb{R}^n$ allows us to suppose that there is a positive number ρ such that $\overline{B(0, \rho)} \cap (K \cup E_1 \cup E_2) = \emptyset$. As before, we may suppose that $u_2 = 0$. The number r and the sets W_1, W_2, V are as in the proof of Theorem 5.1, and we now arrange also that $V \cap \overline{B(0, \rho)} = \emptyset$. The pole-pushing argument in the second paragraph of the proof of Theorem 5.1 is easily modified to show that we may again suppose that u has no singularities in $\partial W_1 \cup \partial W_2$; we simply replace the exponent $2 - n$ by $-a$ throughout the argument. With ϕ and ψ defined as before it is easy to see that (5.4) and (5.5) hold with the exponent $-a$ in place of $2 - n$ (and a revised value for C_2). We choose a non-negative integer p such that $p + n - 1 \ge a$ and replace U by U_p in the definitions of w_1, s_2, w_3 and I. It is easy to see that with these changes (5.6), (5.8), (5.9), and hence (5.10) remain valid, provided that we also replace U by U_p in each of the integrals occurring in these formulae. Our estimate for I is now

$$|I(x)| \leq C_3 \varepsilon \int_V (|U_p(x,y)| + ||\nabla_2 U_p(x,y)||)\, d\lambda(y).$$

Since $V \subset B(0,r)$, it follows from Lemma 6.3 (iii) that $|I(x)| \leq C_8 \varepsilon ||x||^{-n-p+1}$ when $||x|| > 4r$. Also, if $\rho \leq ||x|| \leq 4r$, then, since Q_p and $||\nabla_2 Q_p||$ are bounded on $A(0;\rho,4r) \times B(0,r)$, we have

$$|I(x)| \leq C_9 \varepsilon \left(1 + \int_{B(0,r)} (|U(x,y)| + ||\nabla_2 U(x,y)||)\, d\lambda(y) \right) \leq C_{10} \varepsilon.$$

Hence

$$|\psi(x) - (w_1(x) + s_2(x) + w_3(x))| = |I(x)| \leq C_{11} \varepsilon (1 + ||x||)^{-a} \quad (||x|| > \rho).$$

Again $s_2 \in \mathcal{S}(\mathbb{R}^n) \cap \mathcal{H}(K)$ and we wish to approximate $w_1 + w_3$ by an element of $\mathcal{S}(\mathbb{R}^n)$. If, in the definition (5.14) of F, we replace U by U_p, then $w_1 + w_3$ is again given by (5.13). With ω as in the proof of Theorem 5.1, we can approximate the modified integral in (5.13) by a Riemann sum s_1 so that $|w_1 + w_3 - s_1| < \varepsilon$ on $\partial \omega$. Then $s_1 \in \mathcal{S}(\mathbb{R}^n) \cap \mathcal{H}(K)$ and $w_1 + w_3 - s_1 \in \mathcal{H}(\mathbb{R}^n \setminus \overline{\omega})$. It follows from Lemma 6.3 (iii) that $(w_1 + w_3 - s_1)(x) = O(||x||^{-a})$ as $x \to \infty$, and we conclude from Corollary 2.10 that (5.15) holds with the exponent $-a$ in place of $2 - n$ (and a revised value for C_7). We complete the proof, as before, by collecting together inequalities.

We can now strengthen Theorem 5.2 and extend it to include the case $n = 2$.

6.5 Theorem *Let E be a relatively closed proper subset of Ω and let a, ε be positive numbers. If $u \in \mathcal{H}(E)$, then there exists v in $\mathcal{S}(\Omega) \cap \mathcal{H}(E)$ such that*

$$|(u - v)(x)| < \varepsilon (1 + ||x||)^{-a} \quad (x \in E). \tag{6.1}$$

Since E is a relatively closed proper subset of Ω, we have $\overline{E} \neq \mathbb{R}^n$, so we may suppose that there is a positive number ρ such that $\overline{E} \cap B(0,\rho) = \emptyset$. Let (Ω_m) be a sequence of bounded open sets such that $\overline{\Omega}_m \subseteq \Omega_{m+1}$ $(m \in \mathbb{N})$, $\bigcup_{m=1}^{\infty} \Omega_m = \Omega \setminus \{0\}$ and $\Omega_1 \cap E \neq \emptyset$. Repeating the proof of Theorem 5.2, using Theorem 6.4 in place of Theorem 5.1, we find that there exists a function $v' \in \mathcal{S}(\Omega \setminus \{0\}) \cap \mathcal{H}(E)$ such that

$$|(u - v')(x)| < \frac{5}{8} \varepsilon (1 + ||x||)^{-a} \quad (x \in E). \tag{6.2}$$

(We chose (Ω_m) to cover $\Omega \setminus \{0\}$ rather than Ω in order to ensure that the hypothesis $K \cup E_1 \cup E_2 \neq \mathbb{R}^n$ in Theorem 6.4 is always satisfied by the assignments $E_1 = \overline{\Omega}_m$, $E_2 = \overline{E} \setminus \Omega_{m+1}$ and $K = \overline{\Omega}_m \cap E$.) If $0 \notin \Omega$, then we simply define $v = v'$ and the proof is complete. However, if $0 \in \Omega$, then we need to approximate v' on E by a function v in $\mathcal{S}(\Omega) \cap \mathcal{H}(E)$. We may assume that the singularities of v' in $B(0,\rho/2)$ form a sequence (x_m) converging to 0. Let p_m denote the principal part of v' at x_m. By Lemma 3.1, there exists a function q_m in $\mathcal{H}(\mathbb{R}^n \setminus \{0\})$ such that

$$|(p_m - q_m)(x)| < 2^{-m-2} \varepsilon (1 + ||x||)^{-a} \quad (||x|| > 2||x_m||).$$

We define $v = v' + \sum_{m=1}^{\infty}(p_m - q_m)$. Then $v \in S(\Omega) \cap \mathcal{H}(E)$ and

$$|(v - v')(x)| \leq \sum_{m=1}^{\infty} |(p_m - q_m)(x)| < \frac{1}{4}\varepsilon(1 + \|x\|)^{-a} \quad (x \in \Omega \setminus B(0, \rho))$$

which, together with (6.2), yields (6.1).

The following example shows that the closeness of the approximation in Theorem 6.5 cannot be improved, at least in the case where $\Omega = \mathbb{R}^n$; see, however, the remarks at the end of Section 8.

6.6 Example Let $u \in \mathcal{H}(E)$, where $E = \mathbb{R}^n \setminus B(0, 1)$. There does not necessarily exist v in $S(\mathbb{R}^n)$ such that

$$(u - v)(x) = O(\|x\|^{-a}) \quad (x \to \infty) \tag{6.3}$$

for all positive numbers a.

To verify this, suppose that u, v are as described. Then, by Theorem 2.9, we have

$$(u - v)(x) = \sum_{m=0}^{\infty} \|x\|^{2-n-2m} I_m(x) \quad (\|x\| \geq 1),$$

where $I_m \in \mathcal{H}_m$, and the series is Weierstrass convergent on $\mathbb{R}^n \setminus B(0, 1)$. Hence by the orthogonality property (2.6)

$$\int_{S(0,r)} (u - v)^2 \, d\sigma = \sum_{m=0}^{\infty} r^{4-2n-4m} \int_{S(0,r)} I_m^2 \, d\sigma \quad (r \geq 1),$$

and therefore, for each m,

$$\int_{S(0,r)} I_m^2 \, d\sigma \leq r^{4m+2n-4} \int_{S(0,r)} (u - v)^2 \, d\sigma \to 0 \quad (r \to +\infty),$$

which implies that $I_m = 0$. Thus $u = v$ on $\mathbb{R}^n \setminus B(0, 1)$ and hence u has a harmonic continuation to $\mathbb{R}^n \setminus S$, where S (the set of singularities of v) is finite. This provides a contradiction, since u need have no such continuation. (Consider for example, $u = \sum_{j=1}^{\infty} 2^{-j} U(\cdot, x_j)$, where $x_j = (2^{-j}, 0, \ldots, 0)$.)

7 Tangential approximation near the boundary by harmonic functions with isolated singularities

Let E be a relatively closed subset of Ω and let $u \in \mathcal{H}(E)$. If $a > 0$ and $z \in \partial E \cap \partial \Omega$, then Theorem 6.5 with an application of the Kelvin transform shows that there is a function v in $S(\Omega)$ such that $(u - v)(x) = O(\|x - z\|^a)$ as $x \to z$ along E. In general, we cannot expect to achieve such strong tangential approximation near all points of $\partial E \cap \partial \Omega$ simultaneously. However, with some mild restrictions on Ω, we might hope to have $(u - v)(x) \to 0$ as $x \to z$ along E for all z in $\partial E \cap \partial \Omega$. In this section we show that such close approximation is indeed possible for a large class of open sets Ω. The work described here is based on [10].

The central idea is again to rework the proofs in Section 5 with a modified kernel in place of U—in this case, the Green kernel G of Ω. First we recall the definition of G and some of its properties; for proofs, we refer to [6, Chapter 4]. We suppose throughout this section that Ω is a domain; that is to say, Ω is connected (as well as being non-empty and open). For a fixed point y in Ω, the function $U(\cdot, y)$ is superharmonic on Ω (in fact, on $\mathbb{R}^n$) and therefore if there is a function h in $\mathcal{H}(\Omega)$ such that $h \leq U(\cdot, y)$ on Ω, then there is a (pointwise) greatest such function, which we denote by h_y. Thus, h_y, if it exists, is the *greatest harmonic minorant* of $U(\cdot, y)$ on Ω. It can be shown that if h_y exists for some y in Ω, then it exists for all y in Ω, and we then say that Ω is *Greenian* and define the *Green kernel* of Ω by

$$G(x, y) = U(x, y) - h_y(x) \quad (x, y \in \Omega).$$

Thus Ω is Greenian if and only if $U(\cdot, y)$ has a harmonic minorant on Ω for some (equivalently, for all) y in Ω. In the case $n \geq 3$, we have $U > 0$, so all domains are Greenian. In the plane, however, there are non-Greenian domains, for example $\mathbb{R}^2$ itself. (If $h \in \mathcal{H}(\mathbb{R}^2)$ and $h \leq U(\cdot, 0)$, then $h(0) = \mathcal{M}(h; 0, r) \leq \mathcal{M}(U(\cdot, 0); 0, r) \to -\infty$ as $r \to +\infty$, which is impossible.) It can be shown that a domain Ω in $\mathbb{R}^2$ is Greenian if and only if $\mathbb{R}^2 \setminus \Omega$ is non-polar or, equivalently, $\mathbb{R}^2 \setminus \Omega$ has positive logarithmic capacity (see [6, Theorem 5.3.8]). In particular, Ω is Greenian if $\mathbb{R}^2 \setminus \overline{\Omega} \neq \emptyset$.

We assume now that Ω is a Greenian domain and note the main properties of G. Obviously $G > 0$ on $\Omega \times \Omega$ and G is increasing in the sense that if Ω_0 is a domain contained in Ω, then Ω_0 is Greenian and $G_0 \leq G$ on $\Omega_0 \times \Omega_0$, where G_0 is the Green kernel of Ω_0. The Green kernel is symmetric: $G(x, y) = G(y, x)$ for all points x, y in Ω; see [6, Theorem 4.1.9]. The boundary behaviour of G is important for us. We define a point z of $\partial^\infty \Omega$ to be *regular* (*for* Ω) if $G(x, y) \to 0$ as $x \to z$ for all points y in Ω; otherwise z is called *irregular*. (This definition is unorthodox; usually regularity is defined in another way and the vanishing of $G(\cdot, y)$ at regular boundary points then appears as a theorem; see [6, Sections 6.6, 6.8].) It can be shown that z is regular if $G(x, y) \to 0$ as $x \to z$ for *some* y in Ω. The set of irregular boundary points in $\partial \Omega$ is small: in fact it is a set of zero capacity (logarithmic if $n = 2$, Newtonian if $n \geq 3$). If Ω is unbounded, then ∞ is regular in the case $n \geq 3$ but may be irregular in the case $n = 2$. We say that Ω is *regular* if every point of $\partial^\infty \Omega$ is regular. We note that if Ω is regular, then the classical solution of the Dirichlet problem exists; that is to say, if $f : \partial^\infty \Omega \to \mathbb{R}$ is a continuous function, then there exists a harmonic function H_f on Ω such that $H_f(x) \to f(z)$ as $x \to z$ for each z in $\partial^\infty \Omega$. Even if Ω is not regular, it is still true that corresponding to each continuous function $f : \partial^\infty \Omega \to \mathbb{R}$ there exists a unique bounded harmonic function H_f on Ω such that $H_f(x) \to f(z)$ as $x \to z$ for every regular point z in $\partial^\infty \Omega$. This function H_f is the Perron-Wiener-Brelot (PWB) solution of the Dirichlet problem, and it always satisfies the inequality $\sup_\Omega |H_f| \leq \sup_{\partial^\infty \Omega} |f|$. For the PWB approach to the Dirichlet problem, we refer to [6, Chapter 6]. A sufficient condition for the regularity of a point z in $\partial \Omega$ is the existence of a truncated cone in $\mathbb{R}^n \setminus \Omega$ with vertex z. If there exists a ball B such that $B \subseteq \mathbb{R}^n \setminus \Omega$ and $\partial B \cap \partial \Omega = \{z\}$, then we have not only $G(x, y) \to 0$ but $G(x, y) = O(\|x - z\|)$ as $x \to z$ for each y in Ω. This is proved by noting that $G \leq G'$ on $\Omega \times \Omega$, where G' denotes the Green kernel of $\mathbb{R}^n \setminus \overline{B}$, and using an explicit formula for G'.

Fixing a reference point x_0 in Ω once and for all, we define

$$g(x) = \min\{1, G(x, x_0)\} \quad (x \in \Omega).$$

Our aim in this section is to indicate a proof of the following result [10].

7.1 Theorem *Let E be a relatively closed proper subset of Ω. If $u \in \mathcal{H}(E)$ and $\varepsilon > 0$, then there exists v in $S(\Omega) \cap \mathcal{H}(E)$ such that*

$$|(u - v)(x)| < \varepsilon g(x) \quad (x \in E).$$

We note that, like Theorem 6.5, this theorem is a strengthening of Theorem 5.2. To see this, suppose the hypotheses of Theorem 5.2 are satisfied. In the case where Ω is connected, we have

$$g(x) \le \min\{1, U(\cdot, x_0)\} < C(1 + \|x\|)^{2-n} \quad (x \in \Omega).$$

Theorem 7.1 implies that there exists v in $S(\Omega) \cap \mathcal{H}(E)$ such that $|u - v| < C^{-1}\varepsilon g$ on E, and hence (5.16) holds. If Ω is disconnected, then we apply Theorem 7.1 in the same way to each component of Ω.

The proof of Theorem 7.1 requires a fusion theorem, in preparation for which we give the following lemmas.

7.2 Lemma *Let K be a compact subset of Ω and let ε be a positive number. If h is a bounded harmonic function on $\Omega \setminus K$ such that*

$$\limsup_{x \to z} |h(x)| \le \varepsilon \quad (z \in \partial K)$$

and

$$\lim_{x \to z} h(x) = 0$$

for every regular point z in $\partial^\infty \Omega$, then

$$|h(x)| \le C\varepsilon g(x) \quad (x \in \Omega \setminus K),$$

where $C = C(\Omega, K, x_0)$.

We first indicate a proof of this lemma under the additional assumption that Ω is regular. Let $\omega = \Omega \setminus (K \cup \{x_0\})$. Then $G(\cdot, x_0)$ is harmonic on ω and has a positive lower bound C_1 on K. By considering separately the cases where $z \in \partial^\infty \Omega, z \in \partial K$ and $z = x_0$, we see that

$$\liminf_{x \to z} \left(\frac{\varepsilon}{C_1} G(x, x_0) \pm h(x) \right) \ge 0 \quad (z \in \partial^\infty \omega). \tag{7.1}$$

It follows from the minimum principle that

$$|h(x)| \le \frac{\varepsilon}{C_1} G(x, x_0) \quad (x \in \Omega \setminus K). \tag{7.2}$$

Since, also, by the maximum principle, $|h| \le \varepsilon$ on $\Omega \setminus K$, we can take $C = \max\{1, 1/C_1\}$.

In the general case there exists a positive harmonic function w on Ω such that $w(x) \to +\infty$ as $x \to z$ for each irregular point z in $\partial^\infty \Omega$. (The proof of this uses the fact that the set of such points z has harmonic measure zero; for harmonic measure, see [6, Section 6.4].) If δ is a positive number, then (7.1) holds with $\delta w + G(\cdot, x_0)$ in place of $G(\cdot, x_0)$, and hence so does (7.2). Since δ can be arbitrarily small, (7.2) holds and the lemma follows as before.

7.3 Lemma *Let K be a compact set and let ω be an open set such that $K \subset \omega \subset \overline{\omega} \subset \Omega$. There exists $C = C(K, \omega, \Omega, x_0)$ such that*

$$G(x, y) + \|\nabla_2 G(x, y)\| \le Cg(x) \quad (x \in \Omega \setminus \overline{\omega}; \ y \in K).$$

In proving this, we may assume that ω is bounded. Then G is finite-valued and continuous on the compact set $\partial \omega \times K$. Hence $G \le C_1$ on $\partial \omega \times K$. Also, if $y \in K$, then $G(\cdot, y)$ is harmonic and bounded on $\Omega \setminus \overline{\omega}$; see [6, Lemma 4.1.8]. Since, also, $G(x, y) \to 0$ as $x \to z$ for every regular z in $\partial^\infty \Omega$, we can apply Lemma 7.2 (with $\overline{\omega}$ in place of K) to conclude that

$$G(x, y) \le C_2 C_1 g(x) \quad (x \in \Omega \setminus \overline{\omega}; \ y \in K). \tag{7.3}$$

Let $C_3 = \operatorname{dist}(K, \partial \omega)$. If $y \in K$ and $x \in \Omega \setminus \overline{\omega}$, then $G(x, \cdot)$ is positive and harmonic on $B(y, C_3)$ and hence by (7.3) and Corollary 2.5 (ii),

$$\|\nabla_2 G(x, y)\| \le nC_3^{-1}G(x, y) \le nC_2 C_1 C_3^{-1}g(x). \tag{7.4}$$

The lemma follows from (7.3) and (7.4).

7.4 Lemma *Suppose that $0 < r_1 < r_2$ and $\overline{B(y, r_2)} \subset \Omega$. If $h \in \mathcal{H}(\Omega \setminus \overline{B(y, r_1)})$ and $\varepsilon > 0$, then there exists h' in $\mathcal{H}(\Omega \setminus \{y\})$ such that*

$$|(h - h')(x)| < \varepsilon g(x) \quad (x \in \Omega \setminus \overline{B(y, r_2)}).$$

The Laurent expansion of h in $A(y; r_1, r_2)$ gives a decomposition $h = h_1 + h_2$, where $h_1 \in \mathcal{H}(\mathbb{R}^n \setminus \overline{B(y, r_1)})$ and $h_2 \in \mathcal{H}(\Omega)$. We define $\varepsilon' = \varepsilon/(2C_1)$, where C_1 is the constant of Lemma 7.2 corresponding to the assignment $K = \overline{B(y, r_2)}$. By Lemma 3.1, there exists h_1' in $\mathcal{H}(\mathbb{R}^n \setminus \{y\})$ such that

$$|(h_1 - h_1')(x)| < \varepsilon'(1 + \|x\|)^{-1} \quad (\|x\| \ge r_2).$$

We extend the function $h_1 - h_1'$ by assigning it the value 0 at ∞. Then the restriction of $h_1 - h_1'$ to $\partial^\infty \Omega$ is finite-valued and continuous, so the PWB solution of the Dirichlet problem on Ω with boundary function $h_1' - h_1$ exists; we denote this solution by H. Then $H \in \mathcal{H}(\Omega)$ and, since $|h_1' - h_1| < \varepsilon'$ on $\partial^\infty \Omega$, we have $|H| < \varepsilon'$ on Ω. Hence

$$|h_1' - h_1 - H| \le |h_1' - h_1| + |H| < 2\varepsilon' \quad \text{on} \quad S(y, r_2).$$

Defining $h' = h_2 + h_1' - H$ on $\Omega \setminus \{y\}$, we see that $h' \in \mathcal{H}(\Omega \setminus \{y\})$ and $h' - h = h_1' - h_1 - H$, so that $(h' - h)(x) \to 0$ as $x \to z$ for each regular point z in $\partial^\infty \Omega$ and $|h' - h| < 2\varepsilon'$ on $S(y, r_2)$. It follows from Lemma 7.2 that $|h - h'| < 2\varepsilon' C_1 g = \varepsilon g$ on $\Omega \setminus \overline{B(y, r_2)}$.

7.5 Theorem (Fusion—second modification) *Let ε be a positive number, let K, E_1 be compact subsets of Ω with $K \ne \emptyset$ and let E_2 be a relatively closed subset of Ω such that $E_1 \cap E_2 = \emptyset$. There is a constant C with the following property: if $u_1, u_2 \in S(\Omega) \cap \mathcal{H}(K)$ and $|u_1 - u_2| < \varepsilon$ on K, then there exists u in $S(\Omega) \cap \mathcal{H}(K)$ such that*

$$|(u - u_k)(x)| < C\varepsilon g(x) \quad (x \in K \cup E_k; \ k = 1, 2).$$

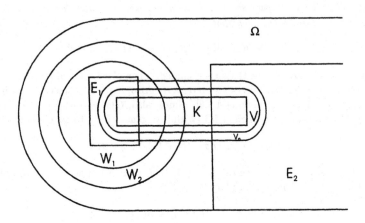

Figure 3

As in the proof of Theorem 5.1, we may suppose that $u_2 = 0$. Let W_1, W_2 be bounded admissible open sets such that

$$E_1 \subset W_1, \qquad \overline{W}_1 \subset W_2, \qquad \overline{W}_2 \subset \Omega \setminus E_2.$$

Also let V_0 be a bounded open set such that $K \subset V_0$ and $\overline{V}_0 \subset \Omega$, and let V be an open set such that $K \subset V \subseteq V_0, u_1 \in \mathcal{H}(\overline{V})$ and $|u_1| < \varepsilon$ on V; see Figure 3. We note that W_1, W_2, V_0 are chosen independently of u_1. Let $W = W_1 \cup V \cup (\Omega \setminus \overline{W}_2)$. We may assume that u_1 has no singularities in $\partial W_1 \cup \partial W_2$. To see this, suppose that there are such singularities $x_1, \ldots, x_m$ and let p_j be the principal part of u_1 at x_j. Then $p_j \in \mathcal{H}(\mathbb{R}^n \setminus \{x_j\})$ and by Lemma 7.4 there exist points $y_1, \ldots, y_m$ in $\Omega \setminus (K \cup \partial W_1 \cup \partial W_2)$ and functions q_j in $\mathcal{H}(\Omega \setminus \{y_j\})$ such that

$$|(p_j - q_j)(x)| < \frac{\varepsilon}{m} g(x) \quad (x \in K \cup E_1).$$

We define $u' = u - \sum_{j=1}^m (p_j - q_j)$. Then $u' \in \mathcal{S}(\Omega) \cap \mathcal{H}(K)$ and u' has no singularities in $\partial W_1 \cup \partial W_2$. Also $|u - u'| < \varepsilon g$ on $K \cup E_1$. If we can find a function u in $\mathcal{S}(\Omega) \cap \mathcal{H}(K)$ such that $|u - u'| < C_1 \varepsilon g$ on $K \cup E_1$ and $|u| < C_1 \varepsilon g$ on $K \cup E_2$ for some constant C_1 independent of u_1, then it will follow that $|u - u_1| < (C_1 + 1)\varepsilon g$ on $K \cup E_1$ and the proof will be complete.

Still following the proof of Theorem 5.1, we let ϕ be a function in $C^\infty(\mathbb{R}^n)$ such that $0 \le \phi \le 1$ on $\mathbb{R}^n, \phi = 1$ on W_1 and supp $\phi \subset W_2$, and then we define $\psi = \phi u_1$ on W_2 and $\psi = 0$ on $\mathbb{R}^n \setminus W_2$. Thus $\psi = u_1$ on $E_1, \psi = 0$ on E_2 and $|\psi| \le |u_1|$ on K, so that

$$|\psi - u_1| \le C_2 \varepsilon g \quad \text{on} \quad K \cup E_1$$

and

$$|\psi| \le C_2 \varepsilon g \quad \text{on} \quad K \cup E_2,$$

where $C_2 = 2/\inf_K g$.

It is now enough to show that there exists u in $\mathcal{S}(\Omega) \cap \mathcal{H}(K)$ such that

$$|u - \psi| < C_3 \varepsilon g \quad \text{on} \quad K \cup E_1 \cup E_2.$$

For this, we follow the final paragraph in the proof of Theorem 5.1 replacing the Newtonian kernel U by the Green kernel G throughout. Our estimate for the integral $I(x)$ appearing in (5.10) is now

$$|I(x)| < C_4 \varepsilon \int_V (G(x,y) + \|\nabla_2 G(x,y)\|) d\lambda(y) \quad (x \in W \setminus S), \tag{7.5}$$

where S again denotes the set of singularities of u_1 in W. Let V_1 be a bounded open set such that $\overline{V}_0 \subset V_1$ and $\overline{V}_1 \subset \Omega$. By Lemma 7.3, the integrand in (7.5) is less than $C_5 g(x)$ when $x \in \Omega \setminus \overline{V}_1$ and hence $|I(x)| \leq C_4 C_5 \varepsilon \lambda(V_0) g(x)$ for such x. It is easy to see that the integral in (7.5) is bounded when $x \in \overline{V}_1$, and g has a positive lower bound on $\overline{V}_1$. Hence $|I| < C_6 \varepsilon g$ on Ω. Thus

$$|\psi - (w_1 + s_2 + w_3)| < C_6 \varepsilon \quad \text{on} \quad W \setminus S.$$

We have $s_2 \in \mathcal{H}(\Omega \setminus S)$ and $w_1, w_3 \in \mathcal{H}(\Omega \setminus (\partial W \cap \operatorname{supp} \psi))$. Let ω be a bounded open set such that

$$\partial W \cap \operatorname{supp} \psi \subset \omega, \qquad \overline{\omega} \subset \Omega, \qquad \omega \cap (K \cup E_1 \cup E_2) = \emptyset.$$

Approximating the integrals defining w_1 and w_3 (in which G now replaces U) by Riemann sums, we find that there exists s_1 in $\mathcal{S}(\Omega) \cap \mathcal{H}(\Omega \setminus (\partial W \cap \operatorname{supp} \psi))$ such that $|w_1 + w_3 - s_1| < \varepsilon$ on $\partial \omega$. The estimates in Lemma 7.3 show that $|s_1| + |w_1| + |w_3| < C_7 g$ on $\Omega \setminus \overline{\omega}$. Hence $w_1 + w_3 - s_1$ is bounded on $\Omega \setminus \overline{\omega}$ and tends to 0 at every regular point of $\partial^\infty \Omega$. It follows from Lemma 7.2 that $|w_1 + w_3 - s_1| < C_8 \varepsilon g$ on $\Omega \setminus \overline{\omega}$. Defining $u = s_1 + s_2$, we see that $u \in \mathcal{S}(\Omega) \cap \mathcal{H}(K)$ and

$$|u - \psi| \leq |s_1 - (w_1 + w_3)| + |w_1 + s_2 + w_3 - \psi| < (C_6 + C_8) \varepsilon g$$

on $K \cup E_1 \cup E_2$, as required.

We can now indicate the proof of Theorem 7.1, which is modelled closely on that of Theorem 5.2. The sequences of sets (Ω_m) and (F_m) are taken to be as in the proof of Theorem 5.2. Let C_m be the constant of Theorem 7.5 corresponding to the assignments $E_1 = \overline{\Omega}_m, E_2 = E \setminus \Omega_{m+1}$ and $K = F_{m+1}$. We may assume that $1 \leq C_1 \leq C_2 \leq \ldots$. Let $\delta_m = \varepsilon 2^{-m-2} C_m^{-1}$. Since g has a positive lower bound on the compact set F_{m+1}, it follows from Theorem 4.1 that there exists a function w_m such that $|u - w_m| < \delta_m g$ on F_{m+1} and hence $|w_{m+1} - w_m| < 2\delta_m g$ on F_{m+1}. It follows from Theorem 7.5 that there exists v_m in $\mathcal{S}(\Omega) \cap \mathcal{H}(F_{m+1})$ such that $|v_m - w_m| < 2\delta_m C_m g$ on $\overline{\Omega}_m \cup F_{m+1}$ and $|v_m - w_{m+1}| < 2\delta_m C_m g$ on E. Defining v as in (5.22), we follow the proof of Theorem 5.2, with only minor modifications, to show that $v \in \mathcal{S}(\Omega) \cap \mathcal{H}(E)$ and $|u - v| < (5/8)\varepsilon g$ on E.

The following simple example shows that in general the closeness of the approximation in Theorem 7.1 cannot be improved; see, however, the remarks at the end of the following section.

7.6 Example Let $\Omega = B(0,1)$, let $E = B(0,1) \setminus B(0,1/2)$ and let F be a non-empty relatively open subset of $\partial \Omega$. We choose $x_0 = 0$ for our reference point. There exists u in

$\mathcal{H}(E)$ for which there is no v in $\mathcal{S}(\Omega)$ such that

$$\frac{(u-v)(x)}{g(x)} \to 0 \quad (x \to z;\ z \in F). \tag{7.6}$$

To see this, let u be a function in $\mathcal{H}(E)$ and suppose that there exists v in $\mathcal{S}(\Omega)$ such that (7.6) holds. From (7.6) and the formulae

$$g(x) = \begin{cases} \min\{1, -\log \|x\|\} & (n = 2) \\ \min\{1, \|x\|^{2-n} - 1\} & (n \geq 3) \end{cases}$$

it follows easily that both the function $u - v$ and its normal derivative vanish continuously on F, and this implies that $u = v$ on E (cf. [6, Exercises 1.4, 1.12]). Thus u has a continuation belonging to $\mathcal{S}(\Omega)$. However, it is easy to construct elements of $\mathcal{H}(E)$ that have no such continuation; for example, $u = \sum_{m=2}^{\infty} 2^{-m} G(\cdot, x_m)$, where $x_m = (2^{-m}, 0, \ldots, 0)$.

8   Tangential approximation by harmonic functions

Let K be a compact subset of Ω. If $u \in \mathcal{H}(K)$, then by Theorem 4.1 we can approximate u uniformly on K by functions in $\mathcal{S}(\Omega)$ or, indeed, $\mathcal{S}(\mathbb{R}^n)$. Theorem 4.2 says that under suitable topological conditions the approximating functions can be taken to be free of singularities. It is natural to ask for conditions that allow *tangential* approximation by harmonic functions free of singularities. In this section we give extensions of Theorems 6.5 and 7.1 in which the approximating functions belong to $\mathcal{H}(\Omega)$. The distinction between approximation by harmonic functions with singularities and approximation by harmonic functions without singularities is important: most applications of the theory depend upon approximation by harmonic functions with no singularities (see Sections 9, 10, 11 below and also Gardiner's article [25] in this volume).

A topological space is called *locally connected* if for each point x in the space and each neighbourhood ω of x there exists a connected neighbourhood ω' of x such that $\omega' \subseteq \omega$. Our extensions of Theorems 6.5 and 7.1, first proved in [7], and [10] respectively, are as follows. Recall that Ω^* denotes the one-point compactification of Ω.

8.1 Theorem *Let E be a relatively closed proper subset of Ω such that $\Omega^* \setminus E$ is connected and locally connected. If $u \in \mathcal{H}(E)$ and ε, a are positive numbers, then there exists v in $\mathcal{H}(\Omega)$ such that*

$$|(u-v)(x)| < \varepsilon(1 + \|x\|)^{-a} \quad (x \in E). \tag{8.1}$$

8.2 Theorem *Let E be a relatively closed proper subset of a Greenian domain Ω such that $\Omega^* \setminus E$ is connected and locally connected. If $\varepsilon > 0$, then there exists v in $\mathcal{H}(\Omega)$ such that*

$$|(u-v)(x)| < \varepsilon g(x) \quad (x \in E),$$

where g is as in Section 7.

Before giving the proofs of these results, we briefly discuss the hypotheses on $\Omega^* \setminus E$. Since $\Omega \setminus E$ is obviously locally connected, the local connectedness of $\Omega^* \setminus E$ is equivalent to the condition that every neighbourhood of $\mathcal{A}$ in $\Omega^* \setminus E$ contains a connected neighbourhood of $\mathcal{A}$. Thus $\Omega^* \setminus E$ is locally connected if and only if for each compact subset K of Ω there is a compact subset L of Ω containing every component of $\Omega \setminus (E \cup K)$ whose closure is a compact subset of Ω. In the case where E is compact this condition is clearly satisfied and the topological hypotheses in Theorem 8.1 and 8.2 reduce to those in Theorem 4.2.

8.3 Example Let $E = \partial F$, where

$$F = \bigcup_{m=1}^{\infty} \left((-\infty, m) \times (2^{-m-1}, 2^{-m}) \right);$$

see Figure 4. Then $(\mathbb{R}^2)^* \setminus E$ is connected but not locally connected. To see that $(\mathbb{R}^2)^* \setminus E$ is not locally connected, note that if $K = S(0,1)$, then no compact set L contains all the bounded components of $\mathbb{R}^2 \setminus (E \cup K)$. The set F is sometimes called an *Arakelyan glove*.

Theorems 8.1 and 8.2 will be deduced from Theorems 6.5 and 7.1 by a pole-pushing argument using the following lemma.

8.4 Lemma *Let E be a relatively closed proper subset of Ω such that $\Omega^* \setminus E$ is connected and locally connected and let (y_m) be a sequence in $\Omega \setminus E$ having no accumulation point in Ω. There exist tracts T_m contained in $\Omega \setminus E$ such that T_m connects y_m to $\mathcal{A}$ for each m and no compact subset of Ω meets infinitely many of the tracts T_m.*

To prove this lemma, we first show that there is a sequence (K_j) of compact sets with $K_j \subset K_{j+1}^{\circ}$ and $\bigcup_j K_j = \Omega$ such that both $\Omega^* \setminus K_j$ and $\Omega^* \setminus (E \cup K_j)$ are connected for each j. We will say that a set A is Ω-*bounded* if $\overline{A}$ is a compact subset of Ω. Extending our earlier usage, we write $\hat{F}$ for the union of a relatively closed subset F of Ω with all the Ω-bounded components of $\Omega \setminus F$. To construct the compact sets K_j, we start by letting (L_j) be a sequence of compact sets such that $L_1 \subseteq E, L_j \subset L_{j+1}^{\circ}$ and $\bigcup_j L_j = \Omega$. The sequence (K_j) is defined inductively as follows. Let $K_1 = L_1$. Having chosen K_j, we choose k such that $K_j \subset L_k^{\circ}$ and define $K_{j+1} = \hat{F}$, where F is the union of L_k with the closure of the union of the Ω-bounded subsets of $\Omega \setminus (E \cup L_k)$. Then $K_j \subset L_k^{\circ} \subset K_{j+1}^{\circ}$ and neither $\Omega \setminus K_{j+1}$ nor $\Omega \setminus (E \cup K_{j+1})$ has any Ω-bounded components.

For each m, let $j(m)$ be the largest integer j such that $y_m \in \Omega \setminus (E \cup K_j)$. Since the component of $\Omega \setminus (E \cup K_{j(m)})$ containing y_m is not Ω-bounded, there is a continuous function $f \colon [0,1] \to \Omega \setminus (E \cup K_{j(m)})$ such that $f(0) = y_m$ and $f(1) \in \Omega \setminus (E \cup K_{j(m)+1})$. Proceeding inductively, we obtain a path in $\Omega \setminus (E \cup K_{j(m)})$ connecting y_m to $\mathcal{A}$, and hence there is a tract T_m in $\Omega \setminus (E \cup K_{j(m)})$ connecting y_m to $\mathcal{A}$. Let K be a compact subset of Ω. Then $K \subseteq K_{j_0}$ for some j_0. If $y_m \in \Omega \setminus K_{j_0}$, then $T_m \subseteq \Omega \setminus K_{j_0}$. Since at most finitely many points y_m belong to K_{j_0}, it follows that at most finitely many tracts T_m meet K.

We can now complete the proof of Theorem 8.1. By Theorem 6.5 there exists a function v_0 in $\mathcal{S}(\Omega) \cap \mathcal{H}(E)$ such that

$$|(u - v_0)(x)| < \frac{1}{2}\varepsilon(1 + \|x\|)^{-a} \quad (x \in E).$$

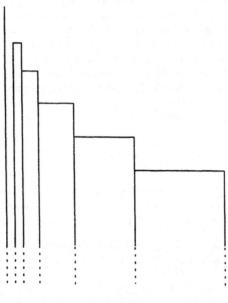

Figure 4

Let the singularities of v_0 be $y_1, y_2, \ldots$. By Lemma 8.4 there exist tracts $T_1, T_2, \ldots$ contained in $\Omega \setminus E$ such that T_m connects y_m to $\mathcal{A}$ and each compact subset of Ω meets at most finitely many of the tracts. Let p_m denote the principal part of v_0 at y_m. Then $p_m \in \mathcal{H}(\mathbb{R}^n \setminus \{y_m\})$ and by Lemma 3.2 there exists q_m in $\mathcal{H}(\Omega)$ such that

$$|(p_m - q_m)(x)| < 2^{-m-1}\varepsilon(1 + ||x||)^{-a} \quad (x \in \Omega \setminus T_m).$$

We write $v = v_0 - \sum_{m=1}^{\infty}(p_m - q_m)$. If ω is a bounded open set such that $\overline{\omega} \subset \Omega$, then there exists k such that $\overline{\omega} \cap T_m = \emptyset$ when $m > k$. The series $\sum_{m=k+1}^{\infty}(p_m - q_m)$ converges uniformly on ω to a harmonic sum, and $v_0 - \sum_{m=1}^{k}(p_m - q_m)$ is harmonic on ω, except possibly for removable singularities. It follows from the arbitrary nature of ω that if v is assigned suitable values at the points y_m, then $v \in \mathcal{H}(\Omega)$. Also, since $E \subset \Omega \setminus \bigcup_m T_m$, we have

$$|(u - v)(x)| \leq |(u - v_0)(x)| + |(v_0 - v)(x)|$$

$$\leq \frac{1}{2}\varepsilon(1 + ||x||)^{-a} + \sum_{m=1}^{\infty}|(p_m - q_m)(x)| < \varepsilon(1 + ||x||)^{-a} \quad (x \in E).$$

To complete the proof of Theorem 8.2, we need the following lemma.

8.5 Lemma *Let T be a tract in Ω connecting a point y to $\mathcal{A}$. If $u \in \mathcal{H}(\Omega \setminus \{y\})$ and $\varepsilon > 0$, then there exists v in $\mathcal{H}(\Omega)$ such that $|u - v| < \varepsilon g$ on $\Omega \setminus T$.*

As in the proof of Lemma 3.2, we choose a sequence $(B_k) = (B(y_k, r_k))$ of open balls such that $y \in B_1$, $y_k \in B_{k+1}$, $B_k \subset T$ and $B_k \to A$ and define $y_0 = y$ and $v_0 = u$. Applying Lemma 7.4 repeatedly, we find that for each non-negative integer k there exists v_k in $\mathcal{H}(\Omega \setminus \{y_k\})$ such that $|v_{k+1} - v_k| < \varepsilon 2^{-k-1}g$ on $\Omega \setminus B_{k+1}$. It follows that (v_k) converges locally uniformly on Ω to a limit function v in $\mathcal{H}(\Omega)$, and we have

$$|u - v| \le \sum_{k=0}^{\infty} |v_{k+1} - v_k| < \varepsilon g \quad \text{on} \quad \Omega \setminus T.$$

The proof of Theorem 8.2 now follows very closely that of Theorem 8.1. By Theorem 7.1, there exists v_0 in $\mathcal{S}(\Omega) \cap \mathcal{H}(E)$ such that $|u - v_0| < (1/2)\varepsilon g$ on E. Let $y_1, y_2, \ldots$ be the singularities of v_0 and let $T_1, T_2, \ldots$ be tracts exactly as in the proof of Theorem 8.1. If p_m is the principal part of v_0 at y_m, then by Lemma 8.5 there exists q_m in $\mathcal{H}(\Omega)$ such that $|p_m - q_m| < \varepsilon 2^{-m-1}g$ on $\Omega \setminus T_m$. Writing $v = v_0 - \sum_{m=1}^{\infty}(p_m - q_m)$, we see, as in the proof of Theorem 8.1 that v is harmonic on Ω, except for removable singularities $y_1, y_2, \ldots$, and $|u - v| < \varepsilon g$ on E.

We conclude this section by mentioning some improvements to Theorems 8.1 and 8.2. In [7] it was shown that, under the hypotheses of Theorem 8.1 or Theorem 6.5, we can arrange that

$$|D^{\alpha}(u - v)(x)| < \varepsilon(1 + ||x||)^{-a} \quad (x \in E)$$

for all multi-indices α with $|\alpha|$ less than some prescribed number. The corresponding result for uniform approximation was earlier proved in [27] (see also [28]). The main modification required to prove this extension of Theorem 8.1 is to replace the fusion result, Theorem 6.4, by the following theorem. The proof given in [7] was inspired by the proof of the uniform version in [27].

8.6 Theorem (Fusion) *Let E_1, E_2 be sets in $\mathbb{R}^n$ such that E_1 is compact, E_2 is closed and $E_1 \cup E_2 \ne \mathbb{R}^n$. Let V be a non-empty open neighbourhood of $E_1 \cap E_2$ and let a be a positive number. For each multi-index α there exists a constant C_{α} with the following property: if $u_1, u_2 \in \mathcal{S}(\mathbb{R}^n)$, then there exists u in $\mathcal{S}(\mathbb{R}^n)$ such that*

$$|D^{\alpha}(u - u_k)(x)| \le C_{\alpha} \sup_V |u_1 - u_2|(1 + ||x||)^{-a} \quad (x \in E_k;\ k = 1, 2).$$

We mentioned in Section 4 that, even in Walsh's theorem, the topological hypotheses are sufficient but not necessary. It is therefore not surprising that the same is true in Theorems 8.1 and 8.2. Gardiner has given necessary and sufficient conditions on Ω and E for the validity of the conclusion in Theorem 8.2 and has similarly given necessary and sufficient conditions for the conclusion of Theorem 8.1 in the important case where $\Omega = \mathbb{R}^n$. These conditions involve thinness and also the so-called (K, L)-condition: a pair (Ω, E), where E is a relatively closed subset of Ω is said to satisfy the (K, L)-*condition* if for each compact subset K of Ω, there exists a compact subset L of Ω containing every Ω-bounded component of $\Omega \setminus (E \cup K)$ whose closure meets K. We note that if (Ω, E) satisfies the (K, L)-condition then $\Omega^* \setminus \hat{E}$ is locally connected but $\Omega^* \setminus E$ need not be. Gardiner's improvement of Theorem 8.2 in the case where $\Omega = \mathbb{R}^n$ is as follows ([24, Theorem 5.9]).

8.7 Theorem *Let E be a closed subset of $\mathbb{R}^n$. The following are equivalent:*

(i) *for each u in* $\mathcal{H}(E)$ *and each positive number* ε *there exists v in* $\mathcal{H}(\mathbb{R}^n)$ *such that* $|u - v| < \varepsilon$ *on E;*

(ii) *for each u in* $\mathcal{H}(E)$ *and each pair of positive numbers* a, ε *there exists v in* $\mathcal{H}(\mathbb{R}^n)$ *such that*
$$|(u - v)(x)| < \varepsilon(1 + ||x||)^{-a} \quad (x \in E);$$

(iii) $\mathbb{R}^n \setminus \hat{E}$ *and* $\mathbb{R}^n \setminus E$ *are thin at the same points of E, and* $(\mathbb{R}^n, E)$ *satisfies the* (K, L)- *condition.*

Next we state Gardiner's improvement of Theorem 8.1 (see [24, Theorem 3.15]).

8.8 Theorem *Let E be a relatively closed proper subset of a domain* Ω. *The following are equivalent:*

(i) *for each u in* $\mathcal{H}(E)$ *and each positive number* ε *there exists v in* $\mathcal{H}(\Omega)$ *such that* $|u-v| < \varepsilon$ *on E;*

(ii) *for each u in* $\mathcal{H}(E)$ *and each function s that is positive and superharmonic on some open set containing E there exists v in* $\mathcal{H}(\Omega)$ *such that* $0 < v - u < s$ *on E;*

(iii) $\Omega \setminus \hat{E}$ *and* $\Omega \setminus E$ *are thin at the same points of E, and* (Ω, E) *satisfies the* (K, L)- *condition.*

Since the function εg in Theorem 8.2 is positive and superharmonic on an open set (namely Ω) containing E, the approximation in Theorem 8.8 (ii) is at least as strong as that in Theorem 8.2. In some cases it is much stronger, as the following example shows.

8.9 Example Let $\Omega = \mathbb{R}^{n-1} \times (0, +\infty)$ and $E = \mathbb{R} \times [-c, c]^{n-2} \times (0, 2c]$, where $0 < c < \pi/2$. If $u \in \mathcal{H}(E)$ and $\varepsilon > 0$, then there exists v in $\mathcal{H}(\Omega)$ such that
$$0 < (v - u)(x) < \varepsilon x_n \exp(-|x_1|\sqrt{n-1}) \quad (x \in E). \tag{8.2}$$

If $a > 0$, then Theorem 8.1 shows that in this example there exists v_1 in $\mathcal{H}(\Omega)$ such that (8.1) holds, and Theorem 8.2 shows that there exists v_2 in $\mathcal{H}(\Omega)$ such that
$$|(u - v_2)(x)| < \varepsilon x_n (1 + ||x||)^{-n} \quad (x \in E) \tag{8.3}$$

(see [6, p. 98] for estimates of the Green kernel of Ω). We note in passing that if $a > n$, then (8.3) is neither stronger nor weaker than (8.1). To see that there is a function v in $\mathcal{H}(\Omega)$ satisfying (8.2), we first define
$$h_\pm = \exp(\pm x_1\sqrt{n-1}) \cos x_2 \cdots \cos x_{n-1} \sin x_n.$$

Then $h_+, h_- \in \mathcal{H}(\mathbb{R}^n)$, so the function $s = \varepsilon \min\{h_+, h_-\}$ is superharmonic on $\mathbb{R}^n$. Also, $s > 0$ on an open set containing E and $s(x) < \varepsilon x_n \exp(-|x_1|\sqrt{n-1})$ for all x in E. Thus the required conclusion follows from Theorem 8.8.

9 Radial growth and decay of harmonic functions

It has been known for a very long time that there exist non-constant entire functions f with the property that $f(re^{i\theta}) \to 0$ as $r \to +\infty$ for all θ; see Lindelöf's book [35, p. 122] of 1905. The real part of such a function f provides an example of a non-constant harmonic function on the plane with the same radial decay property. Modern approximation results make the construction of such functions very easy, as we shall see. By an *algebraic curve* we mean the image $\phi([0, +\infty))$ of a function $\phi \colon [0, +\infty) \to \mathbb{R}^n$ of the form $\phi = (\phi_1, \dots, \phi_n)$, where $\phi_1, \dots, \phi_n$ are real-valued polynomials on $\mathbb{R}$, not all of which are constant.

9.1 Example For each positive number a there exists a non-constant harmonic function v on $\mathbb{R}^n$ such that

$$\lim_{x \to \infty,\ x \in A} \|x\|^a v(x) = 0$$

for every algebraic curve A.

To show this, we take the tract T in Example 3.3 to be the set of points at a distance less than 1 from the path

$$\{(t, e_1(t), e_2(t), \dots, e_{n-1}(t)) \colon t \geq 0\},$$

where $e_1(t) = e^t$ and $e_j(t) = \exp(e_{j-1}(t))$ when $j \geq 2$. Example 3.3 shows that there is a function v in $\mathcal{H}(\mathbb{R}^n) \setminus \{0\}$ such that $|v(x)| < (1 + \|x\|)^{-a-1}$ when $x \in \mathbb{R}^n \setminus T$. Since every algebraic curve has bounded intersection with T, it follows that v has the stated decay property or every such curve.

An example concerning the growth of harmonic functions along rays is as follows.

9.2 Example For each positive number a there exists a harmonic function v on $\mathbb{R}^n$ such that

$$r^{-a} v(x_0 + ry) \to +\infty \quad (r \to +\infty)$$

for all x_0 in $\mathbb{R}^n$ and all y in $S(0,1)$.

We first demonstrate this in the case $n = 2$, identifying $\mathbb{R}^2$ with $\mathbb{C}$ in the usual way. Let m be a positive integer larger than a and let

$$F = \bigcup_{j=1}^{2m} \{re^{ij\pi/m} \colon r \geq 0\}$$

and

$$E = \{z \in F \colon |z| \geq 1\} \cup \{z \in \mathbb{C} \colon \inf_{w \in F} |w - z| \geq (1 + |z|)^{-1}\};$$

see Figure 5. It is easy to see that there exists u in $\mathcal{H}(E)$ such that $u(z) > 1 + |z|^b$ for all z in E, where $b = (1/2)(a + m)$; for example, if E_0 is a component of E, then we can take u to be given on some neighbourhood of E_0 by $u(re^{i\theta}) = 1 + Cr^b \sin(b\theta + \delta)$, where C and δ are suitably chosen constants. By Theorem 8.1 there exists v in $\mathcal{H}(\mathbb{R}^n)$ such that $v(z) > u(z) - 1 > |z|^b$ when $z \in E$. Since $b > a$ and every line has bounded intersection with $\mathbb{C} \setminus E$, we see that v has the stated growth property.

We now proceed by induction on n. Suppose that $n \geq 3$ and that there is a function v_0 in $\mathcal{H}(\mathbb{R}^{n-1})$ with the property that $v_0(x) > \|x\|^b$ on $\mathbb{R}^{n-1} \setminus E$, where $b > a$ and E is some

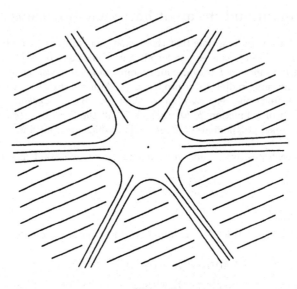

Figure 5

set whose complement has bounded intersection with every line in $\mathbb{R}^{n-1}$. (We saw in the preceding paragraph that there is such a function on $\mathbb{R}^2$.) Let

$$E_1 = \{x \in \mathbb{R}^n : |x_n| \geq 1 + x_1^2 + \cdots + x_{n-1}^2\}$$

and

$$E_2 = \{x \in \mathbb{R}^n : \operatorname{dist}(x, E_1) \geq 1\},$$

and let ω_1, ω_2 be disjoint open neighbourhoods of E_1, E_2. Let v_1 be a function in $\mathcal{H}(\mathbb{R}^n)$ such that $v_1(x) \geq ||x||^{a+1}$ when $x \in E_1$. (We could take v_1 to be of the form $C_1 + C_2 J_k(y_0, \cdot)$, where $y_0 = (0, \ldots, 0, 1)$ and k is an even integer larger than or equal to $a+1$.) Now we define

$$u(x) = \begin{cases} v_1(x) & (x \in \omega_1) \\ v_0(x_1, \ldots, x_{n-1}) & (x \in \omega_2) \end{cases}.$$

By Theorem 8.1 there exists v in $\mathcal{H}(\mathbb{R}^n)$ such that $v > u - 1$ on $E_1 \cup E_2$. Hence there exist numbers $r > 0$ and $c > a$ such that $v(x) > ||x||^c$ when $x \in (E_1 \cup E_2) \setminus B(0, r)$. Since $B(0, r) \cup (\mathbb{R}^n \setminus (E_1 \cup E_2))$ has bounded intersection with every line, the induction is complete.

9.3 Example For each positive number a there exists a non-constant harmonic function v on $\mathbb{R}^n$ such that

$$\lim_{r \to +\infty} \exp(r^a) v(ry) = 0 \qquad (9.1)$$

for every y in $S(0, 1)$.

The construction given for Example 9.2 in the case $n = 2$ shows that there is a function u in $\mathcal{H}(\mathbb{R}^2)$ such that $u(z) > |z|^{a+1}$ when z lies in a set E whose complement has bounded intersection with every line. Let f be an entire function such that $\mathrm{Re}\, f = u$ and let

$$v(x) = \mathrm{Re}\{ze^{-f(z)}\} \quad (x = (x_1, \ldots, x_n) \in \mathbb{R}^n; \ z = x_1 + ix_2).$$

Then $v \in \mathcal{H}(\mathbb{R}^n)$ and it is easy to see that (9.1) holds for every y in $S(0, 1)$.

Examples 9.1 and 9.2 are given in [9], where it is also shown that these examples are extreme in the following sense. Let F be a second category subset of $S(0, 1)$; that is to say F is not contained in any countable union of closed subsets of $S(0, 1)$ each with empty interior relative to $S(0, 1)$. Then there is no harmonic function u on $\mathbb{R}^n$ such that

$$r^{-a}u(ry) \to +\infty \quad (r \to +\infty)$$

for all y in F and all a in $(0, +\infty)$, and there is no v in $\mathcal{H}(\mathbb{R}^n) \setminus \{0\}$ such that (9.1) holds for all y in F and all a in $(0, +\infty)$. The first of these results was originally proved by Schneider [38] in the case $n = 2$; for a refinement of Schneider's theorem see [4, Theorem 1]. Bonilla [17] has recently used Theorem 8.1 to show that for each a there is a large class of harmonic functions u such that

$$\lim_{x \to \infty, x \in \Sigma} \|x\|^a D^\alpha u(x) = 0$$

for every multi-index α and every strip Σ. For some results concerning the growth of harmonic functions on sets of parallel lines, see [11].

By the minimum principle a non-null harmonic function on $B(0, 1)$ cannot have limit 0 at every point of $S(0, 1)$. Our final example in this section shows that in this statement "limit" cannot be replaced by "non-tangential limit". We recall that a function f on $B(0, 1)$ has *non-tangential limit* ℓ at a point y of $S(0, 1)$ if $f(x) \to \ell$ as $x \to y$ along any set of the form $B(0, 1) \cap \{x \colon \langle x - y, -y \rangle > \|x - y\| \cos \alpha\}$, where $0 < \alpha < \pi/2$.

9.4 Example Let $y_1 = (1, 0, \ldots, 0) \in \mathbb{R}^n$. There exists a non-null harmonic function v on $B(0, 1)$ such that $v(x) \to 0$ as $x \to y$ for each y in $S(0, 1) \setminus \{y_1\}$ and v has non-tangential limit 0 at y_1.

In the case $n = 2$ this was shown by Ash and Brown [12]. Bonilla [17], using Theorem 8.2 showed that there is a large class of such functions v. Here we use Lemma 8.5 to show that there is a function v as described. Let $y_t = (t, 0, \ldots, 0)$ and $T = B(y_{1/4}, 3/4) \setminus \overline{B(y_{3/4}, 1/4)}$. Also let $u = G(0, \cdot)$, where G denotes the Green function of $B(0, 1)$. By Lemma 8.5 there exists a function v in $\mathcal{H}(B(0, 1))$ such that

$$|(u - v)(x)| < G(0, x) \quad (x \in B(0, 1) \setminus T),$$

and hence $|v| < 2G(0, \cdot)$ on $B(0, 1) \setminus T$. Clearly v has the stated limiting properties.

10 The Radon transform

Let $\Pi^{(n)}$ denote the collection of all $(n-1)$-dimensional hyperplanes in $\mathbb{R}^n$, and let λ' denote $(n-1)$-dimensional Lebesgue measure on an element Π of $\Pi^{(n)}$. If f is a real-valued (or complex-valued) function on $\mathbb{R}^n$ and the restriction of f to Π is λ'-integrable for each Π in $\Pi^{(n)}$, then we write $f \in \mathcal{R}$ and define

$$\hat{f}(\Pi) = \int_\Pi f \, d\lambda' \quad (\Pi \in \Pi^{(n)}).$$

The function $\hat{f} \colon \Pi^{(n)} \to \mathbb{R}$ (or $\mathbb{C}$) is called the *Radon transform* of f. If $f \in C(\mathbb{R}^n) \cap \mathcal{R}$ and $\hat{f} \equiv 0$, can we conclude that $f \equiv 0$? The answer is affirmative if a further hypothesis is imposed. If we suppose, for example, that f is λ-integrable on $\mathbb{R}^n$, then consideration of the Fourier transform shows that $f \equiv 0$; see, e.g., [2, p. 319]. Without some extra hypothesis, however, the answer is negative. In the case $n = 2$ this was first shown by Zalcman [42], who proved that there exists an entire function f with zero integral on every line in $\mathbb{C}$. Zalcman's construction uses a theorem of Arakelian [1] on holomorphic approximation. Inspired by this work, Armitage and Goldstein [8] proved the following result, using Theorem 8.1.

10.1 Theorem *There exists a function v in $\mathcal{H}(\mathbb{R}^n) \cap \mathcal{R}$ such that $\hat{v} \equiv 0$ but $v \not\equiv 0$.*

Here we indicate an elementary proof of this result. An analogous construction yielding a non-null entire function f on $\mathbb{C}$ (equivalently $\mathbb{R}^2$) with $\hat{f} \equiv 0$ was given in [3]. We use Example 3.3, which depends only upon simple pole-pushing arguments. We take the tract T in that example to be the set of points lying at a distance less than 1 from the path $\{(t, t^2, \ldots, t^n) \in \mathbb{R}^n : t \geq 0\}$. Then there exists a function v in $\mathcal{H}(\mathbb{R}^n) \setminus \{0\}$ such that

$$|v(x)| < (1 + \|x\|)^{-n-1} \quad (x \in \mathbb{R}^n \setminus T). \tag{10.1}$$

We fix a point y in $S(0, 1)$ and define hyperplanes $\Pi(y, t)$ by

$$\Pi(y, t) = \{x \in \mathbb{R}^n : \langle x, y \rangle = t\} \quad (t \in \mathbb{R}).$$

It is not hard to see that, for each positive number b, the strip $\{x \in \mathbb{R}^n : |\langle x, y \rangle| < b\}$ has bounded intersection with T. Hence, by (10.1), the function

$$t \mapsto \int_{\Pi(y,t)} |v| \, d\lambda' \quad (t \in \mathbb{R})$$

is locally bounded on $\mathbb{R}$. A standard result (see e.g. [6, Theorem 1.5.12]) now implies that the function

$$t \mapsto \hat{v}(\Pi(y, t)) = \int_{\Pi(y,t)} v \, d\lambda' \quad (t \in \mathbb{R})$$

is a polynomial of degree at most 1. If $\Pi(y, t) \cap T = \emptyset$, then, by (10.1),

$$\begin{aligned}
|\hat{v}(\Pi(y, t))| &\leq \int_{\Pi(y,t)} |v| \, d\lambda' \\
&< \int_{\Pi(y,t)} (1 + \|x\|)^{-n-1} \, d\lambda'(x) \\
&= \int_{\mathbb{R}^{n-1}} (1 + \sqrt{t^2 + \|x'\|^2})^{-n-1} d\lambda'(x') \\
&< (1 + |t|)^{-1} \int_{\mathbb{R}^{n-1}} (1 + \|x'\|)^{-n} \, d\lambda'(x') \\
&= C(1 + |t|)^{-1}.
\end{aligned}$$

It is easy to see that $\Pi(y, t) \cap T = \emptyset$ either for all sufficiently large positive values of t or else for all sufficiently large negative values of t. Hence $\hat{v}(\Pi(y, t)) \to 0$ either as $t \to +\infty$ or as $t \to -\infty$. Since $\hat{v}(\Pi(y, \cdot))$ is a polynomial, it follows that $\hat{v}(\Pi(y, t)) = 0$ for all t, and since y is an arbitrary point of $S(0, 1)$, we find that $\hat{v}(\Pi) = 0$ for all Π in $\Pi^{(n)}$.

The above proof can easily be modified to show that there are non-null functions v in $\mathcal{H}(\mathbb{R}^n) \cap \mathcal{R}$ such that $\widehat{D^\alpha v} \equiv 0$ for every multi-index α. Moreover, Bonilla [17] has shown that there is a large class of such functions.

In connection with Theorem 10.1, Zalcman [43] has raised the following question: if $k \in \{1, \ldots, n - 2\}$ and f is a continuous function on $\mathbb{R}^n$ which is integrable with respect to k-dimensional Lebesgue measure on every k-dimensional hyperplane in $\mathbb{R}^n$ and whose integral over every such hyperplane is 0, do we necessarily have $f \equiv 0$? He remarks that if f is integrable with zero integral on every k-dimensional hyperplane and every ℓ-dimensional hyperplane, where $1 \leq k < \ell < n$, then f must vanish identically. Another open question concerns the growth rate of functions whose Radon transforms vanish identically. It seems not to be known, for example, whether the function v in Theorem 10.1 can be of finite order.

11 Universal harmonic functions

Let $\{F_\alpha : \alpha \in J\}$ be a collection of continuous mappings from a topological space X into a topological space Y. An element ξ of X is called *universal (with respect to $\{F_\alpha\}$)* if the set $\{F_\alpha(\xi) : \alpha \in J\}$ is dense in Y. For a bibliographic survey of the extensive literature on universality we refer to Grosse-Erdmann's article [29]. We now mention an early example of universality due to Birkhoff [15]. Let $\mathcal{E}$ denote the vector space of all entire functions on $\mathbb{C}$ equipped with the topology of local uniform convergence, and for each a in $\mathbb{C}$ let $F_a : \mathcal{E} \to \mathcal{E}$ be the mapping given by $F_a(f)(z) = f(a + z)$. Birkhoff showed that there exist functions g that are universal with respect to $\{F_a : a \in \mathbb{C}\}$. Thus if g is such a universal function, then for every positive number ε, every compact subset K of $\mathbb{C}$, and every function f in $\mathcal{E}$ there exists a point a in $\mathbb{C}$ such that $|f(z) - g(a + z)| < \varepsilon$ for all z in K.

Dzagnidze [22] proved an analogue of Birkhoff's result in the space $\mathcal{H}(\mathbb{R}^n)$, also with the topology of local uniform convergence. We will use Theorem 8.1 to prove the following version of Dzagnidze's result.

11.1 Theorem *Let (a_m) be an unbounded sequence in $\mathbb{R}^n$. There exists a function H in $\mathcal{H}(\mathbb{R}^n)$ with the following property: for every positive number ε, every compact subset K of $\mathbb{R}^n$, and every function h in $\mathcal{H}(\mathbb{R}^n)$ there exists m such that $|h(x) - H(a_m + x)| < \varepsilon$ for all x in K.*

By working with a subsequence, if necessary, we may assume that $\|a_{m+1}\| > \|a_m\| + 2^{m+3}$ for each m. Let (h_m) be a sequence in $\mathcal{H}(\mathbb{R}^n)$ that is dense in $\mathcal{H}(\mathbb{R}^n)$. (We could, for example, let $\mathcal{B}_k$ be a basis for $\mathcal{H}_k$ for each k and take (h_m) to be an enumeration of the set of all harmonic polynomials of the form $b_1 H_1 + \cdots + b_p H_p$, where $p \in \mathbb{N}, b_1, \ldots, b_p \in \mathbb{Q}$ and $H_1, \ldots, H_p \in \bigcup_{k=0}^\infty \mathcal{B}_k$). We define a function u by

$$u(x) = h_m(x - a_m) \quad (x \in B(a_m, 2^{m+1}); \ m \in \mathbb{N}).$$

Then $u \in \mathcal{H}(E)$, where E is the closed set $\bigcup_{m=1}^\infty \overline{B(a_m, 2^m)}$. Since the topological conditions of Theorem 8.1 are clearly satisfied (with $\Omega = \mathbb{R}^n$), it follows that there is a function H in

$\mathcal{H}(\mathbb{R}^n)$ such that

$$|(H - u)(x)| < (1 + \|x\|)^{-1} \quad (x \in E).$$

To show that H has the stated property, let ε, K and h be as in the statement of the theorem. Then there exists m such that $|h - h_m| < \varepsilon/2$ on $K, K \subset B(0, 2^m)$ and $2^{-m} < \varepsilon$. When $x \in K$, we have

$$
\begin{aligned}
|H(x + a_m) - h(x)| &\leq |H(x + a_m) - u(x + a_m)| + |u(x + a_m) - h(x)| \\
&\leq \sup\{(1 + \|y\|)^{-1} : y \in B(a_m, 2^m)\} + |h_m(x) - h(x)| \\
&< 2^{-m-1} + 2^{-1}\varepsilon < \varepsilon.
\end{aligned}
$$

Although it is convenient to use Theorem 8.1 in the above proof, it is certainly not necessary; see [24, pp. 116, 117] for a recursive construction using Walsh's theorem.

If $H \in \mathcal{H}(\mathbb{R}^n)$ and the set of translates $\{x \mapsto H(a + x) : a \in \mathbb{R}^n\}$ is dense in $\mathcal{H}(\mathbb{R}^n)$, then we call H a *universal harmonic function*. Since, by Theorem 11.1, such functions are known to exist, it follows immediately from their definition that they form a dense subset of $\mathcal{H}(\mathbb{R}^n)$. In fact a much stronger result concerning the abundance of universal harmonic functions is true. We recall that $\mathcal{H}(\mathbb{R}^n)$ with the topology of local uniform convergence is a complete metric space: an explicit metric d is given by

$$d(u, v) = \sum_{k=1}^{\infty} 2^{-k} \frac{\sup_{B(0,k)} |u - v|}{1 + \sup_{B(0,k)} |u - v|} \quad (u, v \in \mathcal{H}(\mathbb{R}^n)).$$

Therefore, by the Baire category theorem, $\mathcal{H}(\mathbb{R}^n)$ is of the second category as a subset of itself. (A subset of a topological space is of the *first category* if it can be expressed as a countable union of nowhere dense sets; otherwise it is of the *second category*.) It is a striking fact that the universal harmonic functions form a second category subset of $\mathcal{H}(\mathbb{R}^n)$. Even more is true:

11.2 Theorem *The functions in $\mathcal{H}(\mathbb{R}^n)$ that are not universal harmonic functions form a first category subset of $\mathcal{H}(\mathbb{R}^n)$.*

We indicate a proof based on work of Duios-Ruis [20], who proved the corresponding result for Birkhoff's universal entire functions. If $a \in \mathbb{R}^n$ and $u \in \mathcal{H}(\mathbb{R}^n)$, then we write u_a for the function $x \mapsto u(x + a)$. For each v in $\mathcal{H}(\mathbb{R}^n)$ and each positive number ε, we define

$$N(v, \varepsilon) = \{u \in \mathcal{H}(\mathbb{R}^n) : d(u_a, v) < \varepsilon \text{ for some } a \text{ in } \mathbb{R}^n\}.$$

It is easy to see that $N(v, \varepsilon)$ is open. Again let (h_m) be a dense sequence in $\mathcal{H}(\mathbb{R}^n)$. An element u of $\mathcal{H}(\mathbb{R}^n)$ is universal if and only if for each pair (m, k) of positive integers there exists a in $\mathbb{R}^n$ such that $d(u_a, h_m) < 1/k$. Thus the set of all universal harmonic functions is $\bigcap_{m=1}^{\infty} \bigcap_{k=1}^{\infty} N(h_m, 1/k)$, a countable intersection of open sets. Thus the non-universal harmonic functions can be expressed as a countable union of closed sets, and each of these closed sets has empty interior, since the universal harmonic functions are dense in $\mathcal{H}(\mathbb{R}^n)$.

We note that Bonilla [18] has shown that $\mathcal{H}(\mathbb{R}^n)$ contains a dense vector subspace all of whose elements, except 0, are universal. In studies of universality in other contexts it has often been shown that universal elements form a large class (their complement is of first category), and there are general theorems to this effect (see [29, Section 1b]).

We conclude with an open question. Duios-Ruis [21] has shown that Birkhoff's universal entire functions can be of arbitrarily slow transcendental growth, and permissible growth rates have been determined for other kinds of universal functions; some general theorems are given in [30]. However, it appears that nothing is known about how slowly the universal harmonic functions discussed above can grow.

12 Acknowledgements

I am very grateful to Sheila O'Brien for her expert typing of these notes. I also thank Michael Gormley for producing the diagrams and Michael Elliott for his technical assistance.

References

[1] N. U. Arakelyan, Uniform and tangential approximation by analytic functions (Russian), *Izv. Akad. Nauk Armyan. SSR Ser. Mat.* **3** (1968), 273–286; English translation *Amer. Math. Soc. Transl. Ser. 2* **122** (1984), 85–97.

[2] D. H. Armitage, A new proof of a uniqueness theorem for harmonic functions in half-spaces, *Bull. London Math. Soc.* **9** (1977), 317–320.

[3] D. H. Armitage, A non-constant continuous function on the plane whose integral on every line is zero, *Amer. Math. Monthly* **101** (1994), 892–894.

[4] D. H. Armitage, Radial limits of superharmonic functions in the plane, *Colloq. Math.* **67** (1994), 245–252.

[5] D. H. Armitage, An entire holomorphic function associated to an entire harmonic function, *J. Approx. Theory* **99** (1999), 325–343.

[6] D. H. Armitage and S. J. Gardiner, *Classical Potential Theory*, Springer, London, 2000.

[7] D. H. Armitage and M. Goldstein, Better than uniform approximation on closed sets by harmonic functions with singularities, *Proc. London Math. Soc. (3)* **60** (1990), 319–343.

[8] D. H. Armitage and M. Goldstein, Nonuniqueness for the Radon transform, *Proc. Amer. Math. Soc.* **117** (1993), 175–178.

[9] D. H. Armitage and M. Goldstein, Radial limiting behaviour of harmonic functions in cones, *Complex Variables Theory Appl.* **22** (1993), 267–276.

[10] D. H. Armitage and M. Goldstein, Tangential harmonic approximation on relatively closed sets, *Proc. London Math. Soc. (3)* **68** (1994), 112–136.

[11] D. H. Armitage and C. S. Nelson, The growth of superharmonic functions on sets of parallel lines, *Analysis* **14** (1994), 127–138.

[12] J. M. Ash and R. Brown, Uniqueness and nonuniqueness for harmonic functions with zero nontangential limits, in: *Harmonic Analysis (Sendai, 1990), ICM-90 Satell. Conf. Proc.* (S. Igari, ed.), Springer, 1991; 30–40.

[13] S. Axler, P. Bourdon, and W. Ramey, *Harmonic Function Theory*, Grad. Texts in Math. **137**, Springer, New York, 1992.

[14] T. Bagby and N. Levenberg, Bernstein theorems for harmonic functions, in: *Methods of Approximation Theory in Complex Analysis and Mathematical Physics* (A. A. Gonchar and E. A. Saff, eds.) Lecture Notes in Math. **1550**, Springer, New York, 1993; 7–18.

[15] G. D. Birkhoff, Démonstration d'un théorème élémentaire sur les fonctions entières, *C. R. Acad. Sci. Paris* **189** (1929), 473–475.

[16] J. Bliedtner and W. Hansen, Simplicial cones in potential theory II (Approximation theorems), *Invent. Math.* **46** (1978), 255–275.

[17] A. Bonilla, "Counterexamples" to the harmonic Liouville theorem and harmonic functions with zero nontangential limits, *Colloq. Math.* **83** (2000), 155–160.

[18] A. Bonilla, Universal harmonic functions, manuscript.

[19] J. B. Conway, *Functions of One Complex Variable*, Grad. Texts in Math. **11**, Springer, New York, 1973.

[20] S. M. Duios Ruis [S. M. Duyos-Ruiz], On the existence of universal functions (Russian), *Dokl. Akad. Nauk SSSR* **268** (1983), 18–22; English translation *Soviet Math. Dokl.* **27** (1983), 9–13.

[21] S. M. Duios Ruis [S. M. Duyos-Ruiz], Universal functions and the structure of the space of entire functions (Russian), *Dokl. Akad. Nauk SSSR* **279** (1984), 792–795; English translation *Soviet Math. Dokl.* **30** (1984), 713–716.

[22] O. P. Dzagnidze, The universal harmonic function in the space E_n (Russian), *Soobšč. Akad. Nauk. Gruzin. SSR* **34** (1964), 525–528.

[23] S. J. Gardiner, Superharmonic extension and harmonic approximation, *Ann. Inst. Fourier (Grenoble)* **44** (1994), 65–91.

[24] S. J. Gardiner, *Harmonic Approximation*, Cambridge University Press, Cambridge, 1995.

[25] S. J. Gardiner, Harmonic approximation and its applications, *These proceedings*.

[26] P. M. Gauthier, M. Goldstein, and W. Ow, Uniform approximation on unbounded sets by harmonic functions with logarithmic singularities, *Trans. Amer. Math. Soc.* **261** (1980), 169–183.

[27] P. M. Gauthier, M. Goldstein, and W. Ow, Uniform approximation on closed sets by harmonic functions with Newtonian singularities, *J. London Math. Soc. (2)* **28** (1983), 71–82.

[28] P. M. Gauthier and W. Hengartner, *Approximation uniforme qualitative sur des ensembles non bornés*, Sém. Math. Sup. **82**, Les Presses de l'Université de Montréal, Montréal, 1982.

[29] K.-G. Grosse-Erdmann, Universal families and hypercylic operators, *Bull. Amer. Math. Soc.* **36** (1999), 345–381.

[30] K.-G. Grosse-Erdmann, Rate of growth of hypercyclic entire functions, to appear in *Indag. Math.*

[31] W. K. Hayman, Power series expansions for harmonic functions, *Bull. London Math. Soc.* **2** (1970), 152–158.

[32] W. K. Hayman and P. B. Kennedy, *Subharmonic Functions*, vol. I, Academic Press, London, 1976.

[33] L. L. Helms, *Introduction to Potential Theory*, Wiley, New York, 1969.

[34] C. O. Kiselman, Prolongement des solutions d'une équation aux dérivées partielles à coefficients constants, *Bull. Soc. Math. France* **97** (1969), 329–356.

[35] E. Lindelöf, *Le calcul des résidus*, Gauthier-Villars, Paris, 1905.

[36] A. Roth, Uniform and tangential approximations by meromorphic functions on closed sets, *Canad. J. Math.* **28** (1976), 104–111.

[37] C. Runge, Zur Theorie der eindeutigen analytischen Funktionen, *Acta Math.* **6** (1885), 228–244.

[38] W. J. Schneider, On the growth of entire functions along half-rays, in: *Entire functions and related parts of analysis*, Amer. Math. Soc., Providence, RI, 1968; 377–385.

[39] J. Siciak, Holomorphic continuation of harmonic functions, *Ann. Polon. Math.* **29** (1974), 67–73.

[40] G. Szegö, *Orthogonal Polynomials*, 3rd ed., Amer. Math. Soc., Providence, RI, 1967.

[41] J. L. Walsh, The approximation of harmonic functions by harmonic polynomials and by harmonic rational functions, *Bull. Amer. Math. Soc. (2)* **35** (1929), 499–544.

[42] L. Zalcman, Uniqueness and non-uniqueness for the Radon transform, *Bull. London Math. Soc.* **14** (1982), 241–245.

[43] L. Zalcman, Problems on uniqueness for k-plane transforms, in: *Approximation by Solutions of Partial Differential Equations* (B. Fuglede et al., eds.), NATO ASI Ser. C **365**, Kluwer Academic Publishers, Dordrecht, 1992; 201.

Sobolev spaces and approximation problems for differential operators

Thomas BAGBY

Department of Mathematics
Indiana University
Bloomington, IN 47405
USA

Nelson CASTAÑEDA

Department of Mathematical Sciences
Central Connecticut State University
New Britain, CT 06050
USA

Abstract

In this paper we give an introduction to the theory of Sobolev spaces, and discuss the connection between these spaces and certain approximation problems associated with elliptic differential operators. In particular, we discuss uniform approximation by harmonic functions, mean approximation by solutions of elliptic equations, and the continuity of Dirichlet eigenvalues for an open set when the set is perturbed.

1 Introduction

The Sobolev spaces are Banach spaces which are related to L^p spaces, but also incorporate a type of differentiability. In this paper we will discuss the definitions and basic properties of these spaces, and the connection between these spaces and certain approximation problems associated with elliptic partial differential operators.

In Section 2 we discuss some of the basic concepts related to the theory of Sobolev spaces. In Section 3 we define the concept of "stability" for compact sets in the sense of Babushka [5, 6], and discuss the connection between the concept of stability and the problem of uniform approximation by harmonic functions. In Section 4 we will discuss the connection between the concept of stability and problems of approximation in the mean by solutions of elliptic partial differential equations. In Section 5 we develop the basic theory of the Dirichlet eigenvalues for the negative of the Laplace operator on a bounded open set. In Section 6 we discuss the connection between the concept of stability and the problem of continuity of the Dirichlet eigenvalues for a bounded open set when the open set is perturbed. Finally, in Section 7 we will introduce "capacities" and discuss criteria for stability of a compact set.

73

N. Arakelian and P.M. Gauthier (eds.), Approximation, Complex Analysis, and Potential Theory, 73–106.
© 2001 *Kluwer Academic Publishers. Printed in the Netherlands.*

2   Sobolev spaces

In this section we introduce the basic concepts related to the theory of Sobolev spaces [1, 2, 16, 32, 50]. We work first on the Euclidean space $\mathbf{R}^N$ of all real N-tuples $(x_1, x_2, \ldots, x_N)$, where $N \geq 2$, and we denote N-dimensional Lebesgue measure by dx. We let $B(a, r)$ denote the open ball with center a and radius r. We will work with real-valued functions and real Banach spaces, except when we indicate otherwise. A *subspace* of a Banach space X will mean a real-linear subspace of X. We will assume that $1 < p < \infty$; we use the notation $L^p(E) = L^p(E, dx)$ for Lebesgue measurable sets E of positive Lebesgue measure, and we set $L^p = L^p(\mathbf{R}^N)$.

We will use the language of distribution theory in Euclidean spaces. We will work with real-valued distributions, except when we indicate otherwise. Recall that a distribution on an open set $\Omega \subset \mathbf{R}^N$ is a linear function $u : C_0^\infty(\Omega) \to \mathbf{R}$ with certain additional properties [27, Chapter II]. Every locally integrable function h on an open set Ω defines a distribution u_h on Ω by the rule

$$u_h(\varphi) = \int_\Omega h\varphi \, dx \quad \text{for all} \quad \varphi \in C_0^\infty(\Omega).$$

Now let u be a locally integrable function on an open set $\Omega \subset \mathbf{R}^N$. Regarding u as a distribution on Ω, the partial derivative $\partial u/\partial x_i$ is always defined as a distribution on Ω. However, the distribution $\partial u/\partial x_i$ may or may not be defined by a locally integrable function on Ω. If there is a locally integrable function f on Ω such that

$$\int_\Omega u\frac{\partial \varphi}{\partial x_i} \, dx = -\int_\Omega f\varphi \, dx \quad \text{for all} \quad \varphi \in C_0^\infty(\Omega),$$

then we have $\partial u/\partial x_i = f$ on Ω in the sense of distribution theory.

2.1 Definition We define the Sobolev space

$$W^{1,p}(\mathbf{R}^N) = \left\{ u \in L^p(\mathbf{R}^N) : \frac{\partial u}{\partial x_i} \in L^p(\mathbf{R}^N) \text{ for } i = 1, 2, \ldots, N \right\}.$$

2.2 Remark We know that the set

$$B = \{(F_0, F_1, \ldots, F_N) : F_i \in L^p \text{ for } i = 0, 1, 2, \ldots, N\}$$

is a Banach space with operations

$$(F_0, F_1, \ldots, F_N) + (G_0, G_1, \ldots, G_N) = (F_0 + G_0, F_1 + G_1, \ldots, F_N + G_N)$$

and

$$c(F_0, F_1, \ldots, F_N) = (cF_0, cF_1, \ldots, cF_N) \quad \text{if} \quad c \in \mathbf{R},$$

and norm

$$\|(F_0, F_1, \ldots, F_N)\| = \sum_{i=0}^{N} \|F_i\|_{L^p(\mathbf{R}^N)}.$$

We can identify $W^{1,p}(\mathbf{R}^N)$ with the closed subspace S of B consisting of all $(N+1)$-tuples $(F_0, F_1, \ldots, F_N)$ such that

$$\int F_i\varphi \, dx = -\int F_0\frac{\partial \varphi}{\partial x_i} \, dx$$

for all $\varphi \in C_0^\infty(\mathbf{R}^N)$ and $i = 1, 2, \ldots, N$, using the identification mapping

$$S \ni (F_0, F_1, \ldots, F_N) \to F_0 \in W^{1,p}(\mathbf{R}^N).$$

This shows that $W^{1,p}(\mathbf{R}^N)$ is a Banach space with the norm inherited from B, or with the equivalent norm

$$\|u\|_{W^{1,p}(\mathbf{R}^N)} = \left(\int |u|^p \, dx + \sum_{i=1}^N \int \left| \frac{\partial u}{\partial x_i} \right|^p dx \right)^{1/p}.$$

2.3 Remark If $u \in W^{1,p}(\mathbf{R}^N)$ and $\varphi \in C_0^\infty(\mathbf{R}^N)$, then one may verify that the product $v = u\varphi \in L^p(\mathbf{R}^N)$ defines an element of $W^{1,p}(\mathbf{R}^N)$, with the first-order partial derivatives given by the usual product rule

$$\frac{\partial v}{\partial x_i} = u \frac{\partial \varphi}{\partial x_i} + \varphi \frac{\partial u}{\partial x_i} \quad \text{for} \quad i = 1, 2, \ldots, N.$$

We turn now to the definition of the Sobolev spaces on $\mathbf{R}^N$ which involve higher order derivatives. For any multi-index $\alpha = (\alpha_1, \alpha_2, \ldots, \alpha_N)$, where each α_j is a nonnegative integer, we use the standard notations

$$x^\alpha = x_1^{\alpha_1} x_2^{\alpha_2} \cdots x_N^{\alpha_N} \quad \text{if} \quad x = (x_1, x_2, \ldots, x_N),$$
$$D^\alpha = (\partial/\partial x_1)^{\alpha_1} (\partial/\partial x_2)^{\alpha_2} \cdots (\partial/\partial x_N)^{\alpha_N},$$
$$|\alpha| = \alpha_1 + \alpha_2 + \cdots + \alpha_N.$$

Recall that if u and f are locally integrable functions on Ω, then $D^\alpha u = f$ on Ω in the sense of distribution theory provided

$$\int_\Omega u D^\alpha \varphi \, dV = (-1)^{|\alpha|} \int_\Omega f \varphi \, dV \quad \text{for all} \quad \varphi \in C_0^\infty(\Omega).$$

2.4 Definition Let $1 < p < \infty$, and let m be a positive integer. We define the Sobolev space

$$W^{m,p}(\mathbf{R}^N) = \{u \in L^p(\mathbf{R}^N) : D^\alpha u \in L^p(\mathbf{R}^N) \text{ for } |\alpha| \leq m\}.$$

2.5 Remarks Let $1 < p < \infty$, and let m be a positive integer.

(a) As before, $W^{m,p}(\mathbf{R}^N)$ is a Banach space with the norm defined by adding the L^p norms of the functions $D^\alpha u$ for $|\alpha| \leq m$, or with the equivalent norm

$$\|u\|_{W^{m,p}(\mathbf{R}^N)} = \left(\sum_{|\alpha| \leq m} \int |D^\alpha u|^p \, dx \right)^{1/p}.$$

(b) If $u \in W^{m,p}(\mathbf{R}^N)$ and $\varphi \in C_0^\infty(\mathbf{R}^N)$, then the product $u\varphi$ is an element of $W^{m,p}(\mathbf{R}^N)$.

We sometimes use the notation $W^{m,p} = W^{m,p}(\mathbf{R}^N)$.

An element of $W^{m,p}(\mathbf{R}^N)$ is called a *Sobolev function*. Note that a "Sobolev function" is actually an equivalence class of functions defining an element of $L^p(\mathbf{R}^N)$, such that this element of $L^p(\mathbf{R}^N)$ has certain additional properties. If u and v are locally integrable functions on $\mathbf{R}^N$ which define the same element of $L^p(\mathbf{R}^N)$, and if $u \in W^{m,p}(\mathbf{R}^N)$, then $v \in W^{m,p}(\mathbf{R}^N)$, and in fact u and v are the same element of $W^{m,p}(\mathbf{R}^N)$.

2.6 Remarks Suppose that $1 < p < \infty$ and m is a positive integer.

(a) Suppose that u is a locally integrable function on $\mathbf{R}^N$ which defines an element of $W^{m,p}(\mathbf{R}^N)$. If $mp > N$, it follows from a theorem of Sobolev [1, Theorem 5.4], [50, Theorem 2.4.2] that there is a continuous function v on $\mathbf{R}^N$ such that $u = v$ dx-almost-everywhere. In other words, if $mp > N$, then every element of $W^{m,p}(\mathbf{R}^N)$ is represented by a continuous function on $\mathbf{R}^N$.

(b) If $mp \leq N$, then there exist elements of $W^{m,p}(\mathbf{R}^N)$ which cannot be represented by continuous functions on $\mathbf{R}^N$. Nevertheless, it is possible to represent each element of $W^{m,p}(\mathbf{R}^N)$ with a function on $\mathbf{R}^N$ having certain properties related to continuity (see [2, Chapter 6], [16, Chapter 4] and [50], and the references given there). An illustration of this will be given in the proof of Theorem 3.5 in Section 3.

2.7 Remarks Let X be a real Banach space. If $z, z_1, z_2, \ldots$ are elements of X, we say that $\{z_j\}_{j=1}^\infty$ converges *weakly* to z provided that

$$\lim_{j \to \infty} T(z_j) = T(z)$$

for every continuous linear function $T : X \to \mathbf{R}$.

(a) Let S be a closed subspace of X. If $z, z_1, z_2, \ldots$ are elements of X such that $\{z_j\}_{j=1}^\infty$ converges weakly to z, and if $z_j \in S$ for each positive integer j, then $z \in S$. (This follows from the Hahn-Banach theorem.)

(b) If $z \in X$, then $||z|| = \sup_T |Tz|$, where the supremum is taken over all continuous linear functions $T : X \to \mathbf{R}$ satisfying $||T|| \leq 1$. (This follows from the Hahn-Banach theorem; see [39, Theorem 5.20 and Remarks 5.21].)

(c) (Lower semicontinuity of the norm with respect to weak convergence) If $z, z_1, z_2, \ldots$ are elements of X such that $\{z_j\}_{j=1}^\infty$ converges weakly to z, then $||z|| \leq \liminf_{j \to \infty} ||z_j||$. (This follows from (b).)

(d) If X is separable and reflexive, and $\{z_j\}_{j=1}^\infty$ is any sequence in X with $\sup_j ||z_j|| < \infty$, then there is a subsequence of $\{z_j\}_{j=1}^\infty$ which converges weakly to an element of X. (This follows from the Alaoglu theorem [45, Appendix A, Section 4, Proposition 4.4] and a metrizability theorem stated in [45, Appendix A, Section 4, Exercise 4].)

(e) If $1 < p < \infty$ and m is a positive integer, then the Sobolev space $W^{m,p}(\mathbf{R}^N)$ is separable and reflexive. (See [1, Section 3.4 and Theorem 3.5].)

2.8 Remark We recall that for any distribution u on $\mathbf{R}^N$, and any $\varphi \in C_0^\infty(\mathbf{R}^N)$, the convolution $u * \varphi$ is a function of class C^∞ on $\mathbf{R}^N$; and for each multi-index α we have

$$D^\alpha(u * \varphi) = u * (D^\alpha \varphi) = (D^\alpha u) * \varphi,$$

where $D^\alpha u$ denotes the distribution derivative, and the other derivatives are classical. In case u is a locally integrable function, and $\varphi \in C_0^\infty(\mathbf{R}^N)$, we have

$$(u * \varphi)(x) = \int_{\mathbf{R}^N} u(y)\varphi(x - y)\, dy \quad \text{for all} \quad x \in \mathbf{R}^N.$$

2.9 Convention In this paper we let Φ be a fixed function in $C_0^\infty(B(0,1))$ such that $\Phi \geq 0$ and $\int \Phi \, dV = 1$. If $1 < p < \infty$ and $u \in L^p(\mathbf{R}^N)$, then for each $\varepsilon > 0$ we define

$$u_{[\varepsilon]} = u * (\Phi_\varepsilon),$$

where

$$\Phi_\varepsilon(x) = \frac{1}{\varepsilon^N} \Phi\left(\frac{x}{\varepsilon}\right).$$

The functions $u_{[\varepsilon]}$ are called *regularizations* of u because of the following theorem, which is proved in [50, Theorem 1.6.1 and Lemma 2.1.3].

2.10 Theorem *Suppose $1 < p < \infty$.*

(a) *If $u \in L^p(\mathbf{R}^N)$, then $u_{[\varepsilon]} \in L^p(\mathbf{R}^N)$ for each $\varepsilon > 0$, and $u_{[\varepsilon]} \to u$ in L^p as $\varepsilon \downarrow 0$.*

(b) *If m is a positive integer and $u \in W^{m,p}(\mathbf{R}^N)$, then $u_{[\varepsilon]} \in W^{m,p}(\mathbf{R}^N)$ for each $\varepsilon > 0$, and $u_{[\varepsilon]} \to u$ in $W^{m,p}(\mathbf{R}^N)$ as $\varepsilon \downarrow 0$.*

In particular, Theorem 2.10 shows that every element in $W^{m,p}(\mathbf{R}^N)$ can be approximated, in the $W^{m,p}(\mathbf{R}^N)$ norm, by a smooth function in $W^{m,p}(\mathbf{R}^N)$.

2.11 Remark Let $1 < p < \infty$. If $u \in W^{1,p}(\mathbf{R}^N)$, and if $\partial u/\partial x_i$ is the zero element of $L^p(\mathbf{R}^N)$ for each $i \in \{1, 2, \ldots, N\}$, then u is the zero element of $W^{1,p}(\mathbf{R}^N)$.

In fact, using Convention 2.9 and Remark 2.8, we see that for each $\varepsilon > 0$, the regularization $u_{[\varepsilon]}$ is a function in $C^\infty(\mathbf{R}^N)$ whose first-order partial derivatives are each equal to zero on all of $\mathbf{R}^N$, and hence $u_{[\varepsilon]}$ is a constant function on $\mathbf{R}^N$; since we have $u_{[\varepsilon]} \in W^{1,p}(\mathbf{R}^N)$ by Theorem 2.10(b), it follows that $u_{[\varepsilon]}$ is identically equal to zero. Since $u_{[\varepsilon]} \to u$ in $W^{1,p}(\mathbf{R}^N)$ as $\varepsilon \downarrow 0$ by Theorem 2.10(b), we conclude that u is the zero element of $W^{1,p}(\mathbf{R}^N)$.

2.12 Definition Suppose that m is a positive integer, and $1 < p < \infty$. If Ω is a bounded open subset of $\mathbf{R}^N$, we let $W_0^{m,p}(\Omega)$ denote the closure of $C_0^\infty(\Omega)$ in $W^{m,p}(\mathbf{R}^N)$.

Note that each $u \in W_0^{m,p}(\Omega)$ will be regarded as an element of the Sobolev space $W^{m,p}(\mathbf{R}^N)$.

If u is an element of $W^{m,p}(\mathbf{R}^N)$, then we may regard u as a distribution on $\mathbf{R}^N$, and we let $\text{supp}\, u$ denote the support of u in the sense of distribution theory [27, Section 2.2]. If $u \in W^{m,p}(\mathbf{R}^N)$ and $K \subset \mathbf{R}^N$ is closed, then we have $\text{supp}\, u \subset K$ if and only if

$$\int u\varphi \, dx = 0 \quad \text{for all} \quad \varphi \in C_0^\infty(\mathbf{R}^N \setminus K).$$

2.13 Remark Suppose that m is a positive integer, and $1 < p < \infty$. Let K be a compact subset of a bounded open set $\Omega \subset \mathbf{R}^N$. If $u \in W^{m,p}(\mathbf{R}^N)$ satisfies $\text{supp}\, u \subset K$, then $u \in W_0^{m,p}(\Omega)$. (This follows from Theorem 2.10 (b).)

If B_1 and B_2 are Banach spaces, we say that a linear map $T : B_1 \to B_2$ is *compact* provided that the image of every bounded set in B_1 is contained in a compact set in B_2. The following famous theorem of Rellich and Kondrachov will play an important role in the present paper (see [1, 32, 50], and the references given there).

2.14 Theorem (Rellich-Kondrachov theorem) *Let Ω be a bounded open set in $\mathbf{R}^N$, and $1 < p < \infty$. Then the linear mapping*

$$W_0^{1,p}(\Omega) \ni u \to u \in L^p(\mathbf{R}^N)$$

is compact.

A simple, direct proof of Theorem 2.14 may be given using part of the proof of Theorem 1 in Section 4.6 of the text of Evans and Gariepy [16].

In the theory of Sobolev spaces, a basic role is played by the study of inequalities [1, 32, 50]. In this paper we shall need the following Poincaré inequality. In the proof we will use a well-known technique (see [35, proof of Theorem 3.6.5], [50, Chapter 4], and the references given there.)

2.15 Theorem (Poincaré inequality) *Let Ω be a bounded open set in $\mathbf{R}^N$, and $1 < p < \infty$. Then there is a positive constant $C = C(\Omega)$ such that*

$$\int_\Omega |u|^p \, dx \le C \sum_{i=1}^N \int_\Omega \left| \frac{\partial u}{\partial x_i} \right|^p dx \quad \text{for all} \quad u \in W_0^{1,p}(\Omega).$$

Proof If the theorem is not true, then there exists a sequence of functions $u_j \in W_0^{1,p}(\Omega)$ such that

$$\int_{\mathbf{R}^N} |u_j|^p \, dx = 1 \tag{2.1}$$

and

$$\lim_{j \to \infty} \int_{\mathbf{R}^N} \left| \frac{\partial u_j}{\partial x_i} \right|^p dx = 0 \quad \text{for each} \quad i \in \{1, 2, \ldots, N\}. \tag{2.2}$$

Note that the sequence $\{u_j\}$ is bounded in $W^{1,p}(\mathbf{R}^N)$. In view of the Rellich-Kondrachov Theorem 2.14, we may assume without loss of generality that $\{u_j\}$ is a Cauchy sequence in $L^p(\mathbf{R}^N)$; using this and (2.2) we see that $\{u_j\}$ is a Cauchy sequence in $W^{1,p}(\mathbf{R}^N)$. If w denotes the limit of $\{u_j\}$ in $W^{1,p}(\mathbf{R}^N)$, then we conclude from (2.2) that $\partial w/\partial x_i$ is the zero element of $L^p(\mathbf{R}^N)$, for each $i \in \{1, 2, \ldots, N\}$. It then follows from Remark 2.11 that w is the zero element of $W^{1,p}(\mathbf{R}^N)$, whereas from (2.1) we have $\int_{\mathbf{R}^N} |w|^p \, dx = 1$. This contradiction proves the theorem. $\qquad\square$

For readers who are familiar with Riemannian manifolds, we will now define the Sobolev spaces $W^{1,2}$ for such manifolds. (Compare Aubin [4] and Hebey [19, 20].)

For the rest of this section we let M denote a noncompact, oriented, connected C^∞ Riemannian manifold of dimension $N \ge 2$. We use the standard notation (g_{ij}) for the metric tensor, g for the determinant of the metric tensor, and (g^{ij}) for the inverse of the metric tensor, in local coordinates. It will be understood that "coordinate region" refers to a connected coordinate neighborhood.

If u is a continuous function on M whose support is a compact subset of a coordinate region $\mathcal{R}$, and if $x_1, x_2, \ldots, x_N$ are local coordinates on $\mathcal{R}$, then the integral of u over M is the quantity

$$I(u) = \int_{\mathcal{R}} u(x_1, x_2, \ldots, x_N) \sqrt{g(x)} \, dx_1 \cdots dx_N.$$

From the transformation law for multiple integrals under a smooth change of coordinates, one can show that we obtain the same value for $I(u)$ if we use other local coordinates $y_1, y_2, \ldots, y_N$ for the coordinate region $\mathcal{R}$.

If u is an arbitrary continuous function of compact support on M, then we define $I(u)$ in the following way. Let $\{\mathcal{R}_i\}$ be a finite open cover of $\operatorname{supp} u$ by coordinate regions. Let $\{\varphi_i\}$ be a partition of unity subordinate to this covering of $\operatorname{supp} u$, by which we mean that we have $\varphi_i \in C_0^\infty(\mathcal{R}_i)$ and $\varphi_i \geq 0$ for each index i, and $\sum_i \varphi_i \equiv 1$ on an open neighborhood of $\operatorname{supp} u$. Then we define

$$I(u) = \sum_i I(u\varphi_i),$$

and this definition is independent of the choice of the cover $\{\mathcal{R}_i\}$ and the partition of unity $\{\varphi_i\}$. By the Riesz representation theorem [39, Chapter 2] there is a unique regular Borel measure dV on M such that

$$I(u) = \int_M u \, dV \quad \text{for every continuous function } u \text{ of compact support on } M.$$

The completion of this measure is also denoted by dV, and we refer to this completion as *Lebesgue measure* on the *Lebesgue measurable* subsets of M. We use the notation $L^2(E) = L^2(E, dV)$ for Lebesgue measurable sets E of positive Lebesgue measure, and we set $L^2 = L^2(M)$.

If Ω is an open subset of M, we say that a Lebesgue measurable function $u : \Omega \to \mathbf{R}$ is *locally integrable on* Ω provided $\int_K |u| \, dV < \infty$ for every compact set $K \subset \Omega$.

2.16 Remarks Let $\mathcal{R}$ be a coordinate region of M, with local coordinates $x_1, x_2, \ldots, x_N$.

(a) We may view $\mathcal{R}$ as a subset of the manifold M with Lebesgue measure dV, or as a subset of the parameter space $\mathbf{R}^N$ with Lebesgue measure dx. The concept of Lebesgue measurability of a subset of $\mathcal{R}$ is the same for the two viewpoints; and the concept of Lebesgue measurability of a real-valued function on $\mathcal{R}$ is the same for the two viewpoints. A Lebesgue measurable set $E \subset \mathcal{R}$ will have measure zero with respect to dV if and only if it has measure zero with respect to dx.

(b) If u is a Lebesgue measurable function on $\mathcal{R}$, then it is locally integrable on $\mathcal{R}$ if and only if $\int_K |u| \, dx < \infty$ for every compact set $K \subset \mathcal{R}$.

Suppose that u is a locally integrable function on a coordinate region $\mathcal{R}$ with local coordinates $x_1, x_2, \ldots, x_N$. We will then say that u has *locally integrable first-order derivatives on* $\mathcal{R}$ provided that there are locally integrable functions $f_1, f_2, \ldots, f_N$ on $\mathcal{R}$ such that $\partial u / \partial x_i = f_i$ on $\mathcal{R}$ for each $i \in \{1, 2, \ldots, N\}$, in the sense of distribution theory, when we regard $\mathcal{R}$ as a subset of $\mathbf{R}^N$.

2.17 Remarks Let u and v be locally integrable functions on M. The following remarks may be proved using the transformation law for multiple integrals, and the chain rule for distributions [27, Formula (6.1.2)], under a smooth change of coordinates.

(a) Suppose that each point of M is contained in a coordinate region on which u has locally integrable first-order derivatives. Then u has locally integrable first-order derivatives on every coordinate region of M.

(b) Suppose that u has locally integrable first-order derivatives on every coordinate region of M. Then there is a Lebesgue measurable function $|\nabla u|$ on M with the following property: for any coordinate region $\mathcal{R}$ in M, with local coordinates $x_1, x_2, \ldots, x_N$, we have

$$|\nabla u|^2 = \sum_{i,j=1}^{N} g^{ij} \frac{\partial u}{\partial x_i} \frac{\partial u}{\partial x_j} \quad dV\text{-almost-everywhere on } \mathcal{R}.$$

(c) Suppose that u and v each have locally integrable first-order derivatives on every coordinate region of M. Then there is a Lebesgue measurable function $\nabla u \cdot \nabla v$ on M with the following property: for any coordinate region $\mathcal{R}$ in M, with local coordinates $x_1, x_2, \ldots, x_N$, we have

$$\nabla u \cdot \nabla v = \sum_{i,j=1}^{N} g^{ij} \frac{\partial u}{\partial x_i} \frac{\partial v}{\partial x_j} \quad dV\text{-almost-everywhere on } \mathcal{R}.$$

2.18 Definition We define the Sobolev space $W^{1,2} = W^{1,2}(M)$ to be the set of all functions $u \in L^2(M)$ which have locally integrable first-order derivatives on each coordinate region in M and satisfy

$$\int_M |\nabla u|^2 \, dV < \infty.$$

In this paper we will find it convenient to reserve the term "parametric ball" for the following situation.

2.19 Definition If we have local coordinates carrying an open subset of M onto an open subset of $\mathbf{R}^N$ containing a closed ball $\overline{B(a,r)}$ with $r > 0$, we refer to the preimage of the open ball $B(a,r)$ as a *parametric ball* in M.

2.20 Remark Suppose B is a parametric ball in M. In local coordinates for a coordinate region containing $\overline{B}$, let (g_{ij}) be the metric tensor, g the determinant of the metric tensor, and (g^{ij}) the inverse of the metric tensor. Then there are positive constants $a < b$ such that the interval $[a, b]$ contains all the values $g(x)$, and all the eigenvalues of the matrices $(g^{ij}(x))$, for $x \in \overline{B}$.

2.21 Theorem $W^{1,2}(M)$ *is a Hilbert space with the inner product*

$$(u, v) = \int_M (uv + \nabla u \cdot \nabla v) \, dV \quad for \quad u \in W^{1,2}(M) \quad and \quad v \in W^{1,2}(M). \tag{2.3}$$

Proof In this proof we will use the notation $\|\nabla w\|_{L^2(E)}$ as an abbreviation for $\| \, |\nabla w| \, \|_{L^2(E)}$, when $w \in W^{1,2}(M)$ and E is a Lebesgue measurable subset of M.

From our earlier discussion we see that (2.3) does define an inner product on $W^{1,2}(M)$, and we must show that this inner-product space is complete. To do this we assume that $\{u_j\}$ is an arbitrary Cauchy sequence in $W^{1,2}(M)$, and we will prove that the sequence converges in $W^{1,2}(M)$. Note that our assumption implies that there is a positive constant C such that

$$\|u_j\|_{W^{1,2}(M)} \leq C. \tag{2.4}$$

Our assumption also implies that $\{u_j\}$ is a Cauchy sequence in $L^2(M)$, and hence converges in $L^2(M)$ to an element $u \in L^2(M)$.

Next let B be any parametric ball, and let $x_1, x_2, \ldots, x_N$ be local coordinates for a coordinate region containing $\overline{B}$. Then from the fact that $\{u_j\}$ is a Cauchy sequence in $W^{1,2}(M)$ we conclude that, for each fixed i, the sequence $\{\partial u_j / \partial x_i\}$ is a Cauchy sequence in $L^2(B, dx)$, and its limit in $L^2(B, dx)$ must be the partial derivative $\partial u / \partial x_i$ on B, in the sense of distribution theory, when we regard B as a subset of $\mathbf{R}^N$.

From Remark 2.17 (a) and the result of the preceding paragraph we see that u has locally integrable first-order derivatives on each coordinate region of M, and

$$\lim_{j \to \infty} \int_B |\nabla(u - u_j)|^2 \, dV = 0 \quad \text{for each parametric ball } B.$$

We conclude that

$$\lim_{j \to \infty} \int_K |\nabla(u - u_j)|^2 \, dV = 0 \quad \text{for each compact set } K \subset M, \tag{2.5}$$

since any compact subset of M can be covered by finitely many parametric balls.

We now show that $\int |\nabla u|^2 \, dV < \infty$. If K is an arbitrary compact subset of M, then by (2.5) we may select a positive integer L such that

$$\|\nabla(u - u_L)\|_{L^2(K)} \le 1,$$

and from this estimate and (2.4) we obtain

$$\|\nabla u\|_{L^2(K)} \le \|\nabla(u - u_L)\|_{L^2(K)} + \|\nabla u_L\|_{L^2(K)} \le 1 + C.$$

Since K is an arbitrary compact subset of M, we conclude that $\|\nabla u\|_{L^2(M)} \le 1 + C$, and hence $u \in W^{1,2}(M)$.

To complete the proof of Theorem 2.21, we show that $\|\nabla(u - u_j)\|_{L^2(M)} \to 0$ as $j \to \infty$. For this we let $\varepsilon > 0$ be arbitrary. We select a positive integer Q such that

$$\|\nabla(u_j - u_k)\|_{L^2(M)} < \frac{\varepsilon}{4} \quad \text{for} \quad j \ge Q \quad \text{and} \quad k \ge Q.$$

We then select a compact set $K \subset M$ such that

$$\|\nabla u_Q\|_{L^2(M \setminus K)} < \frac{\varepsilon}{4}$$

and

$$\|\nabla u\|_{L^2(M \setminus K)} < \frac{\varepsilon}{4}.$$

We may then apply (2.5) to the compact set K to find a positive integer $R \ge Q$ such that

$$\|\nabla(u - u_j)\|_{L^2(K)} < \frac{\varepsilon}{4} \quad \text{for} \quad j \ge R.$$

For each positive integer $j \ge R$ we then have

$$\|\nabla(u - u_j)\|_{L^2(M)}$$
$$\le \|\nabla(u - u_j)\|_{L^2(K)} + \|\nabla(u_j - u_Q)\|_{L^2(M \setminus K)} + \|\nabla u_Q\|_{L^2(M \setminus K)} + \|\nabla u\|_{L^2(M \setminus K)}$$
$$< \frac{\varepsilon}{4} + \frac{\varepsilon}{4} + \frac{\varepsilon}{4} + \frac{\varepsilon}{4} = \varepsilon,$$

which completes the proof. □

We say that a subset of M is *bounded* if its closure in M is compact.

2.22 Definition If Ω is a bounded open set in M, we define $W_0^{1,2}(\Omega)$ to be the closure of $C_0^\infty(\Omega)$ in $W^{1,2}(M)$.

As before, note that we regard each element of $W_0^{1,2}(\Omega)$ as an element of the Sobolev space $W^{1,2}(M)$.

2.23 Remark Let B be a parametric ball in M, and let M' denote the Riemannian manifold $\mathbf{R}^N$ with the usual flat metric. We may regard the parametric ball B as a bounded open subset of the Riemannian manifold M, or as a bounded open subset of the Riemannian manifold M'. Thus we may regard $C_0^\infty(B)$ as a subspace of $W^{1,2}(M)$ or as a subspace of $W^{1,2}(M')$, and from Remark 2.20 it is clear that this yields two equivalent norms on $C_0^\infty(B)$.

2.24 Remarks (a) If $u \in W^{1,2}(M)$ and $\varphi \in C_0^\infty(M)$, then the product $u\varphi$ is an element of $W^{1,2}(M)$.

(b) If Ω is a bounded open set in M, and $u \in W_0^{1,2}(\Omega)$, and $|\nabla u|$ is the zero element of $L^2(M)$, then u is the zero element of $W^{1,2}(M)$. (In fact, from a result in the theory of distributions [42, Section 2.1.2] we see that on every parametric ball u is represented by a constant function. By a standard connectedness argument one can show that the element $u \in W^{1,2}(M)$ is represented by a constant function on M. But this constant must be zero, since u is in the subspace $W_0^{1,2}(\Omega)$, and hence the restriction of u to the nonempty open set $M \setminus \overline{\Omega}$ is represented by the zero function.)

2.25 Theorem (Rellich-Kondrachov theorem for manifolds) *Let Ω be a bounded open set in M. Then the linear mapping*

$$W_0^{1,2}(\Omega) \ni u \to u \in L^2(M)$$

is compact.

The Rellich-Kondrachov Theorem 2.25 follows from the earlier Rellich-Kondrachov Theorem 2.14 by using a partition of unity subordinate to a finite open cover of $\overline{\Omega}$ by parametric balls, and using Remark 2.20 and Remark 2.23 on each of the parametric balls; see also [4, Theorem 2.34].

A discussion of basic inequalities for Sobolev spaces on manifolds is given by Aubin [4] and Hebey [19, 20]. We will need the following Poincaré inequality for manifolds, which can be derived from the Rellich-Kondrachov Theorem 2.25 by modifying the proof given earlier for the Poincaré inequality in Euclidean spaces.

2.26 Theorem (Poincaré inequality for manifolds) *Let Ω be a bounded open subset of M. Then there is a positive constant $C = C(\Omega)$ such that*

$$\int_\Omega |u|^2 \, dV \le C \int_\Omega |\nabla u|^2 \, dV \quad \text{for all} \quad u \in W_0^{1,2}(\Omega).$$

3   Uniform approximation by harmonic functions

In this paper we will consider several approximation problems, involving elliptic partial differential operators, in which a central role is played by the concept of "stability" in the sense of

Babushka. For the initial work of Babushka we refer to the papers [5, 6], and the references given there. For later work we refer to [2, Chapter 11], [23], [25], [37, Theorem 1.1], [41, Chapter IX, Section 5.2] and [48], and the references given there.

3.1 Definition Let m be a positive integer, and $1 < p < \infty$. We say that a compact set $K \subset \mathbf{R}^N$ is (m,p)-*stable* provided that every function $u \in W^{m,p}$ satisfying

$$\operatorname{supp} u \subset K$$

is in $W_0^{m,p}(\operatorname{int} K)$.

In the present section we will consider the connection between the concept of $(1,2)$-stability and the problem of uniform harmonic approximation.

3.2 Definition If $K \subset \mathbf{R}^N$ is compact, we denote by $h(K)$ the set of continuous functions on K which are harmonic on the interior of K.

3.3 Theorem *Let K be a compact subset of $\mathbf{R}^N$. Then the following statements are equivalent:*

(a) *The functions harmonic near K are uniformly dense in $h(K)$.*

(b) *The set K is $(1,2)$-stable.*

Theorem 3.3 follows from combining results on uniform harmonic approximation by Keldysh and Deny (see [17, Section 1.3]) with the method used to prove Theorem 2 of the paper [18] by Havin. A complete discussion of Theorem 3.3, including historical references, is given in the paper [25] by Hedberg.

To illustrate the usefulness of Theorem 3.3, we will now apply it to the problem of uniform approximation on surfaces.

3.4 Definition A k-*cell* in a topological space X is a subset of X which is homeomorphic to the closed unit ball in $\mathbf{R}^k$.

The following theorem is due to Keldysh and Lavrentiev. We refer to [29, Theorem 5.20] for a discussion of the theorem and references to the papers of Keldysh and Lavrentiev.

3.5 Theorem *If J is a 2-cell in $\mathbf{R}^3$, then every continuous function on J can be appoximated uniformly on J by harmonic polynomials.*

Proof We give the proof of Bagby and Gauthier [11]. In the proof we will say that a topological space is homeomorphic to the letter T provided it is homeomorphic to

$$\{(x,0) \in \mathbf{R}^2 : -1 \le x \le 1\} \cup \{(0,y) \in \mathbf{R}^2 : 0 \le y \le 1\}.$$

A theorem of R. L. Moore [34] shows that the closed unit disc of $\mathbf{R}^2$ cannot contain an uncountable family of mutually disjoint subsets which are homeomorphic to the letter T.

From topology it is known that J is a nowhere dense compact subset of $\mathbf{R}^3$ such that $\mathbf{R}^3 \setminus J$ is connected. Because of a Runge-type theorem for harmonic approximation which was proved by Walsh (see [17, Section 1.1]), it suffices to prove that every continuous function on J can be uniformly approximated on J by functions harmonic near J. Therefore, in view of Theorem 3.3, we may prove Theorem 3.5 in the following way: we assume that u is a locally integrable function on $\mathbf{R}^3$ which defines a nonzero distribution in $W^{1,2}(\mathbf{R}^3)$, with $\operatorname{supp} u \subset J$, and we will obtain a contradiction. In the proof we will use the notation $\mathcal{L}_d$ to denote d-dimensional Lebesgue measure.

For each positive number r we define the continuous function

$$u^{(r)}(x) = \frac{1}{\mathcal{L}_3(B(x,r))} \int_{B(x,r)} u(y)\,dy \quad \text{for} \quad x \in \mathbf{R}^3,$$

and we define

$$\widetilde{u}(x) = \lim_{r\to 0} u^{(r)}(x)$$

at each point $x \in \mathbf{R}^3$ where the limit exists. From standard results in real analysis we know that $\widetilde{u}$ is defined and equal to u at $\mathcal{L}_3$-almost-every point of $\mathbf{R}^3$.

We now let E be the set of all points in $x \in \mathbf{R}^3$ such that $\widetilde{u}(x)$ is defined and $\widetilde{u}(x) \neq 0$; since u is not the zero element of $W^{1,2}(\mathbf{R}^3)$, we must have $\mathcal{L}_3(E) > 0$. Moreover, the hypotheses on u imply that $\widetilde{u}$ is defined and equal to zero at each point of $\mathbf{R}^3 \setminus J$, so E must be a subset of J.

Since $u \in W^{1,2}(\mathbf{R}^3)$, it is known that the function $\widetilde{u}$ has the following property: for $\mathcal{L}_2$-almost-every point (y,z) in $\mathbf{R}^2$, the function $\widetilde{u}$ will be defined and continuous on the entire straight line which passes through $(0,y,z)$ and is parallel to the x-axis; this is proved in [16, Section 4.9, Theorem 2 (i)]. Because of this property, there exists a Lebesgue measurable set $E' \subset E$ such that $\mathcal{L}_3(E \setminus E') = 0$, and every point in E' is the center of a compact line segment of positive length which is parallel to the x-axis and lies in E. Similarly, for $\mathcal{L}_2$-almost-every point (x,z) in $\mathbf{R}^2$, the function $\widetilde{u}$ will be defined and continuous on the straight line which passes through $(x,0,z)$ and is parallel to the y-axis; it follows that there exists a Lebesgue measurable set $E'' \subset E'$ such that $\mathcal{L}_3(E \setminus E'') = 0$, and every point in E'' is an endpoint of a compact line segment of positive length which is parallel to the y-axis and lies in E.

Since $\mathcal{L}_3(E'') > 0$, the projection of E'' onto the z-axis must be uncountable. It follows that the set J must contain an uncountable family of mutually disjoint subsets, each of which is homeomorphic to the letter T. Since J is homeomorphic to the closed unit disc in $\mathbf{R}^2$, we obtain a contradiction to the theorem of R. L. Moore stated above, and the proof of Theorem 3.5 is complete.   □

The preceding proof of Theorem 3.5 has been extended in [11] to prove the following: if k and N are positive integers such that $k < N$, and if $p > \max\{1, k-1\}$, then the only distribution in $W^{1,p}(\mathbf{R}^N)$ with support contained in an k-cell is the zero distribution. More recently, this result has been extended by W. K. Ziemer [49].

In recent years there has been a considerable amount of progress in the study of uniform harmonic approximation on unbounded sets. A discussion of this topic in the setting of Euclidean spaces is given in the book of Gardiner [17], and a discussion in the setting of Riemannian manifolds is given in the papers [10, 12]. These techniques permit Theorem 3.5 to be extended to the following theorem [12, Theorem 3]. In the statement of the theorem, "N-dimensional measure" refers to the Lebesgue measure on M defined in Section 2.

3.6 Theorem *Let M be a noncompact, oriented, connected C^∞ Riemannian manifold of dimension $N \geq 2$. Let J be the image of a continuous, proper injection from the closed or* open unit ball of $\mathbf{R}^k$ into M, where $1 \leq k < N$. In case $k \geq 3$, assume also that the N-dimensional measure of J is zero. If f and ε are arbitrary continuous real-valued functions on J, with $\varepsilon > 0$ everywhere, then there is a harmonic function u on M such that $|u - f| < \varepsilon$ at every point of J.

4 Mean approximation by solutions of differential equations

We will now consider mean approximation problems which are analogous to the uniform approximation problem discussed in the preceding section. Throughout this Section 4 we assume that P is a fixed homogeneous polynomial of degree m in $\mathbf{R}^N$, with real coefficients, which satisfies the ellipticity condition

$$P(x) \neq 0 \quad \text{if} \quad x \in \mathbf{R}^N \setminus \{0\}.$$

We define the associated partial differential operator

$$P(D) \equiv P(\partial/\partial x_1, \partial/\partial x_2, \ldots, \partial/\partial x_N).$$

4.1 Remarks (a) If $1 < p < \infty$, there is a positive constant $C = C(P,p)$ such that

$$\|D^\alpha u\|_{L^p} \leq C\|P(D)u\|_{L^p} \quad \text{if} \quad |\alpha| = m \quad \text{and} \quad u \in C_0^\infty(\mathbf{R}^N).$$

This follows from [44, Chapter III, Section 3.5, Corollary].

(b) If $1 < p < \infty$, and if u is a distribution of compact support on $\mathbf{R}^N$ such that both the distribution u and the distribution $P(D)u$ are functions in L^p, then $u \in W^{m,p}$.

We may prove this, using the notation of Convention 2.9, as follows. Select a radius ρ so that $\operatorname{supp} u \subset B(0,\rho)$. From Remark 2.8 we see that $P(D)(u_{[1/j]}) = (P(D)u)_{[1/j]}$ for each positive integer j. Using this fact, the hypothesis on $P(D)u$, and Theorem 2.10 (a), we see that the sequence $\{P(D)(u_{[1/j]})\}_{j=1}^\infty$ must be a Cauchy sequence in $L^p(\mathbf{R}^N)$. Using this fact, the preceding remark (a), and the Poincaré inequality on the ball $B(0,\rho+1)$, we conclude that the sequence $\{u_{[1/j]}\}_{j=1}^\infty$ is a Cauchy sequence in $W^{m,p}(\mathbf{R}^N)$. Then (b) is clear, since we know from Theorem 2.10 (a) that $\{u_{[1/j]}\}_{j=1}^\infty$ converges to u in $L^p(\mathbf{R}^N)$.

(c) If u is a distribution on an open set $\Omega \subset \mathbf{R}^N$ which satisfies the distribution equation $P(D)u = 0$ on Ω, then u must be defined by a function of class C^∞ on Ω and must satisfy $P(D)u = 0$ there in the classical sense.

This is a special case of Weyl's lemma for elliptic equations [27, Corollary 8.3.2].

(d) We say that a distribution E on $\mathbf{R}^N$ is a *fundamental solution* for $P(D)$ if it satisfies the distribution equation $P(D)E = \delta$ on $\mathbf{R}^N$, where δ is the measure consisting of a unit mass at the origin. In view of the hypotheses on $P(D)$, it follows from [27, Theorem 7.1.20] and Remark (c) above that a fundamental solution for $P(D)$ exists, and that every fundamental solution for $P(D)$ is defined by a locally integrable function on $\mathbf{R}^N$. From this and the standard theory of convolutions (see [2, formula (1.1.6)]) we obtain the following: if E is a fundamental solution for $P(D)$, and if $g \in L^p(\mathbf{R}^N)$ has compact support, where $1 < p < \infty$, then $|E * g|^p$ is a locally integrable function on $\mathbf{R}^N$.

4.2 Examples (a) Consider the Laplace operator

$$\Delta \equiv \sum_{i=1}^{N} \frac{\partial^2}{\partial x_i^2}$$

in $\mathbf{R}^N$. If Ω is an open subset of $\mathbf{R}^N$, then the distributions u solving $P(D)u = 0$ on Ω are precisely the harmonic functions on Ω. A fundamental solution for $P(D)$ on the plane $\mathbf{R}^2$ is given by the locally integrable function

$$E(z) = \frac{1}{2\pi} \log |z|.$$

If $N \geq 3$, there is a fundamental solution for Δ on $\mathbf{R}^N$ given by a locally integrable function of the form

$$E(x) = \frac{C_N}{|x|^{N-2}}$$

for some constant C_N.

(b) If Δ is the Laplace operator of the preceding example (a), then the powers $\Delta^2, \Delta^3, \ldots$ are also examples of differential operators $P(D)$ to which the results of this Section 4 apply.

(c) In this paper we have worked with real-valued functions and distributions, but one can develop an analogous theory for complex-valued functions and distributions. If this is done, and the polynomial P in this section is allowed to have complex coefficients instead of real coefficients, then the results given in this Section 4 are still valid. An important example is given by the Cauchy-Riemann operator

$$P(D) \equiv \frac{1}{2} \left(\frac{\partial}{\partial x} + i \frac{\partial}{\partial y} \right)$$

in $\mathbf{R}^2$; here i denotes the imaginary unit in $\mathbf{C}$. If Ω is an open subset of $\mathbf{R}^2$, then the complex-valued distributions u solving $P(D)u = 0$ on Ω are precisely the holomorphic functions on Ω. A fundamental solution for $P(D)$ is given by the locally integrable function

$$E(z) = \frac{1}{\pi z}.$$

The connection between the concept of stability and problems of mean approximation for elliptic equations was studied by Babushka (see [5] and the references given there) and Polking [37]. In particular, the following Theorem 4.3 is a special case of [37, Theorem 1.1]. An exposition of these ideas is given by Adams and Hedberg [2, Theorem 11.5.3].

4.3 Theorem *Let $p > 1$ and $q > 1$ satisfy $1/p + 1/q = 1$. Let $K \subset \mathbf{R}^N$ be a compact set of positive Lebesgue measure. Then the following are equivalent.*

(a) *Every function $u \in L^q(K, dx)$ satisfying*

$$P(D)u \equiv 0 \quad \text{on int } K \tag{4.1}$$

can be approximated in $L^q(K, dx)$ by functions v satisfying

$$P(D)v \equiv 0 \quad \text{near } K. \tag{4.2}$$

(b) *K is (m, p)-stable.*

Proof In the proof we let E be a fundamental solution for $P(D)$.

(b) $\implies$ (a): Assume that (b) holds. We let g be any element of $L^p(K)$ which satisfies

$$\int_K vg\, dx = 0 \tag{4.3}$$

for every function v satisfying (4.2), and we will prove that

$$\int_K ug\, dx = 0 \tag{4.4}$$

for every function $u \in L^q(K, dx)$ satisfying (4.1). In view of the Hahn-Banach theorem, this will prove (a).

We define

$$w \equiv E * g.$$

Condition (4.3) then implies that $\operatorname{supp} w \subset K$, and we have

$$P(D)w = [P(D)E] * g = \delta * g = g.$$

From Remarks 4.1 (b) and 4.1 (d) we conclude that $w \in W^{m,p}$. We now conclude from the stability hypothesis that $w \in W_0^{m,p}(\operatorname{int} K)$, which means that there is a sequence $\{\varphi_j\}$ in $C_0^\infty(\operatorname{int} K)$ satisfying

$$\varphi_j \to w \quad \text{in } W^{m,p}.$$

Now if u is any function in $L^q(K, dx)$ satisfying (4.1), then

$$\int ug\, dx = \int uP(D)w\, dx = \lim_{j\to\infty} \int uP(D)\varphi_j\, dx = \lim_{j\to\infty} 0 = 0,$$

where the next-to-last equality follows from the fact that $P(D)u = 0$ on $\operatorname{int} K$ in the sense of distribution theory. This proves (4.4), and hence we have proved (a).

(a) $\implies$ (b): We assume that (a) holds, and we let w be a fixed element of $W^{m,p}$ which satisfies $\operatorname{supp} w \subset K$. We will prove the following claim: *there is a sequence $\{\varphi_j\}$ in $C_0^\infty(\operatorname{int} K)$ such that*

$$\|P(D)(\varphi_j - w)\|_{L^p} \to 0 \quad \text{as} \quad j \to \infty.$$

Once this claim is proved, we may apply Remark 2.13, Remark 4.1 (a), and the Poincaré inequality on a fixed open ball containing K, to see that $\|\varphi_j - w\|_{W^{m,p}} \to 0$ as $j \to \infty$; and this will complete the proof of (b).

To prove the claim, let g be any element of $L^q(K)$ such that

$$\int gP(D)\varphi\, dx = 0 \quad \text{for all} \quad \varphi \in C_0^\infty(\operatorname{int} K), \tag{4.5}$$

and we will prove that

$$\int gP(D)w\, dx = 0. \tag{4.6}$$

In view of the Hahn-Banach theorem, this will prove the claim.

From (4.5) we see that $P(D)g = 0$ on int K in the sense of distribution theory, so from (a) we can write $g = \lim_j v_j$ in $L^q(K)$, where $P(D)v_j = 0$ on an open neighborhood G_j of K. For each index j we then have

$$\int v_j P(D)w \, dx = 0; \tag{4.7}$$

this is proved by using Remark 2.13 to approximate w in $W^{m,p}$ by functions which lie in $C_0^\infty(G_j)$, and using the fact that $P(D)v_j = 0$ on G_j in the sense of distribution theory. We conclude that

$$\int g P(D)w \, dx = \lim_{j \to \infty} \int v_j P(D)w \, dx = \lim_{j \to \infty} 0 = 0,$$

where the next-to-last equality follows from (4.7). This proves (4.6), which completes the proof of the theorem. $\square$

5   Dirichlet eigenvalues

We now turn to the study of the Dirichlet eigenvalues of the negative of the Laplace operator [14, 15, 28], [46, Chapter 8].

In this section, M denotes a noncompact, oriented, connected C^∞ Riemannian manifold of dimension $N \geq 2$, with the conventions adopted in Section 2. The reader who is interested only in the applications in Euclidean spaces may interpret M as Euclidean space $\mathbf{R}^N$ with the usual flat metric. We let Ω denote a fixed bounded open set in M.

According to Remark 2.24 (b) we may define an inner product on the vector space $W_0^{1,2}(\Omega)$ by the formula

$$[u, v] = \int_\Omega \nabla u \cdot \nabla v \, dV \quad \text{for} \quad u \in W_0^{1,2}(\Omega) \quad \text{and} \quad v \in W_0^{1,2}(\Omega),$$

and we let $H(\Omega)$ denote this inner-product space. In view of the Poincaré inequality for manifolds, the bijection

$$W_0^{1,2}(\Omega) \ni f \to f \in H(\Omega)$$

is an isomorphism of normed spaces; thus $H(\Omega)$ is a Hilbert space, and the Hilbert spaces $H(\Omega)$ and $W_0^{1,2}(\Omega)$ are isomorphic as Banach spaces.

We wish to describe the spectrum of the negative of the Laplace operator acting on $H(\Omega)$. Our development incorporates ideas from the references [6, 41, 45, 46].

For each fixed $f \in L^2(\Omega)$ we define a linear functional

$$T_f : H(\Omega) \to \mathbf{R}$$

by the rule

$$T_f(\varphi) = \int \varphi f \, dV \quad \text{for all} \quad \varphi \in H(\Omega),$$

and note that by the Cauchy-Schwarz inequality and the Poincaré inequality on manifolds we have

$$|T_f(\varphi)| \leq \|f\|_{L^2(\Omega)} \|\varphi\|_{L^2(\Omega)} \leq C \|f\|_{L^2(\Omega)} \|\varphi\|_{H(\Omega)} \quad \text{for all} \quad \varphi \in H(\Omega),$$

where C is a positive constant depending on Ω. Thus the functional $T_f : H(\Omega) \to \mathbf{R}$ is bounded, with norm

$$\|T_f\|_{[H(\Omega)]^*} \le C\|f\|_{L^2(\Omega)}.$$

From the Riesz representation theorem we see that there is an element $w_f \in H(\Omega)$ such that

$$\|w_f\|_{H(\Omega)} = \|T_f\|_{[H(\Omega)]^*} \le C\|f\|_{L^2(\Omega)}$$

and

$$\int \varphi f \, dV = [\varphi, w_f] = \int \nabla\varphi \cdot \nabla(w_f) \, dV \quad \text{for all} \quad \varphi \in H(\Omega).$$

It follows that

$$L^2(\Omega) \ni f \to w_f \in H(\Omega) \tag{5.1}$$

is a bounded linear mapping.

According to the Rellich-Kondrachov Theorem 2.25, the imbedding

$$H(\Omega) \ni \varphi \to \varphi \in L^2(\Omega) \tag{5.2}$$

is a compact linear mapping. We now define the linear mapping

$$S : L^2(\Omega) \to L^2(\Omega)$$

as the composition of the bounded linear mapping (5.1) and the compact linear mapping (5.2), so that S is a compact linear operator on $L^2(\Omega)$.

5.1 Remarks We will need the following facts about the map $S : L^2(\Omega) \to L^2(\Omega)$.

(a) *S is symmetric.*

In fact, if f and f' are elements of $L^2(\Omega)$, then

$$\int_\Omega f(Sf') \, dV = \int_\Omega f w_{f'} \, dV = [w_{f'}, w_f] = \int_\Omega f' w_f \, dV = \int_\Omega f'(Sf) \, dV.$$

(b) *S is positive; that is,* $\int_\Omega fSf \, dV \ge 0 \quad \text{for all} \quad f \in L^2(\Omega).$

In fact, if $f \in L^2(\Omega)$, then

$$\int_\Omega fSf \, dV = \int_\Omega f w_f \, dV = [w_f, w_f] \ge 0.$$

(c) *S is one-to-one.*

In fact, if $f \in L^2(\Omega)$ satisfies $Sf = 0$, then $w_f = 0$. Thus

$$\int_\Omega \varphi f \, dV = [\varphi, w_f] = 0 \quad \text{for all} \quad \varphi \in C_0^\infty(\Omega),$$

which means that f is the zero distribution on Ω.

We will say that a real number σ is an *eigenvalue* for S provided that there exist nonzero elements $f \in L^2(\Omega)$ satisfying

$$Sf = \sigma f. \tag{5.3}$$

If σ is an eigenvalue for S, then the *eigenspace* of S corresponding to σ is the set of all $f \in L^2(\Omega)$ satisfying (5.3).

If $L^2_{\mathbf{C}}(\Omega, dV)$ denotes the usual complex Hilbert space of complex-valued Lebesgue measurable functions w on Ω such that $\int |w|^2 \, dV < \infty$, we may define a function $\widetilde{S}: L^2_{\mathbf{C}}(\Omega, dV) \to L^2_{\mathbf{C}}(\Omega, dV)$ by setting $\widetilde{S}(F + iG) = SF + iSG$ for all real-valued functions F and G in $L^2(\Omega)$. Then $\widetilde{S}$ is a complex-linear, one-to-one, compact, positive self-adjoint operator on $L^2_{\mathbf{C}}(\Omega, dV)$, and from the spectral theorem for such operators one may prove the following theorem for decomposition of the real Hilbert space $L^2(\Omega)$.

5.2 Lemma *The set of all eigenvalues for S can be written as an infinite sequence of positive numbers approaching zero. For each eigenvalue σ, the eigenspace*

$$\{f \in L^2(\Omega) : Sf = \sigma f\}$$

is finite-dimensional; and these finite-dimensional spaces form an orthogonal decomposition of $L^2(\Omega)$.

5.3 Remarks (a) In the following development we will sometimes say that an element $f \in W^{1,2}_0(\Omega)$ satisfies

$$-\Delta f = \lambda f \quad \text{on } \Omega. \tag{5.4}$$

Statement (5.4) is to be interpreted in the sense of the theory of distributions; it means that

$$\int_\Omega (-f\Delta\varphi - \lambda f\varphi) \, dV = 0 \quad \text{for every} \quad \varphi \in C^\infty_0(\Omega).$$

However, in view of Weyl's lemma for elliptic equations, any distribution on Ω which satisfies (5.4) in this sense is automatically of class C^∞ on Ω and satisfies (5.4) in the classical sense.

(b) If $f \in W^{1,2}_0(\Omega)$ and $\varphi \in C^\infty_0(\Omega)$, then

$$\int_\Omega f\Delta\varphi \, dV = -\int_\Omega \nabla f \cdot \nabla\varphi \, dV.$$

In fact, this equation follows from a well-known Green formula in case $f \in C^\infty_0(\Omega)$, and hence the equation holds for arbitrary $f \in W^{1,2}_0(\Omega)$ by a limit argument.

5.4 Lemma *Let $f \in L^2(\Omega)$ and λ a nonzero real number. Then the following are equivalent:*

(a) $Sf = (1/\lambda)f$;

(b) f *is in* $W^{1,2}_0(\Omega)$ *and*

$$-\Delta f = \lambda f \quad \text{on } \Omega.$$

Before giving the proof of Lemma 5.4, we recall that in this paper we regard each element of $W_0^{1,2}(\Omega)$ as an element of the Sobolev space $W^{1,2}(M)$; thus in the statement of part (b), we are identifying each function in $W_0^{1,2}(\Omega)$ with its restriction to Ω, and we will continue this identification later in this section. In order to make this identification we note that the restriction operation $f \to f|_\Omega$ defines a one-to-one mapping from $W_0^{1,2}(\Omega)$ into $L^2(\Omega)$.

Proof (a) $\implies$ (b): Suppose that (a) holds. Then $f = \lambda w_f$ on Ω, so f is the restriction to Ω of the function $\lambda w_f \in W_0^{1,2}(\Omega)$. In view of Remark 5.3 (b) it follows that for any $\varphi \in C_0^\infty(\Omega)$ we have

$$\int_\Omega (-f\Delta\varphi - \lambda f\varphi)\, dV = \int_\Omega \nabla f \cdot \nabla\varphi\, dV - \lambda \int_\Omega f\varphi\, dV$$
$$= \lambda \int_\Omega (\nabla w_f) \cdot (\nabla\varphi)\, dV - \lambda \int_\Omega f\varphi\, dV$$
$$= \lambda \int_\Omega f\varphi\, dV - \lambda \int_\Omega f\varphi\, dV$$
$$= 0,$$

which proves (b).

(b) $\implies$ (a): Suppose that (b) holds. From Remark 5.3 (b) we see that for any $\varphi \in C_0^\infty(\Omega)$ we have

$$0 = \int_\Omega (-f\Delta\varphi - \lambda f\varphi)\, dV$$
$$= \int_\Omega (\nabla f \cdot \nabla\varphi - \lambda \nabla\varphi \cdot \nabla w_f)\, dV$$
$$= [f - \lambda w_f, \varphi].$$

Since $C_0^\infty(\Omega)$ is dense in $W_0^{1,2}(\Omega)$, it follows that $f - \lambda w_f = 0$, which proves (a). $\square$

5.5 Definition We say that a real number λ is a *Dirichlet eigenvalue for* $-\Delta$ *on* Ω provided that the vector space

$$\mathcal{E}(\lambda, \Omega) \equiv \{f \in W_0^{1,2}(\Omega) : -\Delta f = \lambda f \text{ on } \Omega\}$$

has dimension at least one. More briefly, a Dirichlet eigenvalue for $-\Delta$ on Ω will be called a *Dirichlet eigenvalue* for Ω. If λ is a Dirichlet eigenvalue for Ω, then we refer to $\mathcal{E}(\lambda, \Omega)$ as the *eigenspace* corresponding to λ. The set of all Dirichlet eigenvalues for Ω is the *Dirichlet spectrum* for Ω.

5.6 Remark The number $\lambda = 0$ is not a Dirichlet eigenvalue for Ω. (This follows from the proof of the implication (b) $\implies$ (a) in Lemma 5.4.)

5.7 Lemma *Let λ be a nonzero real number.*

(a) *λ is a Dirichlet eigenvalue for Ω if and only if $1/\lambda$ is an eigenvalue for S.*

(b) *If λ is a Dirichlet eigenvalue for Ω, then the eigenspace of S corresponding to $1/\lambda$ consists exactly of the functions in the eigenspace $\mathcal{E}(\lambda, \Omega)$.*

In part (b), we are identifying the functions in $\mathcal{E}(\lambda, \Omega)$ with their restrictions to Ω. Lemma 5.7 follows from Lemma 5.4

From Lemma 5.2 and Lemma 5.7 we immediately obtain the following theorem.

5.8 Theorem *The Dirichlet eigenvalues for Ω can be written as an infinite sequence of positive numbers approaching plus infinity. For each of these eigenvalues λ, the eigenspace $\mathcal{E}(\lambda, \Omega)$ is finite-dimensional; and these finite-dimensional spaces form an orthogonal decomposition of $L^2(\Omega)$.*

As before, we we are identifying the functions in $\mathcal{E}(\lambda, \Omega)$ with their restrictions to Ω.

We will say that the Dirichlet eigenvalues for Ω are "listed by multiplicity" in the sequence $\lambda_1 \leq \lambda_2 \leq \ldots$ provided that the number of times each real number λ occurs in the sequence is exactly the dimension of $\mathcal{E}(\lambda, \Omega)$.

5.9 Definition Let $\lambda_1 \leq \lambda_2 \leq \ldots$ be the Dirichlet eigenvalues for Ω listed by multiplicity. We will then say that elements $e_1, e_2, \ldots$ of $L^2(\Omega)$ form a *special basis* for $L^2(\Omega)$ provided that they form an orthonormal basis for $L^2(\Omega)$, and

$$e_i \in \mathcal{E}(\lambda_i, \Omega) \quad \text{for each positive integer } i.$$

From Theorem 5.8 we see that a special basis for $L^2(\Omega)$ does exist.

5.10 Remarks Let $\lambda_1 \leq \lambda_2 \leq \ldots$ be the Dirichlet eigenvalues for Ω listed by multiplicity, and let $e_1, e_2, \ldots$ be a special basis for $L^2(\Omega)$.

(a) If $u \in W_0^{1,2}(\Omega)$, then

$$[u, e_i] = \lambda_i \int_\Omega u e_i \, dV \quad \text{for each positive integer } i.$$

In fact, it suffices to prove this for $u \in C_0^\infty(\Omega)$, and in this case we see from Remark 5.3 that

$$0 = \int_\Omega (-e_i \Delta u - \lambda_i u e_i) \, dV = \int_\Omega (\nabla u \cdot \nabla e_i - \lambda_i u e_i) \, dV.$$

This proves (a).

(b) $[e_i, e_j] = \begin{cases} \lambda_i & \text{if } i = j \\ 0 & \text{if } i \neq j \end{cases}$. This is a special case of (a).

We now give a well-known proof [14, Chapter I, Section 5] of Rayleigh's variational principle for the Dirichlet eigenvalues.

5.11 Lemma *Let $\lambda_1 \leq \lambda_2 \leq \ldots$ be the Dirichlet eigenvalues for Ω listed by multiplicity, and let $e_1, e_2, \ldots$ be a special basis for $L^2(\Omega)$. If $u \in W_0^{1,2}(\Omega)$, and*

$$c_i = \int_\Omega u e_i \, dV \quad \text{for each positive integer } i,$$

then

$$\sum_{i=1}^\infty c_i^2 \lambda_i \leq [u, u].$$

Proof If k is any positive integer, we clearly have

$$0 \leq \left[u - \sum_{i=1}^{k} c_i e_i, u - \sum_{j=1}^{k} c_j e_j\right];$$

expanding the right side, and using the hypotheses and Remarks 5.10, we obtain

$$\sum_{i=1}^{k} c_i^2 \lambda_i \leq [u, u].$$

The lemma follows from this inequality. □

5.12 Theorem (Rayleigh theorem) *Let $\lambda_1 \leq \lambda_2 \leq \ldots$ be the Dirichlet eigenvalues for Ω listed by multiplicity, and let $e_1, e_2, \ldots$ be a special basis for $L^2(\Omega)$.*

(a) $\lambda_1 = \inf\limits_{u} \dfrac{\int_{\Omega} |\nabla u|^2 \, dV}{\int_{\Omega} u^2 \, dV}$, *where the infimum is taken over all $u \neq 0$ in $W_0^{1,2}(\Omega)$.*

(b) *If $n \geq 2$, then $\lambda_n = \inf\limits_{u} \dfrac{\int_{\Omega} |\nabla u|^2 \, dV}{\int_{\Omega} u^2 \, dV}$, where the infimum is taken over all $u \neq 0$ in $W_0^{1,2}(\Omega)$ satisfying*

$$\int_{\Omega} u e_i \, dV = 0 \quad \text{for all} \quad i \leq n - 1.$$

Proof This theorem follows from the preceding lemma and Remark 5.10 (b). □

5.13 Definition If Ω is a bounded open set in M, and $\lambda_1 \leq \lambda_2 \leq \ldots$ are the Dirichlet eigenvalues for Ω listed by multiplicity, then for each $n \in \{1, 2, \ldots\}$ we will use the notation $\lambda_n(\Omega)$ for λ_n.

5.14 Theorem *Let Ω be a bounded open set in M, and let C and D be constants. Suppose that $n \geq 2$, and that $h_1, h_2, \ldots, h_n$ are elements of $W_0^{1,2}(\Omega)$ such that*

$$\int_{\Omega} |\nabla h_i|^2 \, dV \leq C \quad \text{for} \quad i = 1, 2, \ldots, n \tag{5.5}$$

and

$$\left| \int_{\Omega} \nabla h_i \cdot \nabla h_j \, dV \right| \leq D \quad \text{for} \quad 1 \leq i < j \leq n. \tag{5.6}$$

Let

$$I_{ij} = \int_{\Omega} h_i h_j \, dV \quad \text{for} \quad 1 \leq i, j \leq n, \tag{5.7}$$

and suppose that the symmetric matrix (I_{ij}) has all eigenvalues $\geq L$, where $L > 0$. Then

$$\lambda_n(\Omega) \leq \frac{C + n(n-1)D}{L}.$$

Proof Let $e_1, e_2, \ldots$ be a special basis for $L^2(\Omega)$, and let $n \geq 2$ be a fixed integer. We note that there exist real numbers $a_1, a_2, \ldots, a_n$ such that $a_1^2 + a_2^2 + \cdots + a_n^2 = 1$ and

$$\int_\Omega (a_1 h_1 + a_2 h_2 + \cdots + a_n h_n) e_k \, dV = 0 \quad \text{for all positive integers} \quad k \leq n-1; \qquad (5.8)$$

this is clear if we regard (5.8) as a homogeneous system of $n-1$ equations in n unknowns $a_1, a_2, \ldots, a_n$. Writing $h \equiv a_1 h_1 + a_2 h_2 + \cdots + a_n h_n$, it now follows from our hypothesis on the matrix (I_{ij}) that

$$\|h\|_{L^2(\Omega)}^2 = \sum_{i,j=1}^n I_{ij} a_i a_j \geq L. \qquad (5.9)$$

We now have

$$\lambda_n(\Omega) \leq \frac{[h,h]}{\|h\|_{L^2(\Omega)}^2}$$

$$\leq \frac{1}{L} \sum_{1 \leq i,j \leq n} [h_i, h_j] a_i a_j$$

$$\leq \frac{1}{L} \left((a_1^2 + a_2^2 + \cdots + a_n^2) C + \sum_{1 \leq i,j \leq n;\ i \neq j} |a_i a_j| D \right)$$

$$\leq \frac{C + n(n-1)D}{L},$$

where the first inequality follows from (5.8) and the Rayleigh Theorem 5.12; the second inequality follows from (5.9); the third inequality follows from (5.5) and (5.6); and the fourth inequality follows from the fact that $a_1^2 + a_2^2 + \cdots + a_n^2 = 1$. This proves the theorem. $\qquad\square$

5.15 Theorem (Monotonicity of eigenvalues) *Suppose Ω_1 and Ω_2 are bounded open subsets of M such that $\Omega_1 \subset \Omega_2$. Then*

$$\lambda_n(\Omega_2) \leq \lambda_n(\Omega_1) \quad \text{for every positive integer } n. \qquad (5.10)$$

Proof If $n = 1$, then (5.10) follows from part (a) of the Rayleigh Theorem 5.12, since $W_0^{1,2}(\Omega_1) \subset W_0^{1,2}(\Omega_2)$.

If $n \geq 2$, we let $e_1, e_2, \ldots$ be a special basis for $L^2(\Omega_1)$; then (5.10) follows from applying Theorem 5.14 with $\Omega = \Omega_2$ and

$$h_i = e_i \quad \text{for} \quad i = 1, 2, \ldots, n.$$

$\qquad\square$

6   Continuity of Dirichlet eigenvalues

In this section, M denotes a noncompact, oriented, connected C^∞ Riemannian manifold of dimension $N \geq 2$. As before, the reader who is interested only in the applications in Euclidean spaces may interpret M as Euclidean space $\mathbf{R}^N$ with the usual flat metric.

In this Section 6 we will present theorems of Babushka [6] concerning the dependence of the Dirichlet eigenvalues $\lambda_n(\Omega)$ on the open set Ω. For further work on related topics we refer to [3, 47], and to the references given in these papers and in [6].

The following Theorem 6.1 is due to Babushka [6], and the theorem is also part of Theorem 6.2.3 in the book by E. B. Davies, *Spectral Theory and Differential Operators*, Cambridge University Press, 1995. Both of these authors work in the setting of Euclidean spaces, but their proofs also apply in the setting of Riemannian manifolds. Our proof incorporates ideas used by these writers, but makes use of Theorem 5.14.

6.1 Theorem (Continuity of eigenvalues from the inside) *If $\Omega_1 \subset \Omega_2 \subset \ldots$ are non-empty open subsets of M such that the union $\Omega \equiv \bigcup_{\ell=1}^{\infty} \Omega_\ell$ is bounded, then*

$$\lim_{\ell \to \infty} \lambda_n(\Omega_\ell) = \lambda_n(\Omega) \quad \text{for any positive integer } n.$$

Proof Let $\lambda_1 \leq \lambda_2 \leq \ldots$ be the Dirichlet eigenvalues for Ω listed by multiplicity, and let $e_1, e_2, \ldots$ be a special basis for $L^2(\Omega)$. In this proof all integrals are understood to be taken over M.

From the monotonicity of eigenvalues we know that, for each positive integer n, the limit in the statement of Theorem 6.1 exists and

$$\lim_{\ell \to \infty} \lambda_n(\Omega_\ell) \geq \lambda_n. \tag{6.1}$$

We must prove the reverse inequality.

We now let n be a fixed positive integer, and let $\varepsilon \in (0,1)$ be arbitrary. By approximating the functions $e_1, e_2, \ldots, e_n$ by functions in $C_0^{\infty}(\Omega)$, in the norm of $W_0^{1,2}(\Omega)$, we may obtain functions $h_1, h_2, \ldots, h_n$ in $C_0^{\infty}(\Omega)$ with the following properties: the $n \times n$ symmetric matrix (I_{ij}) with entries

$$I_{ij} = \int h_i h_j \, dV \quad \text{for} \quad 1 \leq i, j \leq n$$

has all its eigenvalues $\geq 1 - \varepsilon$; and

$$\int |\nabla h_i|^2 \, dV \leq \lambda_i + \varepsilon \quad \text{for} \quad i = 1, 2, \ldots, n;$$

and if $n \geq 2$, then

$$\left| \int \nabla h_i \cdot \nabla h_j \, dV \right| \leq \varepsilon \quad \text{for} \quad 1 \leq i < j \leq n.$$

Note that there is an index ℓ_0 such that for all $\ell > \ell_0$ we have

$$h_i \in C_0^{\infty}(\Omega_\ell) \quad \text{for} \quad i = 1, 2, \ldots, n.$$

In case $n = 1$, we may apply part (a) of the Rayleigh Theorem 5.12 for the open sets Ω_ℓ, for $\ell > \ell_0$, to obtain

$$\lim_{\ell \to \infty} \lambda_1(\Omega_\ell) \leq \frac{\int |\nabla h_1|^2 \, dV}{\int h_1^2 \, dV} \leq \frac{\lambda_1 + \varepsilon}{1 - \varepsilon}. \tag{6.2}$$

From (6.1), (6.2) and the fact that $\varepsilon \in (0,1)$ is arbitrary, we obtain Theorem 6.1 for $n = 1$.

In case $n \geq 2$, we may apply Theorem 5.14 to the functions $h_1, h_2, \ldots, h_n$ on the open sets Ω_ℓ, for $\ell > \ell_0$, to obtain

$$\lim_{\ell \to \infty} \lambda_n(\Omega_\ell) \leq \frac{\lambda_n + \varepsilon + n(n-1)\varepsilon}{1 - \varepsilon} \tag{6.3}$$

From (6.1), (6.3) and the fact that $\varepsilon \in (0,1)$ is arbitrary, we obtain Theorem 6.1 for $n \geq 2$. □

6.2 Definition Let K be a compact subset of M. We let H_K be the set of elements $u \in W^{1,2}(M)$ such that

$$\int_M u\varphi \, dV = 0 \quad \text{for all} \quad \varphi \in C_0^\infty(M \setminus K).$$

In other words, H_K consists of all elements of $W^{1,2}(M)$ which are supported on K.

6.3 Remarks (a) If K is a compact subset of M, then H_K is a closed subspace of $W^{1,2}(M)$.

(b) Let K be a compact subset of a bounded open set $\Omega \subset M$. Then $H_K \subset H(\Omega)$. (This is proved by using a partition of unity subordinate to a finite cover of K by parametric balls whose closures lie in Ω, and using Remark 2.13 and Remark 2.23.)

6.4 Remark Suppose that K is a compact subset of M, and that $\Omega_1, \Omega_2, \ldots$ are bounded open subsets of M such that $\Omega_{\ell+1} \subset\subset \Omega_\ell$ for each positive integer ℓ, and

$$K = \bigcap_{\ell=1}^\infty \Omega_\ell.$$

Then

$$H_K = \bigcap_{\ell=1}^\infty H(\Omega_\ell).$$

To prove Remark 6.4, we first note that by Remark 6.3 (b) the left side of this equality is a subset of the right side. To prove the reverse inclusion, suppose that $u \in \bigcap_{\ell=1}^\infty H(\Omega_\ell)$. If $\varphi \in C_0^\infty(M \setminus K)$ is arbitrary, then the sets $\{M \setminus \overline{\Omega_\ell}\}_{\ell=1}^\infty$ form an open cover of the compact set $\operatorname{supp} \varphi$. It follows that there must be some index L such that $\operatorname{supp} \varphi$ is disjoint from $\overline{\Omega_L}$; since $u \in H(\Omega_L)$, we conclude that

$$\int u\varphi \, dV = 0.$$

Since $\varphi \in C_0^\infty(M \setminus K)$ is arbitrary, this proves that $u \in H_K$. This completes the proof of the remark.

6.5 Definition Let K be a compact subset of M. We say that K is $(1,2)$-*stable* provided that $H_K = H(\operatorname{int} K)$.

The following theorem is due to Babushka [6]. Our proof makes use of the ideas of Babushka and Theorem 5.14.

6.6 Theorem (Continuity of eigenvalues from the outside) *Suppose that K is a compact subset of M with nonempty interior Ω. Suppose that $\Omega_1, \Omega_2, \ldots$ are bounded open subsets of M such that $\Omega_{\ell+1} \subset\subset \Omega_\ell$ for each positive integer ℓ, and*

$$K \equiv \bigcap_{\ell=1}^{\infty} \Omega_\ell.$$

Then the following are equivalent.

(a) $\lim_{\ell \to \infty} \lambda_n(\Omega_\ell) = \lambda_n(\Omega)$ *for each positive integer n .*

(b) K *is $(1,2)$-stable.*

Proof For each of the bounded open sets U which appear in the statement of the theorem, we let $e_1(U), e_2(U), \ldots$ be a special basis for $L^2(U)$. In this proof all integrals are understood to be taken over M. Recalling Remarks 6.3, we will regard H_K as a closed subspace of the Hilbert space $H(\Omega_1)$, with the inner product inherited from $H(\Omega_1)$.

From the monotonicity of eigenvalues we see that for each positive integer n, the limit

$$\mu_n \equiv \lim_{\ell \to \infty} \lambda_n(\Omega_\ell)$$

exists, and

$$0 < \lambda_n(\Omega_1) \le \mu_n \le \lambda_n(\Omega). \tag{6.4}$$

Note that $\mu_1 \le \mu_2 \le \ldots$. From (6.4) and Theorem 5.8 for the bounded open set Ω_1 we see that

$$\lim_{n \to \infty} \mu_n = \infty. \tag{6.5}$$

We remark that the Hilbert space $H(\Omega_1)$ is separable (this can be proved by using a partition of unity subordinate to a cover of $\overline{\Omega_1}$ by parametric balls, and applying Remark 2.7 (e) and Remark 2.23.) Now by applying Remarks 2.7 and the Rellich-Kondrachov Theorem 2.25 repeatedly, with a diagonal argument, and Remark 6.4 above, we see that there is an increasing sequence of indices $\ell_1 < \ell_2 < \ldots$, and there are functions $w_1, w_2, \ldots$ in H_K, such that the following property holds: for each positive integer n, the sequence $\{e_n(\Omega_{\ell_s})\}_{s=1}^{\infty}$ converges to w_n weakly in $H(\Omega_1)$ and in the norm topology of $L^2(M)$. From the norm convergence in $L^2(M)$ we conclude that

$$\int w_i w_j \, dV = \begin{cases} 1 & \text{if } i = j \\ 0 & \text{if } i \ne j. \end{cases} \tag{6.6}$$

From Remark 5.10 (a) we see that for each positive integer s we have

$$[v, e_i(\Omega_{\ell_s})] = \lambda_i(\Omega_{\ell_s}) \int v e_i(\Omega_{\ell_s}) \, dV \quad \text{for} \quad v \in H(\Omega_{\ell_s}) \quad \text{and} \quad i = 1, 2, \ldots \tag{6.7}$$

Using the fact that for each positive integer i, the sequence $\{e_i(\Omega_{\ell_s})\}_{s=1}^{\infty}$ converges to w_i weakly in $H(\Omega_1)$ and in the norm topology of $L^2(M)$, we conclude from (6.7) and Remark 6.3 (b) that

$$[v, w_i] = \mu_i \int v w_i \, dV \quad \text{for} \quad v \in H_K \quad \text{and} \quad i = 1, 2, \ldots, \tag{6.8}$$

and in particular,

$$[w_i, w_j] = \begin{cases} \mu_i & \text{if } i = j \\ 0 & \text{if } i \neq j. \end{cases} \tag{6.9}$$

Claim 1: *If n is a positive integer, and $h \in H_K$ satisfies*

$$\int \nabla h \cdot \nabla w_i \, dV = 0 \quad \text{for} \quad i = 1, 2, \ldots, n, \tag{6.10}$$

then

$$\mu_{n+1} \int h^2 \, dV \leq \int |\nabla h|^2 \, dV.$$

Proof of Claim 1: For each positive integer s, we see from Remark 6.3 (b) that we may regard h as an element of $H(\Omega_{\ell_s})$, and we let q_s be the projection of h onto the subspace Y_s of $H(\Omega_{\ell_s})$ spanned by $\{e_i(\Omega_{\ell_s}) : i \leq n\}$, so that

$$\int |\nabla q_s|^2 \, dV = \sum_{i=1}^{n} \left(\frac{1}{\sqrt{\lambda_i(\Omega_{\ell_s})}} \int \nabla h \cdot \nabla e_i(\Omega_{\ell_s}) \, dV \right)^2.$$

Using the fact that for each positive integer i, the sequence $\{e_i(\Omega_{\ell_s})\}_{s=1}^{\infty}$ converges to w_i weakly in $H(\Omega_1)$, we conclude from (6.4) and (6.10) that each summand on the right side of this equation approaches zero as $s \to \infty$, and hence the left side approaches zero as $s \to \infty$; from this fact, and the Poincaré inequality on the bounded open set Ω_1, we conclude that

$$\lim_{s \to \infty} \int q_s^2 \, dV = 0$$

and therefore

$$\lim_{s \to \infty} \int (h - q_s)^2 \, dV = \int h^2 \, dV. \tag{6.11}$$

Since $h - q_s$ is the projection of h onto the orthogonal complement of Y_s in $H(\Omega_{\ell_s})$, we have

$$\int |\nabla h|^2 \, dV \geq \int |\nabla(h - q_s)|^2 \, dV \geq \lambda_{n+1}(\Omega_{\ell_s}) \int (h - q_s)^2 \, dV, \tag{6.12}$$

where the second inequality follows from the Rayleigh Theorem 5.12 for the bounded open set Ω_{ℓ_s} and Remark 5.10 (a). Claim 1 follows from (6.11) and (6.12).

Claim 2: *The sequence $\{w_i/\sqrt{\mu_i}\}_{i=1}^{\infty}$ is an orthonormal basis for H_K.*

Proof of Claim 2: We know from (6.9) that $\{w_i/\sqrt{\mu_i}\}_{i=1}^{\infty}$ is an orthonormal set in H_K. Moreover, if h is any function in H_K which satisfies

$$\int \nabla h \cdot \nabla w_i \, dV = 0 \quad \text{for} \quad i = 1, 2, \ldots,$$

then from Claim 1 we have

$$\mu_{n+1} \int h^2 \, dV \leq \int |\nabla h|^2 \, dV,$$

for every positive integer n, and hence from (6.5) we conclude that $h = 0$. This proves Claim 2.

We now prove the equivalence of the conditions in Theorem 6.6.

(b) $\implies$ (a): Suppose that (b) is true. Then for each positive integer i we have $w_i \in W_0^{1,2}(\Omega)$. In view of (6.6) and (6.9), we may apply part (a) of the Rayleigh Theorem 5.12 to obtain

$$\lambda_1(\Omega) \leq \frac{\int |\nabla w_1|^2 \, dV}{\int w_1^2 \, dV} = \mu_1,$$

and for $n \geq 2$ we may apply Theorem 5.14 to the functions $w_1, w_2, \ldots, w_n$ to obtain

$$\lambda_n(\Omega) \leq \mu_n.$$

From these facts and (6.4) we obtain (a).

(a) $\implies$ (b): We list all the *distinct* numbers of the sequence $\{\mu_i\}_{i=1}^{\infty}$ in the strictly increasing sequence

$$\nu_1 < \nu_2 < \ldots,$$

and for each positive integer k we let $\mathcal{I}_k$ be the set of all the positive integers i such that $\mu_i = \nu_k$. If k is any positive integer, we define $\mathcal{E}_k$ to be the set of all real linear combinations of the functions $e_i(\Omega)$ with $i \in \mathcal{I}_k$, and we define $\mathcal{W}_k$ to be the set of all real linear combinations of the functions w_i with $i \in \mathcal{I}_k$; from (6.6) and (6.9) we obtain

$$[u, u] = \nu_k \int u^2 \, dV \quad \text{for all} \quad u \in \mathcal{W}_k. \tag{6.13}$$

We will now prove that

$$\mathcal{E}_k = \mathcal{W}_k \quad \text{for every positive integer } k. \tag{6.14}$$

Since the finite-dimensional subspaces $\mathcal{W}_1, \mathcal{W}_2, \ldots$ form an orthogonal decomposition of H_K by Claim 2, it will follow from (6.14) that H_K is the orthogonal direct sum of finite-dimensional subspaces lying in $H(\Omega)$, and hence $H_K = H(\Omega)$, which means that K is $(1, 2)$-stable. This will complete the proof of Theorem 6.6.

We now give the proof of (6.14) by mathematical induction. For this purpose we assume that k is a fixed positive integer, and in case $k \geq 2$ we also assume that

$$\mathcal{E}_t = \mathcal{W}_t \quad \text{for all positive integers } t < k. \tag{6.15}$$

We will then prove that $\mathcal{E}_k = \mathcal{W}_k$, which will complete the proof of (6.14).

We next let $i \in \mathcal{I}_k$ be arbitrary, and we will prove that

$$e_i(\Omega) \in \mathcal{W}_k. \tag{6.16}$$

Once this is proved, it will be clear that $\mathcal{E}_k \subset \mathcal{W}_k$; since we have

$$\dim \mathcal{E}_k = [\text{cardinality of } \mathcal{I}_k] = \dim \mathcal{W}_k,$$

it will follow that $\mathcal{E}_k = \mathcal{W}_k$, as required.

We turn now to the proof of (6.16). Regarding $e_i(\Omega)$ as an element of the Hilbert space H_K, we see from our assumption (6.15) that $e_i(\Omega)$ is orthogonal to all the spaces $\mathcal{W}_t = \mathcal{E}_t$ with $t < k$; thus we can find an orthogonal decomposition in H_K,

$$e_i(\Omega) = F + G,$$

where $F \in \mathcal{W}_k$ and G is orthogonal in H_K to all the spaces $\mathcal{W}_t$ with $t \le k$. It then follows from (6.8) that

$$\int FG \, dV = 0,$$

and hence

$$\int F^2 \, dV + \int G^2 \, dV = \int e_i(\Omega)^2 \, dV = 1. \qquad (6.17)$$

We also have

$$\nu_k = \mu_i = \lambda_i(\Omega) = [e_i(\Omega), e_i(\Omega)] = [F, F] + [G, G] \ge \nu_k \int F^2 \, dV + \nu_{k+1} \int G^2 \, dV, \quad (6.18)$$

where the second equality follows from hypothesis (a), the fourth equality follows from the fact that $[F, G] = 0$, and the inequality follows from (6.13) and Claim 1. From (6.17) and (6.18) we conclude that $\int G^2 \, dV = 0$, and hence $G = 0$. Thus $e_i(\Omega) = F \in \mathcal{W}_k$, which completes the proof of (6.16) and the proof of Theorem 6.6. $\qquad \square$

Babushka [6] has also proved results analogous to Theorems 6.1 and 6.6 for other elliptic equations.

7   Criteria for stability

We have now seen that the concept of "(m, p)-stability of a compact set" unifies several approximation problems. In this section we will state without proofs some criteria for (m, p)-stability for compact subsets of $\mathbf{R}^N$, where $N \ge 2$.

7.1 Theorem *If $p > N$, and m is any positive integer, then every compact subset of $\mathbf{R}^N$ is (m, p)-stable.*

Theorem 7.1 is due to Havin [18], Polking [37, Theorem 2.9 (a)], and Burenkov [13]. In order to understand stability in other situations, we introduce set functions known as "capacities."

7.2 Remark There is a unique integrable function G_1 on $\mathbf{R}^N$ such that the Fourier transform $\widehat{G_1}(\xi) \equiv \int_{\mathbf{R}^N} G_1(x) e^{-i\xi \cdot x} \, dx$ is given by the function

$$\widehat{G_1}(\xi) = \frac{1}{(1 + |\xi|^2)^{1/2}}.$$

The reader can find basic properties of the function G_1 and further references in [33, Section 7] and [2, Section 1.2]. Let us note here that $G_1 \ge 0$; we have $G_1(x) = O(e^{-|x|/2})$ as $x \to \infty$; and there are positive constants A and B such that

$$\frac{A}{|x|^{N-1}} \le G_1(x) \le \frac{B}{|x|^{N-1}} \quad \text{for} \quad |x| \le 1.$$

The following Definition 7.3 and Theorem 7.4 were given by Meyers [33].

7.3 Definition (a) If E is any subset of $\mathbf{R}^N$, we define

$$\gamma_{1,p}(E) = \inf_f \int_{\mathbf{R}^N} f^p \, dx,$$

where the infimum is taken over all non-negative functions f in $L^p(\mathbf{R}^N)$ such that $G_1 * f \geq 1$ on E.

(b) We say that a set $E \subset \mathbf{R}^N$ is a $(1,p)$-*nullset* if $\gamma_{1,p}(E) = 0$.

7.4 Theorem (a) *If $p > N$, then there is a positive constant $C = C(N,p)$ such that $\gamma_{1,p}(E) \geq C$ for every nonempty set $E \subset \mathbf{R}^N$.*

(b) *If $p < N$, there are positive constants A and B such that*

$$Ar^{N-p} \leq \gamma_{1,p}(B(a,r)) \leq Br^{N-p} \quad \text{for} \quad a \in \mathbf{R}^N \quad \text{and} \quad r \leq 1.$$

(c) *There are positive constants A and B such that*

$$A(\log(1/r))^{1-N} \leq \gamma_{1,N}(B(a,r)) \leq B(\log(1/r))^{1-N} \quad \text{for} \quad a \in \mathbf{R}^N \quad \text{and} \quad r \leq 1/2.$$

We now define the concept of $(1,p)$-thinness of a set at a point. For more information about this concept, including historical references, we refer to [2, Chapter 6].

7.5 Definition Let $1 < p \leq N$. We say that a set $E \subset \mathbf{R}^N$ is $(1,p)$-*thin* at the point $a \in \mathbf{R}^N$ provided

$$\int_0^1 \left(\frac{\gamma_{1,p}(E \cap B(a,r))}{r^{N-p}}\right)^{1/(p-1)} \frac{dr}{r} < \infty.$$

We can now state several necessary and sufficient conditions for $(1,p)$-stability.

7.6 Theorem *Let K be a compact subset of $\mathbf{R}^N$, and $1 < p \leq N$. Then the following are equivalent.*

(a) K *is $(1,p)$-stable.*

(b) $\gamma_{1,p}(G \setminus K) = \gamma_{1,p}(G \setminus \text{int } K)$ *for every bounded open set $G \subset \mathbf{R}^N$.*

(c) *There is a positive constant η such that $\gamma_{1,p}(B \setminus K) \geq \eta\gamma_{1,p}(B \setminus \text{int } K)$ for every open ball $B \subset \mathbf{R}^N$.*

(d) *The set $\mathbf{R}^N \setminus K$ and the set $\mathbf{R}^N \setminus \text{int } K$ are $(1,p)$-thin at exactly the same points of $\mathbf{R}^N$.*

(e) *The set of points in ∂K where $\mathbf{R}^N \setminus K$ is $(1,p)$-thin forms a $(1,p)$-nullset.*

We have already mentioned that the paper of Hedberg [25] includes a proof of Theorem 3.3. Hedberg's paper also gives a complete discussion of Theorem 7.6 in the case $p = 2$, including historical references for results originally stated in the context of the uniform approximation problem of Theorem 3.3. The reader who is interested in the connections between Sobolev space theory and the traditional concepts of potential theory will find Hedberg's paper to be a valuable source.

The equivalence of conditions (a) and (b) in Theorem 7.6 is analogous to a characterization of $(1, p)$-stability obtained by Bagby [7, Theorems 4 and 5] using a different set function $\gamma_{1,p}$. With the definition of capacity given here, Theorem 7.6 was proved in Hedberg [21] and Hedberg and Wolff [26]; an exposition is given in [2, Theorem 11.4.1].

We next state a simple geometric condition which is sufficient for a compact set to be (m, p)-stable for every positive integer m and every p in the interval $(1, \infty)$. For each positive number r we define the $(N - 1)$-dimensional disk

$$D_r = \{(x_1, x_2, \ldots, x_{N-1},\ 0) \in \mathbf{R}^N : x_1^2 + x_2^2 + \cdots + x_{N-1}^2 \le r^2\}.$$

If r and h are positive numbers and $T : \mathbf{R}^N \to \mathbf{R}^N$ is a Euclidean motion, we say that the convex hull of

$$T(D_r) \cup T((0, 0, \ldots, 0, h))$$

is a *truncated cone* with vertex $T((0, 0, \ldots, 0, h))$.

7.7 Theorem *Let K be a compact subset of $\mathbf{R}^N$ such that for each point $a \in \partial K$ there is a truncated cone, with vertex a, whose intersection with K is the single point a. If m is any positive integer, and if $1 < p < \infty$, then K is (m, p)-stable.*

Note that in the hypotheses of Theorem 7.7, the sizes of the cones are allowed to depend on the point $a \in \partial K$.

In the case $p > N$, Theorem 7.7 follows from Theorem 7.1. In the remaining cases Theorem 7.7 is due to Hedberg, and follows from the sufficient conditions for (m, p)-stability given in [22, Theorem 4] and [23, Theorem 3.10]. For more recent work on sufficient conditions for (m, p)-stability we refer to the papers of Hedberg and Wolff [24], [26, Theorem 5 and Theorem 6], and the exposition in [2, Theorem 11.5.4].

We next discuss the problem of finding necessary and sufficient conditions for (m, p)-stability when $m \ge 2$.

A necessary and sufficient condition for $(m, 2)$-stability when $N > 2m$ was proved by Saak [40].

The constructive techniques of Vitushkin have been used by Sinanyan [43] and Lindberg [30] to study mean approximation by holomorphic functions. The work of Lindberg [30] was extended by Bagby [8] to obtain necessary and sufficient conditions for mean approximation by solutions of certain elliptic differential equations of order $m \ge 2$, and this gives necessary and sufficient conditions for (m, p)-stability of compact sets in $\mathbf{R}^N$ for certain values of m and N; the conditions are analogous to conditions (b) and (c) of Theorem 7.6, expressed in terms of a family of capacities similar to those defined earlier by Maz'ja [31]. Bagby [9] showed that certain conditions analogous to part (e) of Theorem 7.6 were necessary for mean approximation by harmonic functions in $\mathbf{R}^N$ for $N \ge 3$, and asked whether one could establish the sufficiency of such conditions. Netrusov [36] announced results which give a solution to this problem, and improve the results of [8] in other ways; in particular, Netrusov obtains necessary and sufficient conditions for (m, p)-stability for all positive integers m.

The techniques used by Netrusov, in this and related problems, are of fundamental importance. Netrusov's proof of his stability theorem is given in [2, Theorem 11.5.10].

We conclude with some remarks concerning $(1, 2)$-stability on manifolds.

7.8 Remarks We let M be a noncompact, oriented, connected C^∞ Riemannian manifold of dimension $N \geq 2$. Recall that we use the term "parametric ball" in the technical sense of Section 2.

(a) If K is a compact subset of a parametric ball, then K is a $(1, 2)$-stable subset of M if and only if K is $(1, 2)$-stable when regarded as a subset of the parameter space $\mathbf{R}^N$ with the usual flat metric. (This follows from Remark 2.23.)

(b) Let K be a compact subset of M. Suppose that $\{K_i\}$ is a finite family of compact sets, with each K_i contained in a parametric ball B_i, such that the open sets $\{\operatorname{int} K_i\}$ form an open cover of K. If each of the sets $K \cap K_i$ is $(1, 2)$-stable when regarded as a subset of the parameter space $\mathbf{R}^N$ for B_i with the usual flat metric, then K is a $(1, 2)$-stable subset of M. (This is proved using a partition of unity subordinate to the open cover $\{\operatorname{int} K_i\}$ of K, and using the preceding remark (a).)

Remarks 7.8 show that we may use sufficient conditions for $(1, 2)$-stability in Euclidean spaces to prove $(1, 2)$-stability of certain compact sets on manifolds.

References

[1] R. A. Adams, *Sobolev Spaces*, Academic Press, New York–San Francisco–London, 1975.

[2] D. R. Adams and L. I. Hedberg, *Function Spaces and Potential Theory*, Grundlehren Math. Wiss. **314**, Springer-Verlag, Berlin–Heidelberg–New York, 1996.

[3] W. Arendt and S. Monniaux, Domain perturbation for the first eigenvalue of the Dirichlet Schrödinger operator, *Oper. Theory Adv. Appl.* **78** (1995), 9–19.

[4] T. Aubin, *Some Nonlinear Problems in Riemannian Geometry*, Springer-Verlag, Berlin–Heidelberg–New York, 1998.

[5] I. Babushka, The theory of small changes in the domain of existence in the theory of partial differential equations and its applications, in: *Differential Equations and Their Applications*, Publ. House Czechoslovak Acad. Sci., Prague; Academic Press, New York, 1963, 13–26.

[6] I. Babushka, Continuous dependence of eigenvalues on the domain, *Czechoslovak Math. J.* **15** (1965), 169–178.

[7] T. Bagby, Quasi topologies and rational approximation, *J. Funct. Anal.* **10** (1972), 259–268.

[8] T. Bagby, Approximation in the mean by solutions of elliptic equations, *Trans. Amer. Math. Soc.* **281** (1984), 761–784.

[9] T. Bagby, Approximation in the mean by harmonic functions, in: *Linear and Complex Analysis Problem Book. 199 Research Problems* (V. P. Havin, S. V. Hruščev, and N. K. Nikol'skii, eds.), Lecture Notes in Math. **1043**, Springer-Verlag, Berlin–Heidelberg–New York, 1984, 466–470; also in: *Linear and Complex Analysis Problem Book 3, Part II* (V. P. Havin and N. K. Nikolski, eds.), Lecture Notes in Math. **1574**, Springer-Verlag, Berlin–Heidelberg–New York, 1994, 117–120.

[10] T. Bagby and P. M. Gauthier, Harmonic approximation on closed subsets of Riemannian manifolds, in: *Complex Potential Theory* (P. M. Gauthier, ed.), NATO ASI Ser. C **439** Kluwer, Dortrecht, 1994, 75–87.

[11] T. Bagby and P. M. Gauthier, Note on the support of Sobolev functions, *Canad. Math. Bull.* **41** (1998), 257–260.

[12] T. Bagby, P. M. Gauthier, and J. Woodworth, Tangential harmonic approximation on Riemannian manifolds, in: *Harmonic Analysis and Number Theory* (S. W. Drury and M. Ram Murty, eds.), CMS Conf. Proc. **21**, American Mathematical Society, Providence, RI, 1997, 58–72.

[13] V. I. Burenkov, On the approximation of functions in the space $W_p^r(\Omega)$ by functions with compact support for an arbitrary open set Ω, *Proc. Steklov Inst. Math.* **131** (1974), 51–36.

[14] I. Chavel, *Eigenvalues in Riemannian Geometry*, Academic Press, Orlando, 1984.

[15] B. Davies and Y. Safarov, eds., *Spectral Theory and Geometry*, Cambridge University Press, 1999.

[16] L. C. Evans and R. Gariepy, *Measure Theory and Fine Properties of Functions*, CRC Press, Boca Raton–Ann Arbor–London, 1992.

[17] S. J. Gardiner, *Harmonic Approximation*, Cambridge University Press, 1995.

[18] V. P. Havin, Approximation in the mean by analytic functions, *Soviet Math. Doklady* **9** (1968), 245–248.

[19] E. Hebey, *Sobolev Spaces on Riemannian Manifolds*, Lecture Notes in Math. **1635**, Springer-Verlag, Berlin–Heidelberg–New York, 1996.

[20] E. Hebey, *Nonlinear Analysis on Manifolds: Sobolev Spaces and Inequalities*, Courant Institute of Mathematical Sciences, New York, 1999.

[21] L. I. Hedberg, Non-linear potentials and approximation in the mean by analytic functions, *Math. Z.* **129** (1972), 299–319.

[22] L. I. Hedberg, Approximation in the mean by solutions of elliptic equations, *Duke Math. J.* **40** (1973), 9–16.

[23] L. I. Hedberg, Spectral synthesis and stability in Sobolev spaces, in: *Euclidean Harmonic Analysis* (J. J. Benedetto, ed.), Lecture Notes in Math. **779**, Springer-Verlag, Berlin–Heidelberg–New York, 1980, 73–103.

[24] L. I. Hedberg, Spectral synthesis in Sobolev spaces, and uniqueness of solutions of the Dirichlet problem, *Acta Math.* **147** (1981), 237–264.

[25] L. I. Hedberg, Approximation by harmonic functions, and stability of the Dirichlet problem, *Exposition. Math.* **11** (1993), 193–259.

[26] L. I. Hedberg and T. H. Wolff, Thin sets in nonlinear potential theory, *Ann. Inst. Fourier (Grenoble)* **33** (1983), 161–187.

[27] L. Hörmander, *The Analysis of Linear Partial Differential Operators I*, Springer-Verlag, Berlin–Heidelberg–New York–Tokyo, 1983.

[28] M. Kac, Can one hear the shape of a drum?, *Amer. Math. Monthly* **73** (4), Part II (1966), 1–23.

[29] N. S. Landkof, *Foundations of Modern Potential Theory*, Springer-Verlag, Berlin–Heidelberg–New York, 1972.

[30] P. Lindberg, A constructive method for L^p-approximation by analytic functions, *Ark. Mat.* **20** (1982), 61–68.

[31] V. G. Maz'ja, On (p, l)-capacity, imbedding theorems, and the spectrum of a self adjoint elliptic operator, *Math. USSR-Izv.* **7** (1973), 357–387.

[32] V. G. Maz'ja, *Sobolev Spaces*, Springer-Verlag, Berlin–Heidelberg–New York–Tokyo, 1985.

[33] N. G. Meyers, A theory of capacities for potentials of functions in Lebesgue classes, *Math. Scand.* **26** (1970), 255–292.

[34] R. L. Moore, Concerning triods in the plane and the junction points of plane continua, *Proc. Nat. Acad. Sci. U.S.A.* **14** (1928), 85–88.

[35] C. B. Morrey, Jr., *Multiple Integrals in the Calculus of Variations*, Springer-Verlag, New York, 1966.

[36] Yu. V. Netrusov, Spectral synthesis in spaces of smooth functions, *Russian Acad. Sci. Dokl. Math.* **46** (1993), 135–137.

[37] J. C. Polking, Approximation in L^p by solutions of elliptic partial differential equations, *Amer. J. Math.* **94** (1972), 1231–1244.

[38] J. C. Polking, A Leibniz formula for some differentiation operators of fractional order, *Indiana Univ. Math. J.* **21** (1972), 1019–1029.

[39] W. Rudin, *Real and Complex Analysis*, 3rd ed., McGraw-Hill, New York, 1987.

[40] E. M. Saak, A capacity criterion for a domain with stable Dirichlet problem for higher order elliptic equations, *Math. USSR Sbornik* **29** (1976), 177–185.

[41] B.-W. Schulze and G. Wildenhain, *Methoden der Potentialtheorie für elliptische Differentialgleichungen beliebiger Ordnung*, Birkhäuser, Basel–Stuttgart, 1977.

[42] G. E. Shilov, *Generalized Functions and Partial Differential Equations*, Gordon & Breach, New York, 1968.

[43] S. O. Sinanyan, Approximation by polynomials and analytic functions in the areal mean, *Amer. Math. Soc. Transl. Ser. 2* **74** (1968), 91–124.

[44] E. M. Stein, *Singular Integrals and Differentiability Properties of Functions*, Princeton University Press, 1970.

[45] M. Taylor, *Partial Differential Equations I, Basic Theory*, Appl. Math. Sci. **115**, Springer-Verlag, New York, 1996.

[46] M. Taylor, *Partial Differential Equations II, Qualitative Studies of Linear Equations*, Appl. Math. Sci. **116**, Springer-Verlag, New York, 1996.

[47] J. Weidmann, Stetige Abhängigkeit der Eigenwerte und Eigenfunktionen elliptischer Differentialoperatoren vom Gebiet, *Math. Scand.* **54** (1984), 51–69.

[48] G. Wildenhain, Potential theory methods for higher order elliptic equations, in: *Potential Theory–Surveys and Problems*, Lecture Notes in Math. **1344** (1988), 181–195.

[49] W. K. Ziemer, A topological argument concerning the support of a Sobolev function on a *k*-cell, to appear.

[50] W. P. Ziemer, *Weakly Differentiable Functions. Sobolev Spaces and Functions of Bounded Variation*, Grad. Texts in Math. **120**, Springer-Verlag, Berlin–Heidelberg–New York, 1989.

Holomorphic and harmonic approximation on Riemann surfaces

André BOIVIN

Department of Mathematics
University of Western Ontario
London, Ont., N4G 5B7
Canada

Paul M. GAUTHIER[*]

Département de mathématiques et de statistique
Université de Montréal
C.P. 6128, Succ. Centre-ville
Montréal, Qué., H3C 3J7
Canada

Abstract

The most important result on open Riemann surfaces is the analog of Runge's theorem on approximation by complex polynomials, due to Behnke and Stein. In this course, we begin by a discussion of Runge's theorems on polynomial and rational approximation on compact sets. This theory is refined and extended in various ways to Riemann surfaces. We also introduce a corresponding theory of harmonic approximation.

1 Holomorphic approximation

1.1 Introduction

Let S denote a (connected) Riemann surface. We shall denote by $S^* = S \cup \{*\}$ the one-point compactification of S except in the case where S is the complex plane $\mathbf{C}$ in which case we shall use the notation $\overline{\mathbf{C}}$. Since neighbourhoods of the ideal point $*$ of S^* are complements of compact subsets of S, it follows that, if S is already compact, then the ideal point $*$ of S^* is isolated. Here and forevermore, X and K will be compact subsets of S, and E and F will be closed (not necessarily compact) subsets of S. Non-compact surfaces are called open surfaces.

For W an arbitrary subset of S, $\mathrm{Hol}(W)$ denotes the set of functions f holomorphic on (a neighbourhood $U = U(f)$ of) W. Similarly, $\mathrm{Mer}(W)$ denotes the set of functions meromorphic on W. Let $C(E)$ denote the (complex-valued) continuous functions on E, and for $f \in C(E)$, $\|f\|_E := \sup_{p \in E} |f(p)|$. The space $H_S(E)$ consists of the uniform limits on E

---

[*]Research supported in part by NSERC (Canada) and FCAR (Québec).

N. Arakelian and P.M. Gauthier (eds.), Approximation, Complex Analysis, and Potential Theory, 107–128.
© *2001 Kluwer Academic Publishers. Printed in the Netherlands.*

of functions holomorphic on $\mathcal{S}$, i.e. uniform limits on E of "entire" functions. Note that we require that the limit be uniform on *all* of E, as opposed to asking "only" uniform convergence on arbitrary compact subsets of E. That is, $f \in H_{\mathcal{S}}(E)$ if there exists a sequence $\{g_n\}$ of functions in $\mathrm{Hol}(\mathcal{S})$ such that $\lim_{n \to \infty} \|f - g_n\|_E = 0$. $M(E)$ consists of the uniform limits on E of the meromorphic functions which are holomorphic on E (i.e. without poles on E). The space $H(E)$ consists of the uniform limits on E of functions holomorphic on E, and finally, we let $A(E) = C(E) \cap \mathrm{Hol}(E^0)$ denote the functions continuous on E which are holomorphic on E^0 (the interior of E).

We naturally have the following inclusions:

$$H_{\mathcal{S}}(E) \subseteq M(E) \subseteq H(E) \subseteq A(E) \subseteq C(E),$$

and are interested to know when equalities hold.

1.2   The case $\mathcal{S} = \mathbf{C}$

One of the great and beautiful success stories of approximation theory is that we are able to tell exactly when equality holds for every possible inclusion above when $\mathcal{S} = \mathbf{C}$.

We now assume that $\mathcal{S} = \mathbf{C}$ and that E and F, respectively X and K, are closed, respectively compact, subsets of $\mathbf{C}$. We shall first restrict our attention to compact sets. Let $R(X)$ denote the closure in $C(X)$ of the rational functions which are holomorphic on X (i.e. without poles on X) and let $P(X)$ consist of the uniform limits on X of (holomorphic) polyno= mials. We note that $P(X) = H_{\mathbf{C}}(X)$ and $R(X) = M(X)$. So for compact sets, our natural inclusions now read:

$$P(X) \subseteq R(X) \subseteq H(X) \subseteq A(X) \subseteq C(X). \tag{1}$$

As an important example of when equality can hold, let us recall the following well-known theorem.

1.1 Theorem (Weierstrass) *If X is a (closed bounded) segment of the real line, then $C(X) = P(X)$; that is, every continuous function on X can be uniformly approximated on X by polynomials.*

It is natural to ask whether there are other sets X for which the above equality holds, and if so, can we give a complete description of them? Corollary 1.8 below will provide the answer to these two questions. We now proceed to describe some of the other milestones in our story. Some of our presentation is inspired by the lecture notes of Pyotr Paramonov [26], namely, our second proof that $R(X) = H(X)$ and most of our applications of Mergelyan's theorems (Spasibo Pyotr!).

1.3   Runge's theorems

The story really begins with Runge's theorems. Considering the spaces $P(X)$, $R(X)$ and $H(X)$, C. Runge obtained in 1885 a complete characterization of the sets X for which equality holds between them. It is interesting to note that both the Runge and Weierstrass theorems were published in 1885.

1.2 Theorem (Rational Runge theorem) *Let X be an arbitrary compact subset of* **C**. *Then $H(X) = R(X)$ (or equivalently* $\mathrm{Hol}(X) \subset R(X)$*).*

1.3 Theorem (Polynomial Runge theorem) $H(X) = P(X)$ *if and only if* **C** $\setminus X$ *is connected.*

One defines the *hull* of X (denoted $\widehat{X}$) as the union of X with the bounded components of its complement, that is:

$$\widehat{X} = X \cup \{\text{the bounded components of } \mathbf{C} \setminus X\}.$$

Note that $X = \widehat{X}$ if and only if **C** $\setminus X$ is connected, and thus, this last theorem is often stated as:

$$H(X) = P(X) \iff X = \widehat{X}.$$

1.4 Proofs of Runge's theorems - I

We shall first give two proofs that $\mathrm{Hol}(X) \subset R(X)$ for *all* compact sets $X \subset \mathbf{C}$.

First proof that $\mathrm{Hol}(X) \subset R(X)$. Assume that f is holomorphic in a neighbourhood of X. We surround X by a union of smooth closed curves Γ, located in the domain of holomorphicity of f, so that the Cauchy integral representation formula is valid:

$$f(\zeta) = \frac{1}{2\pi i} \int_\Gamma \frac{f(z)}{z - \zeta} \, dz, \quad \zeta \in X.$$

At this point, we could follow Runge's original (constructive) proof, which involves approximating the integral by Riemann sums, and noting that the resulting functions are rational functions of ζ which converge uniformly on X to f (see [11]). Instead we present first an (abstract) functional analytic proof. Using the Hahn-Banach theorem and the Riesz representation theorem (see [33]), we have that $f \in R(X)$ if and only if every measure ν on X orthogonal to $R(X)$ annihilates f.

So let ν be a measure on X orthogonal to $R(X)$. Using Fubini's theorem, we obtain

$$\int f(\zeta) \, d\nu(\zeta) = \frac{1}{2\pi i} \int_\Gamma f(z) \int_X \frac{d\nu(\zeta)}{z - \zeta} \, dz.$$

For each fixed $z \in \Gamma$, the function $\zeta \to 1/(z - \zeta)$ belongs to $R(X)$, and thus the inner integral vanishes by our assumption on ν. We conclude that $\int f d\nu = 0$ for all $\nu \perp R(X)$, and thus $f \in R(X)$. □

Before presenting our second proof that $H(X) = R(X)$, let us recall that if f is **R**-differentiable, we define, for $a \in \mathbf{C}$

$$\overline{\partial} f(a) := \left. \frac{\partial f}{\partial \overline{z}} \right|_a = \frac{1}{2} \left. \left(\frac{\partial}{\partial x} + i \frac{\partial}{\partial y} \right) \right|_a.$$

It follows from the Cauchy-Riemann equations that f is **C**-differentiable at $a \in \mathbf{C}$ if and only if it is **R**-differentiable at a and $\overline{\partial} f(a) = 0$.

The operator $\overline{\partial} : f \mapsto \overline{\partial} f$ is called the *Cauchy-Riemann operator*.

1.4 Pompeiu's formula *Let $\varphi \in C_0^1(\mathbf{C})$ and $z \in \mathbf{C}$. Then*

$$\varphi(z) = \frac{1}{\pi} \int_{\mathbf{C}} \frac{\overline{\partial}\varphi(\zeta)}{z - \zeta} \, dm(\zeta),$$

where $m(\cdot)$ denotes the Lebesgue measure on $\mathbf{C}$.

Proof Fix $z \in \mathbf{C}$ and take $R > 0$ such that $\mathrm{supp}(\varphi) \subset B(z, R) := \{\zeta \in \mathbf{C} : |\zeta - z| < R\}$. Let ρ, θ be the polar coordinates centred at z (i.e. $\zeta - z = \rho e^{i\theta}$, $\rho = \sqrt{(\zeta - z)(\overline{\zeta - z})}$, $e^{2i\theta} = (\zeta - z)/(\overline{\zeta - z})$ and $\theta = (1/2i) \log(\zeta - z)/(\overline{\zeta - z}))$, and let $F(\rho, \theta) := \varphi(\zeta) = \varphi(z + \rho e^{i\theta})$. For all $\zeta \neq z$ we have,

$$\overline{\partial}\varphi(\zeta) = F_\rho' \frac{\partial \rho}{\partial \overline{\zeta}} + F_\theta' \frac{\partial \theta}{\partial \overline{\zeta}} \,,$$

where

$$\frac{\partial \rho}{\partial \overline{\zeta}} = \frac{\zeta - z}{2\sqrt{(\zeta - z)(\overline{\zeta - z})}} = \frac{\rho e^{i\theta}}{2\rho} = \frac{e^{i\theta}}{2} \,,$$

$$\frac{\partial \theta}{\partial \overline{\zeta}} = -\frac{1}{2i} \frac{\overline{\zeta - z}}{\zeta - z} \frac{\zeta - z}{(\overline{\zeta - z})^2} = -\frac{1}{2i} \frac{\rho e^{i\theta}}{\rho^2} = \frac{i e^{i\theta}}{2\rho} \,,$$

so

$$\frac{1}{\pi} \int_{\mathbf{C}} \frac{\overline{\partial}\varphi(\zeta)}{z - \zeta} \, dm(\zeta) = \frac{1}{\pi} \int_0^{2\pi} \int_0^R \left(F_\rho' \frac{e^{i\theta}}{2} + F_\theta' \frac{i e^{i\theta}}{2\rho} \right) \frac{1}{-\rho e^{i\theta}} \rho \, d\rho \, d\theta$$

$$= \lim_{\delta \to 0} -\frac{1}{2\pi} \left(\int_0^{2\pi} \int_\delta^R F_\rho' \, d\rho \, d\theta + \frac{i}{\rho} \int_\delta^R \int_0^{2\pi} F_\theta' \, d\rho \, d\theta \right).$$

Since F, as a function of θ, has period 2π, it follows that the second term in the limit above vanishes and thus

$$\frac{1}{\pi} \int_{\mathbf{C}} \frac{\overline{\partial}\varphi(\zeta)}{z - \zeta} \, dm(\zeta) = \lim_{\delta \to 0} \frac{1}{2\pi} \int_0^{2\pi} F(\delta, \theta) \, d\theta = \varphi(z).$$

$\square$

Remark 1. Pompeiu's formula and Cauchy's integral formula are just special cases of the "complex" Green formula. See [6, Theorem 2.1] or [16].

Remark 2. At $z = 0$, Pompeiu's formula becomes

$$\varphi(0) = -\frac{1}{\pi} \int \frac{1}{\zeta} \overline{\partial}\varphi(\zeta) \, dm(\zeta).$$

In terms of distributions, this says that $\overline{\partial}(1/\pi\zeta)$ is the Dirac δ-"function", i.e. $1/\pi\zeta$ is a fundamental solution of the equation $\overline{\partial}f = 0$.

Runge's theorem is an easy consequence of Pompeiu's formula and the first part of the following lemma.

1.5 Lemma *Let K be a compact subset of $\mathbf{C}$ and $h \in L_\infty(K, m(\cdot))$. Define*

$$f(z) = \int_K \frac{h(\zeta)}{z - \zeta} dm(\zeta)$$

(the integral converging absolutely for all z—see below). Then

(i) *For an arbitrary compact X with $X \cap K = \emptyset$, we have that $f \in R(X)$. In fact, f can be approximated uniformly abitrarily closely by rational functions of the form*

$$\sum_{n=1}^{N} \frac{\lambda_n}{z - a_n}$$

with $a_n \in K$ and $\lambda_n \in \mathbf{C}$.

(ii) *f is holomorphic outside of K, continuous on $\overline{\mathbf{C}}$ and $f(\infty) = 0$. Moreover*

$$\|f\| = \|f\|_{\overline{\mathbf{C}}} \leq 2M \sqrt{\pi m(K)},$$

where $M = \|h\|_{K,m}$, the norm of h in $L_\infty(K, m(\cdot))$.

Remark 3. The function f defined in the lemma above is called the *Cauchy potential* (or *transform*) of the function h.

Second proof that $H(X) = R(X)$. Let $0 < \delta < \varepsilon$, and denote respectively by U_ε and U_δ the ε-neighbourhood and δ-neighbourhood of X. Let f be holomorphic on U_ε, and let $\varphi \in C_0^1(U_\varepsilon)$ be such that $0 \leq \varphi \leq 1$, $\varphi = 1$ on U_δ, and $\varphi = 0$ off U_ε.

Let $g = f\varphi$. Then $g \in C_0^1(\mathbf{C})$ and by Pompeiu's formula, for $z \in X$, we have

$$f(z) = g(z) = \frac{1}{\pi} \int_{U_\varepsilon \setminus U_\delta} \frac{\overline{\partial} g(\zeta)}{z - \zeta} \, dm(\zeta),$$

since $\overline{\partial} g(\zeta) = \overline{\partial} f(\zeta) = 0$ on U_δ. The conclusion follows from the first part of the Lemma above with $h(\zeta) = \overline{\partial} g(\zeta)$ and $K = \overline{U}_\varepsilon \setminus U_\delta$. □

Proof of Lemma 1.5 (i) Let $d = \mathrm{dist}(X, K)$, $d > 0$. For $\mu \in (0, d/2)$, we divide K into a finite number ($n = N(\mu)$) of mutually disjoint Borel sets K_n, $1 \leq n \leq N$, such that $\mathrm{diam}(K_n) < \mu$. Fix $a_n \in K_n$ and let $\lambda_n = \int_{K_n} h(\zeta) dm(\zeta)$. Then for $z \in X$, we get

$$\left| \int_K \frac{h(\zeta) \, dm(\zeta)}{z - \zeta} - \sum_{n=1}^{N} \frac{\lambda_n}{z - a_n} \right| = \left| \sum_{n=1}^{N} \int_{K_n} \frac{h(\zeta) \, dm(\zeta)}{z - \zeta} - \sum_{n=1}^{N} \int_{K_n} \frac{h(\zeta) \, dm(\zeta)}{z - a_n} \right|$$

$$\leq \sum_{n=1}^{N} M \int_{K_n} \left| \frac{(z - a_n) - (z - \zeta)}{(z - \zeta)(z - a_n)} \right| dm(\zeta)$$

$$\leq M \sum_{n=1}^{N} \frac{\mu}{d^2} m(K_n)$$

$$\leq \frac{m(K)M}{d^2} \mu \to 0 \quad \text{as} \quad \mu \to 0.$$

(ii) Because the function $\sum_{n=1}^{N} \lambda_n/(z - a_n)$ is holomorphic off K, it follows from (i) that f is holomorphic off K.

That $f(\infty) = 0$ follows immediately from its definition.

We now estimate $|f(z)|$ for an arbitrary $z \in \mathbf{C}$. Let $r = \sqrt{m(K)/\pi}$ (so that $m(B(z,r)) = m(K)$). In view of the fact that the function $1/|z - \zeta|$ decreases as the distance between ζ from (a fixed) z increases, we get

$$|f(z)| \leq M \int_K \frac{1}{|z - \zeta|}\, dm(\zeta) \leq M \int_{B(z,r)} \frac{1}{|z - \zeta|}\, dm(\zeta)$$
$$= M \int_0^{2\pi} \int_0^r \frac{\rho\, dr\, d\theta}{\rho} = 2M\pi r = 2M\sqrt{\pi m(K)},$$

which together with providing us with the uniform estimate we were seeking, also proves (for all z) the absolute convergence of the integral defining f. $\square$

That f is continuous on all of $\overline{\mathbf{C}}$ is interesting, though not needed in the proof of Runge's theorem. It does prove useful though when one is trying to establish Mergelyan's theorem (stated below). In fact, a slightly stronger result is true.

Recall that for $E \subset \mathbf{C}$, and $\alpha \in (0, 1]$, the space $\mathrm{Lip}_\alpha(E)$ is the class of functions g for which there exists a constant $c = c(g) \in [0, \infty)$ such that

$$|g(z_1) - g(z_2)| \leq c|z_1 - z_2|^\alpha \qquad \text{and} \qquad |g(z_1)| \leq c$$

for all $z_1, z_2 \in E$. A Banach space norm is obtained by letting $\|g\|_{\alpha,E} = \min\{c(g)\}$, where the minimum (which is indeed attained) is taken over all $c(g)$ for which the last two inequalities are satisfied (exercise!). Evidently, $\mathrm{Lip}_\alpha(E) \subset C(E)$ for all $\alpha \in (0, 1]$. That f is continuous is thus a consequence of the following lemma.

1.6 Lemma *Under the conditions of Lemma 1.5, for any $\alpha \in (0, 1)$, we have $f \in \mathrm{Lip}_\alpha(\mathbf{C})$ with $\|f\|_{\alpha,E} \leq Mc(\alpha, K)$. Also, there exists a compact set K, such that if f is the Cauchy potential of the characteristic function $h = \chi_K \equiv 1|_K$, then $f \notin \mathrm{Lip}_1(\mathbf{C})$.*

Proof Fix $z_1 \neq z_2$ and let $\delta = |z_1 - z_2|/2$, $a = (z_1 + z_2)/2$, $D_1 = B(z_1, \delta)$, $D_2 = B(z_2, \delta)$, $D_3 = B(a, 2\delta) \setminus (D_1 \cup D_2)$ and $D_4 = \mathbf{C} \setminus B(a, 2\delta)$. Then we have

$$|f(z_1) - f(z_2)| \leq \sum_{s=1}^4 \int_{D_s \cap K} |h(\zeta)| \frac{|z_1 - z_2|}{|z_1 - \zeta||z_2 - \zeta|}\, dm(\zeta)$$
$$\leq \sum_{s=1}^4 2M\delta \int_{D_s \cap K} \frac{1}{|z_1 - \zeta||z_2 - \zeta|}\, dm(\zeta).$$

The term corresponding to $s = 1$ ($s = 2$ is analogous) is dealt with by changing to polar coordinates centred at z_1 (and integrating over all of D_1).

The term with $s = 3$ is evaluated trivially.

Choose $r > 0$ such that $m(B(a, r) \cap D_4) = m(K)$. For the integral over D_4, we use the inequality $|z_1 - \zeta||z_2 - \zeta| \geq |\zeta - a|^2/4$ and polar coordinates centred at a. This gives

$$\int_{D_4 \cap K} \frac{1}{|z_1 - \zeta||z_2 - \zeta|}\, dm(\zeta) \leq \int_{D_4 \cap K} \frac{4}{|\zeta - a|^2}\, dm(\zeta)$$
$$\leq 8\pi \int_{2\delta}^r \rho^{-1}\, d\rho = 8\pi \log\left(\frac{r}{2\delta}\right).$$

The first part of the lemma follows, since the function $t \log(1/t) \in \mathrm{Lip}_\alpha([0,1])$ for all $\alpha \in (0,1)$, but not for $\alpha = 1$ (exercise!).

For the (counter-)example when $\alpha = 1$, let $K = \{z : |z| \le 1, \Re(z) \ge |\Im(z)|\}$ and take $z_1 = 0$ and $z_2 = -2\delta$ where δ is taken sufficiently small. $\qquad\square$

1.5 Proofs of Runge's theorems - II

We now want to prove that $H(X) = P(X) \iff X = \hat{X}$.

Proof of the necessity ($\Longrightarrow$). Assume, to obtain a contradiction, that $H(X) \subset P(X)$, with $\mathbf{C} \setminus X$ not connected, so that there exists a component Ω_1 of $\mathbf{C} \setminus X$ which is bounded. Note in particular that $\partial\Omega_1 \subset X$.

Fix $a_1 \in \Omega_1$. Since $f(z) = 1/(z - a_1) \in H(X) \subset P(X)$, for all $\varepsilon > 0$, there exists a polynomial $p_\varepsilon(z)$ such that

$$\left| \frac{1}{z - a_1} - p_\varepsilon(z) \right| < \varepsilon$$

for all $z \in X$, and in particular, for all $z \in \partial\Omega_1$. Let $d = \mathrm{diam}(\Omega_1)$. Then, for all $z \in \partial\Omega_1$, we have

$$|1 - p_\varepsilon(z)(z - a_1)| < \varepsilon d.$$

Then the maximum modulus principle, applied to the function $F(z) = 1 - p_\varepsilon(z)(z - a_1)$ at $z = a_1$, says that $1 < \varepsilon d$. Clearly a contradiction! $\qquad\square$

First proof of the sufficiency ($\Longleftarrow$). The (abstract) functional analytic proof goes as follows. Let ν be a measure on X orthogonal to $P(X)$, and define

$$\hat{\nu}(z) = \int_X \frac{d\nu(\zeta)}{\zeta - z}, \quad z \in \mathbf{C} \setminus X.$$

Then $\hat\nu$ is holomorphic on $\mathbf{C} \setminus X$ and at ∞ (see Lemma 1.5). Its Taylor expansion at ∞ is obtained by developing $1/(1 - \zeta/z)$ in a geometric series:

$$\hat\nu(z) = -\sum \frac{1}{z^{n+1}} \int_X \zeta^n \, d\nu(\zeta).$$

Since $\nu \perp P(X)$, each coefficient of this series vanishes near ∞. Since $\mathbf{C} \setminus X$ is connected, $\hat\nu$ vanishes identically on $\mathbf{C} \setminus X$. If f is any function holomorphic in a neighbourhood of X, we represent f as a Cauchy integral. As in the first proof of the rational Runge theorem, we obtain $\int f d\nu = 0$. This places $f \in P(X)$. $\qquad\square$

Second proof of the sufficiency ($\Longleftarrow$). Assume that $\Omega = \mathbf{C} \setminus X$ is connected and $f \in H(X)$. We may assume that f is smoothly defined on all of $\mathbf{C}$ with compact su= pport. By Lemma 1.5, given $\varepsilon > 0$, there exists $\{a_1, \ldots, a_N\} \subset \Omega$ and $\{\lambda_1, \ldots, \lambda_N\} \in \mathbf{C} \setminus \{0\}$ such that

$$\left| f(z) - \sum_{n=1}^N \frac{\lambda_n}{z - a_n} \right| < \frac{\varepsilon}{2}, \quad z \in X.$$

It suffices to prove that $1/(z - a) \in P(X)$ for all $a \in \Omega$ (since then the functions $\lambda_n/(z - a_n)$ can be approximated by polynomials $p_{\varepsilon_n}(z)$ within $\varepsilon_n = \varepsilon/(2N)$ and thus the function f can be approximated within ε).

Remark 4. We could use here a pole-pushing argument, as in the original proof of Runge. (See [11, 16]; see also the lecture notes of Armitage and Gardiner for the harmonic version in the present Proceedings.) Again, we choose a different route.

Let

$$G = \left\{ a \in \Omega : \left. \frac{1}{z-a} \right|_X \in P(X) \right\}.$$

We shall prove that $G = \Omega$. In fact, first we have that $G \neq \emptyset$ since, from the Taylor series representation, G contains all points outside any disc containing X.

Next, G is closed in Ω, for let $\{a_k\}_{k=1}^\infty \subset G$ with $a = \lim_k a_k \in \Omega$. Then $a \in G$ follows directly from the uniform convergence of $1/(z-a_k)$ to $1/(z-a)$ on X as $k \to \infty$.

Finally, we show that G is open in Ω. Let $a \in G$, $d = \mathrm{dist}(a, X)$ and $a_1 \in B(a, d)$. Let us prove that $a_1 \in G$. Elementary properties of the geometric series imply that, for all $\varepsilon > 0$, there exists a natural number L such that

$$\left| \frac{1}{z-a_1} - \sum_{\ell=1}^L \frac{(a_1-a)^{\ell-1}}{(z-a)^\ell} \right| < \varepsilon, \quad \text{for all} \quad z \in X.$$

But $1/(z-a) \in P(X)$, thus $1/(z-a)^\ell \in P(X)$ for each natural number ℓ, and consequently $1/(z-a_1) \in P(X)$.

These three facts together imply that $G = \Omega$.   □

1.6   Mergelyan's theorems

In 1952, S. N. Mergelyan showed that the compact sets for which $A(X) = P(X)$ are exactly the same as those for which $H(X) = P(X)$ (see the polynomial Runge theorem above). That is,

1.7 Theorem (Mergelyan) $A(X) = P(X) \iff X = \widehat{X}$.

As an immediate corollary we obtain the following generalization of Weierstrass's theorem.

1.8 Corollary *The equality* $C(X) = P(X)$ *is equivalent to the condition that* $X = \widehat{X}$ *and* $X^0 = \emptyset$.

One implication in Mergelyan's theorem is immediate. Indeed let $A(X) = P(X)$. Then $H(X) \subset P(X)$ and, by the necessity in the polynomial Runge theorem, $X = \widehat{X}$.

Now assume that $X = \widehat{X}$. Then, from the sufficiency in the polynomial Runge theorem, one need only prove that $A(X) = H(X)$. In fact the following stronger result (also due to Mergelyan) is true.

1.9 Theorem (Mergelyan) *Let* X *be a compact subset of* $\mathbf{C}$, *and let* $\Omega_0 = \mathbf{C} \setminus \widehat{\;}X$, $\Omega_1, \Omega_2, \ldots$ *be the components of the complement of* X, *that is* $\mathbf{C} \setminus X = \bigcup_{s \geq 0} \Omega_s$. *If* $d = \inf_s \{\mathrm{diam}(\Omega_s)\} > 0$, *then* $A(X) = H(X) (= R(X))$.

Remark 5. If $\Omega_0 = \mathbf{C} \setminus X$, i.e. if $\mathbf{C} \setminus X$ is connected and thus the indices $s = 1, \ldots$ are all missing, then $d = \infty$.

We shall not present here a proof of this result. A functional analytic proof of these two theorems of Mergelyan can be found in [12, II.9.1 and II.10.4]. Constructive proofs can be found in [26], [33, Chap. 13] and [12, VIII.7.4].

It is known that there exist sets X for which $R(X) \neq A(X)$. From Mergelyan's theorem, these sets must be infinitely connected i.e. their complements must consist of infinitely many distinct components. Sets of rational approximation will be discussed briefly in the light of Vitushkin's theorem, but first we give some applications of Mergelyan's theorem.

1.7 Some applications of Mergelyan's theorem

Let us define a finitely-Jordan domain to be a bounded domain in $\mathbf{C}$ whose boundary consists of finitely many pairwise disjoint Jordan curves. A particular case of the second theorem of Mergelyan is the following deep theorem of Walsh.

1.10 Theorem (Walsh) *If D is a finitely Jordan domain, then $R(\overline{D}) = A(\overline{D})$.*

As Walsh pointed out [40], this easily yields the Cauchy theorem for smoothly bounded domains.

1.11 Theorem (Cauchy) *If D is a smoothly bounded domain and $f \in A(\overline{D})$, then*

$$\int_{\partial D} f(z)\,dz = 0.$$

For other Cauchy theorems which can be deduced from approximation theorems, see [17].

1.12 Argument Principle *Let D be a finitely-Jordan domain and f a function which is continuous on $\overline{D}$ and holomorphic in D except for (finitely many) poles. If f has no zeros on ∂D, then the difference between the number of zeros and the number of poles of f in D is equal to the variation of the argument of f divided by 2π over the oriented boundary of D:*

$$N_D(f) - P_D(f) = \frac{1}{2\pi} \Delta_{\partial D} \arg(f). \tag{2}$$

Proof We recall the following assertions:

(i) If f_1 and f_2 in $C(\partial D)$ have no zeros on ∂D, then

$$\Delta_{\partial D} \arg(f_1 f_2) = \Delta_{\partial D} \arg(f_1) + \Delta_{\partial D} \arg(f_2)$$

(for division of functions, the + is replaced by a $-$).

(ii) $\Delta_{\partial D} \arg(z - b)|_{b \in D} = 2\pi$, $\Delta_{\partial D} \arg(z - b)|_{b \notin \overline{D}} = 0$.

From this we obtain that (2) holds for each polynomial and each rational function f, having neither zeros nor poles on ∂D (if (2) holds for f_1 and f_2, then it holds also for $f_1 f_2$ and f_1/f_2).

To prove the argument principle, we can reduce the general situation to the case when f has neither zeros nor poles in $\overline{D}$. Let Q be a polynomial whose zeros are the same as the poles of f in D and P a polynomial whose zeros are the same as those of f in D. It is sufficient to establish (2) for $F = fQ/P$ in place of f, since $f = FP/Q$ and for P/Q it is proven. Set $\varepsilon = \min\{|F(z)| : z \in \overline{D}\} > 0$. By the theorem of Walsh stated above, there exist polynomials P_ε and Q_ε ($Q_\varepsilon \neq 0$ in $\overline{D}$), such that

$$\left| F - \frac{P_\varepsilon}{Q_\varepsilon} \right|_{\overline{D}} < \frac{\varepsilon}{2}.$$

Then,

$$N_D(F) = N_D(P_\varepsilon/Q_\varepsilon) = 0 = P_D(F) = P_D(P_\varepsilon/Q_\varepsilon).$$

There remains to remark that

$$\Delta_{\partial D} \arg\left(\frac{P_\varepsilon}{Q_\varepsilon}\right) = \Delta_{\partial D} \arg\left(\frac{P_\varepsilon}{Q_\varepsilon} - F + F\right)$$

$$= \Delta_{\partial D} \arg(F) + \Delta_{\partial D} \arg\left(1 + \frac{P_\varepsilon/Q_\varepsilon - F}{F}\right)$$

$$= \Delta_{\partial D} \arg(F),$$

since, from the inequality

$$\left| \frac{P_\varepsilon/Q_\varepsilon - F}{F} \right| \leq \frac{1}{2}$$

on ∂D, it immediately follows that

$$\Delta_{\partial D} \arg\left(1 + \frac{P_\varepsilon/Q_\varepsilon - F}{F}\right) = 0.$$

This concludes the proof.   □

The following theorem follows from the above argument principle as in the classical situation.

1.13 Theorem (Rouché) *Let D be a finitely-Jordan domain and $f, g \in A(\overline{D})$, with $|g| < |f|$ on ∂D. Then $N_D(f) = N_D(f + g)$.*

By the Riemann mapping theorem, each Jordan domain D is conformally equivalent to the unit disc Δ. More generally, if D is a finitely Jordan domain, there is a conformal mapping f from D to a domain of the form $\Omega = \Delta \setminus (\overline{\Delta}_1 \cup \cdots \cup \overline{\Delta}_n)$, where the $\overline{\Delta}_j$'s are disjoint closed discs inside the unit disc Δ. The boundary correspondence principle of Osgood and Carathéodory asserts that the mapping f extends to the boundary. In fact, if $f : D \to \Omega$ is a conformal mapping between finitely Jordan domains, then f extends to a homeomorphism $f : \overline{D} \to \overline{\Omega}$. This result has a converse which we now give in a form more general than the usual formulation.

1.14 Theorem (Inverse boundary correspondence principle) *Suppose f is in $A(\overline{D})$, where D is a finitely-Jordan domain and denote $\Omega = f(D)$. If the restriction of f to ∂D is injective and $f(\partial D) \subset \partial\Omega$, then f is a homeomorphism, Ω is a finitely-Jordan domain, and the restriction of f to D is conformal.*

Proof Denote $\partial D = \Gamma$ and $f(\Gamma) = \Sigma$. Since f is open on D, $\partial\Omega \subset \Sigma$ and by hypothesis $\Sigma \subset \partial\Omega$. Hence, $\Sigma = \partial\Omega$. Thus, Ω is a finitely-Jordan domain. We consider Γ to be oriented positively with respect to D, and endow Σ with the orientation induced by f. Let $b \in \Omega$. By the argument principle (since $f - b \neq 0$ on Γ), we have

$$N_D(f - b) = \frac{1}{2\pi}\Delta_\Gamma \arg(f - b) = \frac{1}{2\pi}\Delta_\Sigma \arg(w - b).$$

The last expression is the winding number of Σ with respect to the interior point b. Since the winding number is a topological invariant, it follows from the boundary correspondence principle that the winding number of the boundary of a finitely Jordan domain with respect to an interior point is the same as for a domain bounded by finitely many disjoint circles, namely, ± 1. From the previous equality, we then have $N_D(f-b) = \pm 1$. Since $N_D(f-b) \geq 0$, the value -1 is excluded. The proof is complete. □

For a Jordan domain, we can simplify the hypotheses as follows.

1.15 Theorem *Suppose f is in $A(\overline{D})$, where D is a Jordan domain. If the restriction of f to ∂D is injective, then f is a homeomorphism, $f(D)$ is a Jordan domain, and the restriction of f to D is conformal.*

Proof Set $\Gamma = \partial D$, $\Omega = f(D)$ and $\Sigma = f(\Gamma)$. Of course Σ is a Jordan curve. Since f is open on D, it follows that $\partial\Omega \subset \Sigma$, and since Ω is necessarily bounded, $\partial\Omega = \Sigma$. The theorem now follows from the Inverse boundary correspondence principle. □

In this last theorem, we cannot replace the Jordan domains by finitely-Jordan domains. To see this, we first construct a Riemann surface. Let W be the slit disc $(|w| < 2) \setminus [-1, 1]$ and let U^+ and U^- be the intersection of the open disc $(|w| < 1)$ with the closed upper and lower half-planes, respectively. The surface S is constructed by adjoining U^+ and U^- to W in the natural manner. We may think of the surface S as lying over the w-plane and we may form a bordered Riemann surface $\overline{S}$ by adjoining to S, in the natural manner, two circles over $|w| = 2$ and $|w| = 1$, respectively. Let π be the projection mapping from $\overline{S}$ to $\mathbf{C}$. There is a conformal mapping ϕ from some annulus D onto S, and by the boundary correspondence principle, we may extend ϕ to a homeomorphism $\phi : \overline{D} \to \overline{S}$. Now, set $f = \pi \circ \phi : \overline{D} \to (|w| \leq 2)$. Then, f is continuous on the closed annulus, holomorphic on D, and injective on ∂D. But, of course, f is not conformal since it assumes the values $|w| < 1$ twice. This example shows that, in the previous theorem, we may not replace Jordan domain by finitely-Jordan domain.

In fact, as Paul Koosis pointed out to us, we can give an explicit formula for such a function f. By the reflection principle, we may extend the function f of the previous paragraph holomorphically to the closed annulus $\overline{D}$. Hence, by Runge's theorem, we may take f to be a rational function, with poles at most at 0 and ∞. Since f conserves the orientation of one circle but reverses that of another, in fact both 0 and ∞ must be poles. The simplest such rational function is the famous Zhukovskiĭ transformation $w = z + 1/z$, and it does indeed have the properties we seek. Choose $0 < r_1 \neq r_2 < 1$ and set $\overline{D} = (r_1 \leq |z| \leq 1/r_2)$. Then, f is injective on ∂D but assumes every value of the interval $(-2, 2)$ twice and so is not conformal.

1.8 Vitushkin's theorems

Although we wish to concentrate in these lectures on the problem of approximation by entire holomorphic functions, we cannot "passer sous silence" the following theorems of A. G. Vitushkin characterizing the sets of rational approximation.

Given a compact set $K \subset \mathbf{C}$, we say that a function f is K-admissible if f is holomorphic on $\overline{\mathbf{C}} \setminus K$ and $f(\infty) = 0$. Let us denote by B_0 the class of K-admissible functions which belong to the unit ball of the space of all bounded Borel functions on $\mathbf{C} \setminus K$, and by B_1 the class of K-admissible functions which belong to the unit ball of the space of all continuous functions on $\overline{\mathbf{C}}$. Both unit balls are taken with respect to the sup norm.

The *analytic capacity* and *continuous analytic capacity* of K are defined as

$$\gamma(K) = \sup_{f \in B_0} \left\{ \lim_{z \to \infty} |z \cdot f(z)| \right\} = \sup_{f \in B_0} |f'(\infty)|$$

and

$$\alpha(K) = \sup_{f \in B_1} \left\{ \lim_{z \to \infty} |z \cdot f(z)| \right\} = \sup_{f \in B_1} |f'(\infty)|,$$

respectively. For an arbitrary bounded set W, we define

$$\gamma(W) = \sup\{\gamma(K) : K \subset W, \ K \text{ compact}\}$$

and

$$\alpha(W) = \sup\{\alpha(K) : K \subset W, \ K \text{ compact}\}.$$

We have that $\alpha(W) \leq \gamma(W)$; and for U open, $\alpha(U) = \gamma(U)$. Moreover the (continuous) analytic capacity of a ball $B(a, R)$ is its radius R, the analytic capacity of a (bounded) line segment is its length L divided by 4 $(L/4)$, while its continuous analytic capacity is zero.

In 1959 and 1966 respectively, Vitushkin proved the following (see [39]):

1.16 Theorem (Vitushkin) *Let X be a compact subset of $\mathbf{C}$. Then the following are equivalent.*

(i) $R(X) = C(X)$;

(ii) *for every open set D, $\gamma(D) = \gamma(D \setminus X)$.*

1.17 Theorem (Vitushkin) *Let X be a compact subset of $\mathbf{C}$. Then the following are equivalent.*

(i) $R(X) = A(X)$;

(ii) *for every open disk D, $\alpha(D \setminus X^0) = \alpha(D \setminus X)$.*

These are *local* conditions on the boundary of X. Vitushkin gave other (weaker) equivalent local conditions, also in terms of capacities. We refer the reader to the paper of Vitushkin [39] or the book of T. W. Gamelin [12].

1.9   Closed subsets of C

With Vitushkin's theorems, we have completed our task of describing when equalities hold in our list of inclusions (1), that is, describing the *compact* subsets of polynomial and rational approximation in **C**. As mentionned in the introduction, all the theorems above have been extended to closed sets. For example, T. Carleman showed in 1927 that every continuous function on the real line **R** is the uniform limit *on* **R** of entire functions, thus extending the theorem of Weierstrass. In fact, Carleman showed much more. We will not pursue this direction. The interested reader is refered to [16, Section 6].

Let us start with the extension of Runge's theorems obtained by Alice Roth in 1938 [31].

1.18 Theorem (Roth)   (i) $H(E) = M(E)$ *for all closed subsets E of* **C**.

(ii) $H(E) = H_{\mathbf{C}}(E) \iff \overline{\mathbf{C}} \setminus E$ *is connected and locally connected.*

We note that the necessity of the condition in (ii) was in fact established by N. U. Arakelyan, who extended in 1964 the theorem of Mergelyan. Arakelyan's theorem shows that as with the compact case, the closed sets E for which $A(E) = H_{\mathbf{C}}(E)$ are the same as those for which $H(E) = H_{\mathbf{C}}(E)$.

1.19 Theorem (Arakelyan) $A(E) = H_{\mathbf{C}}(E)$ *if and only if* $\overline{\mathbf{C}} \setminus E$ *is connected and locally connected.*

Finally, Vitushkin's theorem was extended by A. H. Nersessian in 1972 [25].

1.20 Theorem (Nersessian) *Let E be a closed subset of* **C**. *Then* $A(E) = M(E)$ *if and only if* $\alpha(D \setminus E^0) = \alpha(D \setminus E)$ *for every open set D.*

We note that in this last theorem, contrary to the previous two theorems, there are no extra conditions at ∞.

1.10   Riemann surfaces

We first note that all the theorems above are valid when $S = \mathbf{C}$ is replaced by a *planar domain* Ω, with E being now a (relatively) closed subset of Ω. Also the theorems of Runge and Mergelyan (properly restated), that is the theorems dealing with approximation on compact sets, are also valid on an arbitrary open Riemann surface S. Recalling that we denote by S^* the one-point compactification of S, we have the following generalizations of the Runge and Mergelyan theorems ([5, 7, 21]):

1.21 Theorem (Behnke-Stein) (i) *For each Riemann surface S and each compact subset K of S, we have* $H(K) = M(K)$.

(ii) *For each Riemann surface S and each compact subset K of S, we have* $H(K) = H_S(K)$ *if and only if* $S^* \setminus K$ *is connected.*

1.22 Theorem (Bishop) *For each Riemann surface S and each compact subset K of S, we have* $A(K) = H_S(K)$ *if and only if* $S^* \setminus K$ *is connected.*

On closed sets, the situation is more complicated. In 1975, Gauthier and Hengartner [19] proved the following.

1.23 Theorem *Let F be a closed subset of an open Riemann surface S. If $H_S(F) = H(F)$ (or a fortiori if $H_S(F) = A(F)$), then $S^* \setminus F$ is connected and locally connected.*

That is, the condition

$$S^* \setminus F \text{ is connected and locally connected} \qquad (*)$$

is necessary for holomorphic approximation. Somewhat surprisingly in view of the theorems of Roth, Arakelyan, Behnke-Stein and Bishop, it is also shown in [19] that the condition $(*)$ is not sufficient in general for $H_S(F) = H(F)$ (or a fortiori for $H_S(F) = A(F)$) to hold. Examples will be provided below.

The complete characterization of the sets for which $H_S(F) = H(F)$ or $H_S(F) = A(F)$ on arbitrary open Riemann surfaces is still an open problem.

Though not sufficient in general, the condition $(*)$ is sufficient in some cases, as shown by Arakelyan's theorem, for example. Our next result will provide other such cases. We first need one more definition. We say that a closed subset F of an open Riemann surface S is *essentially of finite genus* if there exists an open covering $\{U_i;\ i \in I\}$ of F such that $U_i \cap U_j = \emptyset$ if $i \neq j$ and each U_i is of finite genus. Scheinberg [36], and Gauthier and Hengartner (see [20]) obtained the following results.

1.24 Theorem *Let $F \subset S$ be closed and essentially of finite genus. The following are equivalent.*

  (i) $H_S(F) = H(F)$;

  (ii) $H_S(F) = A(F)$;

  (iii) $S^* \setminus F$ is connected and locally connected.

1.25 Corollary *Let S be an open Riemann surface of finite genus and let F be an arbitrary closed subset of S. The following are equivalent.*

  (i) $H_S(F) = H(F)$;

  (ii) $H_S(F) = A(F)$;

  (iii) $S^* \setminus F$ is connected and locally connected.

In [38], Schmieder has introduced an even larger class of closed sets containing all sets essentially of finite genus, and has proved that $(*)$ is sufficient for holomorphic approximation on sets F in that class. This is the best result in this direction to date. Gauthier [15] has also proved the sufficiency of the condition $(*)$ under the assumption that S admits a Cauchy kernel which is bounded at infinity.

1.11 Examples

Denote by O_{AB} the class of Riemann surfaces with the property that every bounded entire function is constant. So, for example, $\mathbf{C} \in O_{AB}$, but the unit disc U is not. Let S_0 be an open Riemann surface with the property that $S_0 \setminus K \in O_{AB}$ whenever $K \subset S_0$ is compact and $S_0 \setminus K$ is connected. Z. Kuramochi ([23, Theorem 5.1]) proved that each surface in $O_{HB} \setminus O_G$ has this property (for the definitions, see [1]). Using this result, Gauthier and Hengartner [19] showed that there exist pairs (S, E) for which $(*)$ neither implies $H_S(F) = H(F)$ nor $H_S(F) = A(F)$. Recall that in this case, S must necessarily be of infinite genus (see Corollary 1.25).

1.26 Example Let $D \subset S_0$ be a parametric disc, and choose two points $p, q \in D$, $q \neq p$. Let $S = S_0 \setminus \{p\}$, $E = (S \setminus D) \cup \{q\}$ and

$$f = \begin{cases} 0 & \text{on } S \setminus D \\ 1 & \text{at the point } q. \end{cases}$$

We have that the pair (S, E) satisfies condition $(*)$ and $f \in \text{Hol}(E) \subset A(E)$. Now assume that there exists $g \in \text{Hol}(S)$ with $|f - g|_E < \varepsilon < 1/2$. It then follows that g must be constant, since it is bounded on $S \setminus D$. This contradicts $|f - g|_E < 1/2$.

Note. We note that we can assume $g \in \text{Mer}(S)$ and still conclude that g is constant, since $S \setminus \overline{D} \in O_{AB}$. This shows that local uniform approximation by meromorphic functions does not imply "global" uniform approximation by meromorphic functions. Compare this with the theorems of Roth and Nersessian above. In fact the example above can be slightly "simplified" in this case, since we can take $S = S_0$ and $E = S_0 \setminus D$. It then suffices to assume that f has an essential singularity in D. See also Example 1.27 below.

We now provide an explicit example of a pair (S, E) for which $(*)$ neither implies $H_S(F) = H(F)$ nor $H_S(F) = A(F)$. (following Scheinberg [37]).

1.27 Example Let S be the surface obtained from two copies of the unit disc U, slicing each disc along the intervals $((2n-2)/(2n-1), (2n-1)/2n)$, $n \geq 1$, and joining the two discs along the corresponding cuts in the usual way. Let

$$U^+ = U \cap \{\Re z \geq 0\} \quad \text{and} \quad E = \pi^{-1}(U^+),$$

where π is the projection of S onto the unit disc. Note that (S, E) satisfies condition $(*)$. For each function $g \in \text{Hol}(E)$, we define the function g_Δ on U^+ as follows:

$$g_\Delta(z) = (g(p_{z,1}) - g(p_{z,2}))^2, \quad z \in U^+,$$

where $\pi^{-1}(z) = \{p_{z,1}, p_{z,2}\}$. It is easy to see that g_Δ is holomorphic on U^+ and that g_Δ vanishes at the points $1 - 1/n$, that is, at the points where $p_{z,1} = p_{z,2}$. It follows that if g_Δ is bounded on U^+, then g_Δ vanishes identically since the Blaschke condition on the zeros of g_Δ is not satisfied.

Now assume that f is a meromorphic function on S with a unique pole at the point p, where $\{p, q\} = \pi^{-1}(-1/2)$. We now claim that if h is holomorphic on S, then $f - h$ must be unbounded on E, and that consequently it is not possible to approximate uniformly f on E by entire functions. Indeed, assuming $f - h$ bounded, we would have $(f - h)_\Delta \equiv 0$ on U^+, and thus on $U \setminus \{-1/2\}$, but this is not possible since $f - h$ is holomorphic at the point q, but has a pole at the point p.

Note. In $(*)$ if we replace the one-point compactification by the ends compactification we obtain a stronger condition. This last example shows that this stronger condition would still not suffice to have $H_S(F) = H(F)$ or $H_S(F) = A(F)$, since each end of S is accessible from $S \setminus E$ (we consider the branch points as being points of the surface).

2   Harmonic approximation

2.1   Introduction

In this section, we study harmonic approximation on an arbitrary (possibly compact) Riemann surface S. Our approach follows [8] and we correct an error therein. In terms of a local variable, a harmonic function h on S, having an isolated singularity at a point P, can be written as the sum $h = u + s$ of a harmonic function

$$u(z) = \Re \left(\sum_{n=0}^{\infty} \alpha_n z^n \right),$$

and a singular part

$$s(z) = \Re \left(\sum_{n=1}^{\infty} \beta_n z^{-n} \right) + \lambda \log |z|,$$

with $\alpha_n, \beta_n \in \mathbf{C}$, $\lambda \in \mathbf{R}$; the two series are assumed to be convergent, the first for z sufficiently small, the second for all $z \neq 0$. We say that the singularity of h is *non-essential* if s is of the form

$$s(z) = \Re \left(\sum_{n=1}^{N} \beta_n z^{-n} \right) + \lambda \log |z|,$$

and *Newtonian* if it is of the form

$$\lambda \log |z|.$$

These definitions are independent of local coordinates. h is said to be an *essentially harmonic* function, respectively a *Newtonian* function, on an open set Ω of S if h is harmonic on Ω except possibly for non-essential (respectively Newtonian) singularities. We say that a function is essentially harmonic (Newtonian) on an arbitrary set $W \subset S$ if it is essentially harmonic (Newtonian) on an open set containing W. We then write $h \in \text{EssHar}(W)$ and $h \in N(W)$ respectively. Similarly, $\text{Har}(W)$ denotes the harmonic functions on W. We shall say that $h \in \text{EssHar}(W)$ is the uniform limit of functions essentially harmonic on S if for all $\varepsilon > 0$, there is $h_\varepsilon \in \text{EssHar}(S)$ such that $|h - h_\varepsilon|_W < \varepsilon$. Thus, h_ε has the same type of singularities as h on W. In other words, $(h - h_\varepsilon) \in \text{Har}(W)$.

2.1 Theorem *Let X be a compact subset of a Riemann surface S and E a subset of $S \backslash X$. If E meets each of the bounded components of $S \setminus X$, then each function harmonic (respectively essentially harmonic) on X is the uniform limit of essentially harmonic functions on S all of whose singularities lie in E (respectively $E \cup X$).*

We have a similar result for Newtonian approximation.

2.2 Theorem *Let X be a compact subset of a Riemann surface S and E an open subset of $S \setminus X$. If E meets each of the bounded components of $S \setminus X$, then each function harmonic (respectively Newtonian) on X is the uniform limit of Newtonian functions on S all of whose singularities lie in E (respectively $E \cup X$).*

Remark. Theorem 2.1 allows us to approximate by essentially harmonic functions, whose pole set E is prescribed by arbitrarily choosing a single point in each hole of X. In Theorem 2.2, however, we must suppose that E is open. A single point in each hole would not suffice (cf. the lemmas below). Thus, in Newtonian approximation the exact location of the poles is fuzzy (uncertain), whereas in essentially harmonic approximation the poles can be precisely described. In [8], in a moment of dyslexia, we stated these two theorems in the opposite way. However, the ensuing remarks and proofs were in accord with the correct versions.

These two theorems are harmonic analogs of the rational Runge theorem of section 1. As an immediate consequence of Theorems 2.1 and 2.2, we also have the following analog of the polynomial Runge theorem of section 1 (see [27] and [28, p. 192]).

2.3 Theorem *Let X be a compact subset of a Riemann surface S such that $S^* \setminus X$ is connected. Then, each function harmonic on X is the uniform limit of harmonic functions on S.*

The above theorems show a striking similarity between holomorphic approximation and harmonic approximation. Now, let us point out a fundamental difference between the two theories. The condition that the compact set X have no holes is not only sufficient in the polynomial Runge theorem—it is also *necessary*. In the harmonic analog (Theorem 2.3), however, this condition, although still sufficient, is no longer necessary. Indeed, let X be the unit circle centered at the origin of the complex plane $\mathbf{C}$, and let f be an arbitrary continuous function defined on X (of course, this includes harmonic functions on X). By solving the Dirichlet problem, we may extend f to a function F which is continuous on the closed unit disc K and harmonic on the open unit disc. For $0 < \rho < 1$ and $z \in K$, set $F_\rho(z) = F(\rho z)$. Then, $F_\rho \to F$ uniformly on K, as $\rho \to 1$. Now, the functions F_ρ are harmonic on K and so by Theorem 2.3 may be uniformly approximated by functions harmonic on all of $\mathbf{C}$. By completing these entire harmonic functions to entire holomorphic functions and then taking the real part of the partial sum of the Taylor series, we may in fact approximate the functions F_ρ uniformly on K by harmonic polynomials. It follows that the function f, which was an arbitrary continuous function on the circle X, can be uniformly approximated on the circle by harmonic polynomials, even though the complement of the circle is of course not connected. The reader interested in knowing more about harmonic approximation should consult the book of Gardiner [13]. For a study of harmonic approximation on Riemannian manifolds, we refer to [4] and the bibliography therein. See also the lecture notes of Armitage, Bagby-Castañeda, and Gardiner in the present Proceedings.

2.2 Preliminaries

Let S be a Riemann surface and let Π be a set of isolated points on S. A *singular function* h associated to the discrete set Π is a function defined on

$$\bigcup_{P \in \Pi} D_P \setminus \{P\},$$

where D_P is a parametric disc centred at P and, in the local coordinate for D_P, the singular part s of h has the representation

$$s(z) = \Re \left(\sum_{n=1}^{\infty} \beta_n(P) z^{-n} \right) + \lambda(P) \log |z|, \quad P \in \Pi.$$

If S is compact, then Π is finite, and in this case we say that the singular function is *admissible* if

$$\sum_{P \in \Pi} \lambda(P) = 0.$$

A harmonic function p on $S \setminus \Pi$ such that $p - h$ is harmonic at each point $P \in \Pi$ is called a *principal function* associated to the singular function h. The following result is well known (see [1, pp. 148–157]).

2.4 Lemma *Let S be a compact Riemann surface and h a singular function associated to a finite set of points on S. Then there exists a principal function associated to h on S if and only if h is admissible.*

Let h and h_i, $i = 1, 2, \ldots$, be singular functions on a compact Riemann surface S, and let h be defined in $\bigcup_{j=1}^{k}(D_j \setminus P_j)$, where the P_j are the singular points of h and the D_j are the parametric discs about the P_j. We say that the sequence $\{h_i\}$ converges to h if the singularities of the h_i converge to those of h, and if the h_i converge uniformly to h on compact subsets of $\bigcup_{j=1}^{k}(D_j \setminus P_j)$. We define analogously the convergence of a sequence p_i of principal functions to a principal function p. Note that in this case the maximum principle implies that the p_i converge to p uniformly on each compact subset of S not intersecting the singular points of p. Thus, the convergence which we have just defined differs from the uniform convergence which is involved in our three theorems. However, uniform convergence implies convergence. The following lemma can be found in [29] and [30, p. 52].

2.5 Lemma *Let S be a compact Riemann surface. If a sequence $\{h_i\}$ of admissible singular functions converges to an admissible singular function h, then there exists a sequence $\{p_i\}$ of principal functions associated to $\{h_i\}$ which converges to a principal function p associated to h.*

Consider a singular function of the form $h(z) = \Re(\beta_n z^{-n})$ and f= ix $0 < \epsilon < 1$. We claim that there exists a singular function

$$h_\epsilon(z) = u(z) + \sum_{j=1}^{J} \lambda_j \log|z - \zeta_j| \tag{3}$$

with

$$|h(z) - h_\epsilon(z)| < \epsilon, \quad \text{for} \quad \epsilon \le |z| \le 1, \tag{4}$$

where u is harmonic on $(|z| \le 1)$, the points ζ_j are inside $(|\zeta| < \epsilon)$, for $j = 1, \ldots, J$, and

$$\lambda_1 + \cdots + \lambda_J = 0. \tag{5}$$

Since, in polar coordinates,

$$\Re(\beta_n z^{-n}) = \Re(\beta_n)r^{-n}\cos(n\theta) + \Im(\beta_n)r^{-n}\sin(n\theta),$$

we may assume that

$$h(z) = r^{-n}\cos(n\theta) \text{ or } h(z) = r^{-n}\sin(n\theta).$$

Let $\psi(r)$ be a smooth function of compact support with $0 \leq \psi$, $\psi(r) = 0$ for $0 \leq r \leq \epsilon/2$, and $\psi(r) = 1$, on a neighbourhood of $[\epsilon, 1]$. Now set $H(z) = 0$ and $H(z) = \psi(|z|)h(z)$ for $z \neq 0$. Since H is a smooth function of compact support, it can be written as a logarithmic potential

$$H(z) = \frac{1}{2\pi} \int \log|z - \zeta| \cdot \Delta H(\zeta)\, dm(\zeta).$$

Any Riemann sum is of the form (3), and if the partition is sufficiently fine, we have the approximation (4). We may suppose the points ζ_j in the Riemann sum are evenly distributed on concentric circles about the origin. Denoting $\phi_j = \arg \zeta_j$, on each such circle we have $\sum e^{in\phi_j} = 0$. Hence $\sum \cos(n\phi_j) = 0$ and $\sum \sin(n\phi_j) = 0$. Thus, $\sum \Delta H(\zeta_j) = 0$. Summing over all such circles yields (5). As an application of this result and the two previous lemmas, we have:

2.6 Lemma *Let S be a compact Riemann surface. Then each function essentially harmonic on S is the limit of Newtonian functions on S.*

In this lemma, the convergence is again uniform on each compact set not meeting the singular points of the limit function.

Proof Let h be essentially harmonic on S. Then we may consider h as an admissible singular function on an appropriate finite family of parametric discs. From the previous paragraph, there is a sequence $\{h_i\}$ of admissible singular functions which converges to h, such that each h_i has only Newtonian singularities. By Lemma 2.5, there is a sequence $\{p_i\}$ of principal functions associated to $\{h_i\}$ which converges to a principal function p associated to h. Since each h_i has only Newtonian singularities, each principal function p_i is Newtonian. Since h is defined on all of S, the difference $p - h$ is harmonic on S and hence constant. Thus, we may consider that the Newtonian functions p_i converge to h itself. By the maximum principle, the functions p_i converge to h on each compact subset of S not meeting the singular points of h. □

We also need a Mittag-Leffler type lemma, this time on open Riemann surfaces (see [27] and [28, p. 194]).

2.7 Lemma *Let S be an open Riemann surface and s a singular function associated to a discrete set of points on S. Then, there exists a principal function associated to s on S.*

2.3 Proofs

We may now present:

Proof of Theorem 2.1 Let $u \in \mathrm{Har}(X)$. Let V be a neighbourhood of X bounded by a finite number of Jordan curves and such that $u \in \mathrm{Har}(V)$. Let $\tilde{X}$ be the union of X and the components of $S \setminus X$ which do not intersect $S \setminus V$. Notice that $S \setminus \tilde{X}$ has only finitely many bounded components $G_1, G_2, \ldots, G_n$, since such a component meets and therefore contains a component of $S \setminus V$. Choose a point $P_i \in E \cap G_i$, $i = 1, \ldots, n$ and set $E_u = \{P_1, \ldots, P_n\}$. Fix $\epsilon > 0$. By [27], [28, p. 192] and Theorem 2.3, we know that there exists a function $\nu_1 \in \mathrm{Har}(S \setminus E_u)$ such that

$$|\nu_1 - u|_{\tilde{X}} < \frac{\epsilon}{2}.$$

Locally, for each point $P_i \in E_u$, we may write

$$\nu_i(z) = \lambda^{(i)} \log |z| + \Re f_i(z),$$

where f_i is holomorphic in a punctured neighbourhood of P_i, and $\lambda^{(i)} \in \mathbf{R}$. Since E_u contains only a finite number of points, we can find a function $g \in \mathrm{Mer}(S \setminus E_u)$ such that $g - f_i$ is meromorphic in a neighbourhood of P_i, $i = 1, \ldots, n$ (see for example [22, pp. 648, 649, 680]). Denote by K_1 the set of poles of g. Following [36, Corollary 2], there exists a function $\ell \in \mathrm{Mer}(S)$ all of whose poles are in $E_u \cup K_1$ and such that $|\ell - g|_{X \cup K_1} < \varepsilon/2$. Set $\nu = \nu_1 - \Re g + \Re \ell$. Then ν is in $\mathrm{EssHar}(S) \cap \mathrm{Har}(S \setminus E_u)$ and

$$|\nu - u|_X \le |\nu_1 - u|_X + |\Re \ell - \Re g|_X < \varepsilon.$$

There remains to consider the case when $u \in \mathrm{EssHar}(X)$. By Lemmas 2.4 and 2.7, there exists $u_1 \in \mathrm{EssHar}(S)$ with all of its singularities in $E \cup X$ and such that $(u - u_1) \in \mathrm{Har}(X)$. To complete the proof, it suffices to apply the reasoning of the previous paragraph to the function $u - u_1$. $\qquad\square$

Remark. If the Riemann surface S is open, we may, in the above proof, choose the function g such that it is holomorphic on $S \setminus E_u$ and $g - f_i$ is holomorphic in a neighbourhood of P_i. This avoids introducing the set K_1.

Proof of Theorem 2.2 If S is a compact Riemann surface, then the proof follows easily from Theorem 2.1 and Lemma 2.5.

If S is open, let $u \in N(X)$ and $\varepsilon > 0$. Let G be a neighbourhood of X bounded by finitely many disjoint Jordan curves such that $\overline{G}$ is compact and $S^* \setminus G$ is connected. If $\mathcal{G}$ is the surface obtained by doubling G, then by the previous paragraph, there exists $u_1 \in N(\mathcal{G})$, all of whose singularities lie in $E \cup X \cup (\mathcal{G} \setminus \overline{G})$, and such that $|u_1 - u|_X < \varepsilon/2$. Let $\widehat{X}$ be the hull of X. Then $\widehat{X}$ is a compact subset of G and $S^* \setminus \widehat{X}$ is connected. Let $p \in N(S) \cap \mathrm{Har}(S \setminus \widehat{X})$ such that $p - u_1$ is harmonic on $\widehat{X}$ (cf. Lemma 2.7). Then, by Theorem 2.1, there exists ν harmonic on S such that $|(p - u_1) - \nu|_{\widehat{X}} < \varepsilon/2$. The function $p - \nu$ performs the desired approximation, which concludes the proof of the theorem. $\qquad\square$

References

[1] L. V. Ahlfors and L. Sario, *Riemann Surfaces*, Princeton University Press, 1960.

[2] N. U. Arakelyan, Uniform approximation on closed sets by entire functions, *Izv. Akad. Nauk SSSR Ser. Mat.* **28** (1964), 1187–1206 (Russian).

[3] N. U. Arakelyan, Uniform and tangential approximation by analytic functions, *Izv. Akad. Nauk Armyan. SSR Ser. Mat.* **3** (1968), 273–286 (Russian).

[4] T. Bagby, P. M. Gauthier, and J. Woodworth, Tangential harmonic approximation on Riemannian manifolds, in: *Harmonic Analysis and Number Theory (Montréal, Qué., 1996)* (S. W. Drury and M. Ram Murty, eds.), CMS Conf. Proc. **21**, Amer. Math. Soc., Providence, RI, 1997, 58–72.

[5] H. Behnke and K. Stein, Entwicklung analytischer Funktionen auf Riemannnschen Flächen, *Math. Ann.* **120** (1949), 430–461.

[6] S. R. Bell, *The Cauchy Transform, Potential Theory, and Conformal Mapping*, CRC Press, Boca Raton–Ann Arbor–London–Tokyo, 1992.

[7] E. Bishop, Subalgebras of functions on Riemann surfaces, *Pacific J. Math.* **8** (1958), 29–50.

[8] A. Boivin and P. M. Gauthier, Approximation harmonique sur les surfaces de Riemann, *Canad. J. Math.* **36** (1984), 1–8.

[9] T. Carleman, Sur un théorème de Weierstrass, *Ark. Mat. Astron. Fys.* **20B** (4) (1927), 1–5.

[10] O. Forster, *Lectures on Riemann Surfaces*, Grad. Texts in Math. **81**, Springer-Verlag, Berlin–Heidelberg–New York, 1981.

[11] D. Gaier, *Lectures on Complex Approximation*, Birkhäuser, Boston, 1987.

[12] T. W. Gamelin, *Uniform Algebras*, Prentice-Hall, Englewood Cliffs, NJ, 1969 (1st ed.); Chelsea, New York, 1984 (2nd ed.).

[13] S. J. Gardiner, *Harmonic Approximation*, Cambridge University Press, 1995.

[14] P. M. Gauthier, Meromorphic uniform approximation on closed subsets of open Riemann surfaces, in: *Approximation Theory and Functional Analysis* (*Proc. Conf. Campinas 1977*) (J. B. Prolla, ed.), North-Holland, Amsterdam, 1979, 139–158.

[15] P. M. Gauthier, Analytic approximation on closed subsets of open Riemann surfaces, in: *Constructive Function Theory '77* (*Proc. Conf. Blagoevgrad*), Bulgarian Acad. Sci., Sofia, 1980, 317–325.

[16] P. M. Gauthier, Uniform approximation, in: *Complex Potential Theory*, NATO ASI Series C439, Kluwer, Dordrecht, 1994, 235–271.

[17] P. M. Gauthier, The Cauchy theorem for arbitrary connectivity, unpublished manuscript.

[18] P. M. Gauthier and W. Hengartner, Approximation sur les fermés par des fonctions analytiques sur une surface de Riemann, *C.R. Acad. Bulgare Sci.* **26** (1973), 731–732.

[19] P. M. Gauthier and W. Hengartner, Uniform approximation on closed sets by functions analytic on a Riemann surface, in: *Approximation Theory* (Z. Cielielski and J. Musielak, eds.), Reidel, Dordrecht, 1975, 63–70.

[20] P. M. Gauthier and W. Hengartner, *Approximation uniforme qualitative sur des ensembles non bornés*, Sém. Math. Sup. **82**, Les Presses de l'Université de Montréal, 1982.

[21] H. Köditz and S. Timmann, Randschlichte meromorphe Funktionen auf endlichen Riemannschen Flächen, *Math. Ann.* **217** (1975), 157–159.

[22] A. Hurwitz, *Vorlesungen über allgemeine Funktionentheorie und elliptische Funktionen*, 4th ed., Grundlehren Math. Wiss. **3**, Springer-Verlag, Berlin–New York, 1964.

[23] Z. Kuramochi, On covering surfaces, *Osaka Math. J.* **5** (1953), 155–201; errata **6** (1954), 167.

[24] S. N. Mergelyan, Uniform approximation to a function of a complex variable, *Amer. Math. Soc. Transl.* **3** (1962), 294–391.

[25] A. H. Nersessian, On the uniform and tangential approximation by meromorphic functions, *Izv. Akad. Nauk Armyan. SSR. Ser. Mat.* **7** (1972), 405–412 (Russian).

[26] P. V. Paramonov, *On Mergelyan's theorem, Selected Chapters of Complex Analysis*, Moscow University, 2000.

[27] A. Pfluger, Ein Approximationsatz für harmonische Funktionen auf Riemannschen Flächen, *Ann. Acad. Sci. Fenn. Ser. A I Math.* **216** (1956).

[28] A. Pfluger, *Theorie der Riemannschen Flächen*, Springer-Verlag, Heidelberg, 1957.

[29] B. Rodin and L. Sario, Convergence of normal operators, *Kōdai Math. Sem. Rep.* **19** (1967), 165–175.

[30] B. Rodin and L. Sario, *Principal Functions*, Van Nostrand, Princeton, NJ, 1968.

[31] A. Roth, Approximationseigenschaften und Strahlengrenzwerte meromorpher und ganzer Funktionen, *Comment. Math. Helv.* **11** (1938), 77–125.

[32] A. Roth, Meromorphe Approximationen, *Comment. Math. Helv.* **48** (1973), 151–176.

[33] W. Rudin, *Real and Complex Analysis*, 3rd ed., McGraw Hill, New York, 1987.

[34] C. Runge, Zur Theorie der eindeutigen analytischen Funktionen, *Acta Math.* **6** (1885), 228–244.

[35] S. Scheinberg, Uniform approximation by functions analytic on a Riemann surface, *Annals of Math.* **108** (1978), 257–298.

[36] S. Scheinberg, Uniform approximation by meromorphic functions having prescribed poles, *Math. Ann.* **243** (1979), 83–93.

[37] S. Scheinberg, Approximation and non-approximation on Riemann surfaces, in: *Complex Approximation (Proc. Conf. Quebec 1978)* (B. Aupetit, ed.), Progr. Math. **4** Birkhäuser, Boston–Basel–Stuttgart, 1980, 110–118.

[38] G. Schmieder, Approximation auf abgeschlossenen Teilen Riemannscher Flächen, *Manuscripta Math.* **46** (1984), 165–192.

[39] A. G. Vitushkin, Analytic capacity of sets in problems of approximation theory, *Russian Math. Surveys* **22** (1967), 139–200.

[40] J. L. Walsh, The Cauchy-Goursat theorem for rectifiable Jordan curves, *Proc. Nat. Acad. Sci. USA* **19** (1933), 540–541.

On the Bloch constant

Huaihui CHEN

Department of Mathematics
Nanjing Normal University
Nanjing, Jiangsu 210097
P. R. China

Abstract

This course presents a survey on Bloch constants for analytic mappings, meromorphic mappings and harmonic mappings of one variable and of several variables, including elementary concepts and theorems, Ahlfors'method, and recent results.

1 Introduction

The earliest result concerned with this topic is due to Valiron [37], who proved that the inverse of every non-constant entire function has holomorphic branches in arbitrarily large Euclidian disks. Then, Bloch improved the arguments in Valiron's proof and obtained a stronger conclusion: Every holomorphic function on the unit disk has an inverse branch on some Euclidian disk of radius $\delta|f'(0)|$, where δ is an absolute constant. Landau defined Bloch's constant $\mathbf{B}$ as the least upper bound of all numbers δ which satisfy the above conclusion. It is a famous and extremely hard problem to find the exact value of Bloch's constant.

Landau's elementary argument [19, 20] gave a lower bound for Bloch's constant: $\mathbf{B} \geq 0.23$. In 1938, Ahlfors [1, 2] generalized the Schwarz-Pick lemma with which he obtained a better lower bound: $\mathbf{B} \geq \sqrt{3}/4$. At the same time, Ahlfors and Grunsky [3] gave an important example to establish an upper bound. They proved that

$$0.4332 \approx \sqrt{3}/4 \leq \mathbf{B} \leq \frac{1}{\sqrt{1+\sqrt{3}}} \frac{\Gamma(1/3)\Gamma(11/12)}{\Gamma(1/4)} \approx 0.4719.$$

It is conjectured that this upper bound is the precise value of Bloch's constant.

Over a long period of time, no essential achievement had been attained in this topic until 1990. Other Bloch constants had been introduced for different classes of mappings and for different radii: for analytic mappings, harmonic mappings and meromorphic mappings; for one variable and for several variables; for hyperbolic radius and spherical radius. However, no exact value was found.

In 1990, Bonk [6] obtained a distortion theorem for Bloch functions from which Ahlfors' lower bound follows directly. Further, he gave a slight improvement of the lower bound:

$$\mathbf{B} > \sqrt{3}/4 + 10^{-14}.$$

N. Arakelian and P.M. Gauthier (eds.), Approximation, Complex Analysis, and Potential Theory, 129–161.
© 2001 *Kluwer Academic Publishers. Printed in the Netherlands.*

Then, in 1996, Chen and Gauthier [10] improved the result of Bonk further by showing

$$\mathbf{B} > \sqrt{3}/4 + 2 \times 10^{-4}.$$

Since Bonk's work, interest in this topic has been rekindled, and a lot of research has appeared. This article is a brief survey. The elementary concepts and theorems are formulated in Sections 1–5. Sections 6–8 are devoted to the Ahlfors lemma (Ahlfors' generaliztion of the Schwarz-Pick lemma) and Julia's lemma for multiple-valued functions, whose proof uses the Ahlfors lemma. Sections 9 and 10 are concerned with distortion theorems.

The upper bounds and lower bounds of different Bloch constants are formulated in Section 11 and 12. In Section 13, recent results on lower bounds of Bonk, Chen and Gauthier, and Yanagihara are presented without proof.

The equality statement in the Ahlfors lemma has been proved by Heins, Minda and Royden [18, 23, 32]. In Section 14 of this article, we prove an estimate for ultrahyperbolic metrics, which is a much stronger result than the equality statement, and whose proof is almost the same as that of Ahlfors himself and is much simpler than the proofs in [18, 23, 32].

Ahlfors' method for the lower bound is introduced in Section 15, and Sections 16 and 17 employ this method to treat the spherical and hyprebolic Bloch constants, respectively.

Finally, in Section 18 and 19, we introduce Bloch constants for harmonic mappings, quasiregular holomorphic mappings, and Bloch mappings in several variables. Many recent results of Liu, Gong, Chen and Gauthier and Hengartner are formulated without proof.

Throughout this article, we use the notation D_r to denote the disk in the complex plane, centered at the origin, with radius r.

2   The Schwarz-Pick lemma

2.1 Theorem (Schwarz-Pick lemma) *Let f be an analytic mapping of the unit disk D into itself. Then:*

(i) *for each $z \in D$, we have*

$$\frac{|f'(z)|}{1 - |f(z)|^2} \leq \frac{1}{1 - |z|^2};$$ (2.1)

(ii) *for any $z_1, z_2 \in D$, we have*

$$\left| \frac{f(z_2) - f(z_1)}{1 - \overline{f(z_1)}f(z_2)} \right| \leq \left| \frac{z_2 - z_1}{1 - \overline{z_1}z_2} \right|.$$ (2.2)

Furthermore, the equality in (2.1) holds for some point z, or the equality in (2.2) holds for some pair of different points z_1 and z_2, only if f is a linear transformation of D onto itself; conversely, if f is a linear transformation of D onto itself, (2.1) and (2.2) hold for any $z \in D$ and any $z_1, z_2 \in D$.

Proof Given $z_0 \in D$, let

$$w_0 = f(z_0), \qquad \phi(z) = \frac{z - z_0}{1 - \overline{z_0}z}, \qquad \psi(w) = \frac{w - w_0}{1 - \overline{w_0}w}, \qquad F = \psi \circ f \circ \phi^{-1}.$$

Then, F is an analytic mapping of D into itself such that $F(0) = 0$, since ϕ and ψ are linear transformations of D onto itself such that $\phi(z_0) = \psi(w_0) = 0$. We have

$$|\phi'(z_0)| = \frac{1}{1 - |z_0|^2}, \quad |\psi'(w_0)| = \frac{1}{1 - |w_0|^2}.$$

By the Schwarz lemma,

$$|F'(0)| = \frac{|f'(z_0)|(1 - |z_0|^2)}{1 - |w_0|^2} \le 1.$$

This shows (2.1).

If equality holds in (2.1) for z_0, then, by the Schwarz lemma, F is a rotation around the origin and, consequently, f is a linear transformation of D onto itself. Conversely, if f is such a mapping, then F is a rotation and $|F'(0)| = 1$. Thus, the equality in (2.1) holds for z_0. Since z_0 can be any point in D, the equality in (2.1) holds for any $z \in D$.

To show (2.2), let

$$\phi(z) = \frac{z - z_1}{1 - \overline{z_1}z}, \quad \psi(w) = \frac{w - w_1}{1 - \overline{w_1}w}, \quad F = \psi \circ f \circ \phi^{-1}.$$

Then, $F(0) = 0$ and $F(\phi(z_2)) = \psi(w_2)$. By the Schwarz lemma, $|\psi(w_2)| \le |\phi(z_2)|$. This shows (2.2). The proof can be completed by the same reasoning as in the previous paragraph. □

For the Schwarz lemma and Schwarz-Pick lemma, see [34, 12, 30, 31].

If we denote $w = f(z)$, (2.1) can be written in the form

$$\frac{|dw|}{1 - |w|^2} \le \frac{|dz|}{1 - |z|^2}. \tag{2.3}$$

We define the pseudo-distance between two points $z_1, z_2 \in D$ by

$$d_p(z_1, z_2) = \left| \frac{z_1 - z_2}{1 - \overline{z_1}z_2} \right|.$$

A pseudo-disk $\Delta(z_0, \rho)$ of center $z_0 \in D$ and radius $\rho < 1$ is defined as

$$\Delta(z_0, \rho) = \{ z \in D : d_p(z, z_0) < \rho \}.$$

If L is a linear mapping of D onto itself, then $L(\Delta(z_0, \rho)) = \Delta(L(z_0), \rho)$ and $L(\partial\Delta(z_0, \rho)) = \partial\Delta(L(z_0), \rho)$. With these definitions and notations, the Schwarz-Pick lemma says that (i) under the action of an analytic mapping of D into itself, the pseudo-distance is decreased, (ii) such a mapping f sends a pseudo-disk $\Delta(z_0, \rho)$ into the pseudo-disk $\Delta(f(z_0), \rho) = L(\Delta(z_0, r))$, where L is a linear mapping of D onto itself such that $L(z_0) = f(z_0)$, (iii) some point on $\partial\Delta(z_0, \rho)$ is sent to a point on $\partial\Delta(f(z_0), \rho)$, if and only if f is a linear mapping of D onto itself.

3 Landau's theorem for bounded analytic functions

Let f be an analytic mapping of a domain G. Let $z_0 \in G$, $w_0 = f(z_0)$. We say that $D_r(w_0) = \{ w : |w - w_0| < r \}$ is a *univalent disk* of radius r and center $f(z_0)$ contained in

$f(G)$, if there is a subdomain G' of G, containing z_0, such that f is univalent on G' and $f(G') = D_r(w_0)$. Here, $f(G)$ is regarded as a covering surface, and $w = f(z)$ is regarded as both a point in the covering surface and a complex number (a point in the complex plane).

The following theorem is due to Landau [19].

3.1 Theorem (Landau) *Let f be a holomorphic mapping of the unit disk D into D_M, $M \geq 1$, such that $f(0) = 0$ and $f'(0) = \alpha > 0$. Then:*

(i) *f is univalent in D_{r_0}, where*

$$r_0 = \frac{\alpha}{M + \sqrt{M^2 - \alpha^2}} > \frac{\alpha}{2M};$$

(ii) *for any positive number $r \leq r_0$, $f(D_r)$ contains the disk D_R, where*

$$R = M \cdot \frac{r(\alpha - Mr)}{M - \alpha r} \geq Mrr_0,$$

(iii) *$f(D)$ contains a univalent disk of radius $R_0 = Mr_0^2$ around $f(0)$.*

Furthermore, the numbers r_0 and R_0 are both best possible..

Proof First assume that $M = 1$. Then, $\alpha = 1$ implies $f(z) = z$, so the conclusions hold obviously. Now, we assume that $0 < \alpha < 1$.

Let $z_1, z_2 \in D$ be such that $z_1 \neq z_2$ and $0 \leq |z_1| \leq |z_2| = \rho < 1$ and $f(z_1) = f(z_2) = w_0$. Then the function

$$g(z) = \frac{f(z) - w_0}{1 - \overline{w_0} f(z)} \cdot \frac{1 - \overline{z}_1 z}{z - z_1} \cdot \frac{1 - \overline{z}_2 z}{z - z_2}$$

is holomorphic in D and, by the maximum principle, $|g(z)| \leq 1$ for $z \in D$. In particular, $|g(0)| = |w_0|/|z_1 z_2| \leq 1$ or $|g(0)| = \alpha/|z_2| \leq 1$ according to whether $w_0 \neq 0$ or $w_0 = 0$. If $w_0 = 0$, we have $\rho \geq \alpha$. Thus $\rho \geq r_0$ since $r_0 < \alpha$. If $w_0 \neq 0$, then

$$|w_0| \leq |z_1 z_2| \leq \rho^2. \tag{3.1}$$

Let $h(z) = f(z)/z$ for $z \neq 0$ and $h(0) = \alpha$. Applying the Schwarz-Pick lemma to the function h, we see that

$$\left| \frac{h(z_2) - \alpha}{1 - \alpha h(z_2)} \right| \leq \rho,$$

i.e., $h(z_2) \in \overline{\Delta}(\alpha, \rho)$. Let $\phi(z) = (z + \alpha)/(1 + \alpha z)$. Then $\phi(\overline{D}_\rho) = \phi(\overline{\Delta}(0, \rho)) = \overline{\Delta}(\alpha, \rho)$. Note that $\phi(\overline{D}_\rho)$ is the closed disk with diameter $[(\alpha - \rho)/(1 - \alpha\rho), (\alpha + \rho)/(1 + \alpha\rho)]$. Thus, by (3.1),

$$\rho \geq |w_0|/\rho = |f(z_2)|/\rho = |h(z_2)| \geq \frac{\alpha - \rho}{1 - \alpha\rho}. \tag{3.2}$$

It follows from (3.2) that

$$\rho \geq \frac{\alpha}{1 + \sqrt{1 - \alpha^2}}.$$

This proves (i).

To prove (ii), let $0 < r \le r_0$ and consider the function $h(z)$. By the Schwarz-Pick lemma,

$$\left| \frac{h(re^{i\theta}) - \alpha}{1 - \alpha h(re^{i\theta})} \right| \le r.$$

So,

$$|h(re^{i\theta})| \ge \frac{\alpha - r}{1 - \alpha r}, \qquad |f(re^{i\theta})| \ge r \cdot \frac{\alpha - r}{1 - \alpha r} = R \ge r \cdot \frac{\alpha - r_0}{1 - \alpha r_0} = r r_0.$$

This proves (ii) for $M = 1$. Thus (i) and (ii) have been shown for $M = 1$. We can prove them for any M by considering the function $f(z)/M$.

Let $f(z) = -z(z - \alpha)/(1 - \alpha z)$, $0 < \alpha < 1$, which satisfies the assumption of the theorem for $M = 1$. Then, $f'(r_0) = 0$ and $f(z) = r_0^2$ if and only if $z = r_0$. This shows the numbers $r_0 = 1/(1 + \sqrt{1 - \alpha^2})$ and $R_0 = r_0^2$ are best possible. $\qquad \square$

4 Bloch's theorem and Bloch's constant

4.1 Theorem (Bloch) *Let f be an analytic function on the unit disk D such that $f'(0) = 1$. Then $f(D)$ contains a univalent disk of radius $1/14$.*

Proof First assume that f is analytic on $\overline{D}$. Let $z' \in D$ be such that

$$A = (1 - |z'|)|f'(z')| = \max\{(1 - |z|)|f'(z)| : z \in D\}.$$

Note that $A \ge 1$ since $f'(0) = 1$. For $\zeta \in D$, define

$$g(\zeta) = \frac{2}{A} f\left(z' + \frac{1}{2}(1 - |z'|)\zeta\right) - \frac{2}{A} f(z').$$

Then,

$$g(0) = 0, \quad |g'(0)| = \frac{1}{A}(1 - |z'|)|f'(z')| = 1.$$

For $\zeta \in D$,

$$
\begin{aligned}
|g'(\zeta)| &= \frac{1}{A}(1 - |z'|)|f'\left(z' + (1 - |z'|)\zeta/2\right)| \\
&= \frac{(1 - |z'|)}{1 - |z' + (1 - |z'|)\zeta/2|} \frac{(1 - |z' + (1 - |z'|)\zeta/2|)|f'(z' + (1 - |z'|)\zeta/2)|}{A} < 2,
\end{aligned}
$$

and, consequently, $|g(\zeta)| < 2$. By Theorem 3.1, $g(D)$ contains a univalent disk of radius $2/(2 + \sqrt{3})^2$. Thus, $f(D)$ contains a univalent disk of radius $1/(2 + \sqrt{3})^2 > 1/14$.

If f is not analytic on the boundary, considering the function $f(rz)/r$ for $r < 1$, we conclude that $f(D)$ contains a univalent disk of radius $r/(2 + \sqrt{3})^2$. Taking $r = (2 + \sqrt{3})^2/14$, we show that $f(D)$ contains a univalent disk of radius $1/14$. $\qquad \square$

There are many proofs of Bloch's theorem. The above proof, which is due to Landau [19], is probably the simplest. For the original proof, see Bloch's paper [4].

We can modify the above proof to obtain a bigger number than $1/14$. Let $z' \in D$ be such that

$$A = (1 - |z'|^2)|f'(z')| = \max\{(1 - |z|^2)|f'(z)| : z \in D\}.$$

Then $A \geq 1$. Define $g(\zeta) = (f(\phi(\zeta)) - f(z'))/A$, where ϕ is a linear transformation of D onto itself such that $\phi(0) = z'$. Then,

$$|g'(0)| = \frac{1}{A}(1 - |z'|^2)|f'(z')| = 1,$$

and, for $\zeta \in D$, we have

$$(1 - |\zeta|^2)|g'(\zeta)| = \frac{1}{A}(1 - |\phi(\zeta)|^2)|f'(\phi(\zeta))| \leq 1,$$

since

$$\frac{|\phi'(\zeta)|}{1 - |\phi(\zeta)|^2} = \frac{1}{1 - |\zeta|^2}.$$

Let $0 < r < 1$. For $\zeta \in D_r$, we have

$$|g(\zeta)| < \int_0^r \frac{dt}{1 - t^2} = \frac{1}{2} \log \frac{1+r}{1-r} = M_r.$$

Let $h(\omega) = g(r\omega)$. By Theorem 3.1, $h(D)$ contains a univalent disk of radius

$$R_r = \frac{M_r r^2}{(M_r + \sqrt{M_r^2 - r^2})^2}.$$

Consequently, $f(D)$ contains a univalent disk of the same radius. Taking $r = 0.6$, we have $R_r > 0.2306$. This shows that $f(D)$ contains a univalent disk of radius 0.2306.

Let f be a non-constant analytic function on D. For $z \in D$, let $r(z, f)$ denote the radius of the largest univalent disk contained in $f(D)$ with center $f(z)$. Let $B_f = \sup\{r(z, f) : z \in D\}$. The greatest lower bound of B_f for all analytic functions f on D such that $f'(0) = 1$ is called *Bloch's constant* and is denoted by $\mathbf{B}$, i.e., $\mathbf{B} = \inf\{B_f : f'(0) = 1\}$. We have proved that $\mathbf{B} \geq 0.2306$. In 1938, Ahlfors [1] proved $\mathbf{B} \geq \sqrt{3}/4 \approx 0.4330$. Also, Ahlfors and Grunsky [3] established the following upper bound

$$\mathbf{B} \leq \frac{1}{\sqrt{1 + \sqrt{3}}} \frac{\Gamma(1/3)\Gamma(11/12)}{\Gamma(1/4)} \approx 0.4719.$$

It is conjectured that the correct value of B is precisely this upper bound. In 1990, Bonk [6] proved $\mathbf{B} > \sqrt{3}/4 + 10^{-14}$, and then, in 1996, Chen and Gauthier [10] improved the result of Bonk further by showing

$$\mathbf{B} > \sqrt{3}/4 + 2 \times 10^{-4}.$$

5   Julia's lemma

For $r > 0$,

$$\Delta_r = \left\{ z \in D : \frac{|1 - z|^2}{1 - |z|^2} < r \right\}$$

is called a horodisk. It is contained in D and internally tangent to ∂D at 1. In Euclidian terms Δ_r is the disk with center $1/(1 + r)$ and radius $r/(1 + r)$.

5.1 Theorem (Julia's lemma, disk version) *Let f be a function analytic on $D \cup \{1\}$. If $f(D) \subset D$ and $f(1) = 1$, then $f'(1) = a > 0$ and $f(\overline{\Delta}_r) \subset \overline{\Delta}_{ar}$ for $r > 0$. Further, a point in $\partial \Delta_r$ other than 1 is mapped into $\partial \Delta_{ar}$ if and only if f is a linear transformation of D onto itself.*

Assume that f is a function as above. It is easy to verify that $a > 0$. Let

$$\zeta = \phi(z) = \frac{1+z}{1-z}, \qquad F = \phi \circ f \circ \phi^{-1}.$$

ϕ is a linear mapping of D onto the right half-plane H, which maps a closed horodisk $\overline{\Delta}_r$ onto the closed half plane $\overline{H}_r = \{\zeta : \mathrm{Re}\,\zeta \geq 1/r\}$. Then, F is analytic on H, $F(H) \subset H$, F has a pole at ∞ and $F(\zeta) = \zeta/a + O(1)$ as $\zeta \to \infty$. So, Julia's lemma has another equivalent version:

5.2 Theorem (Julia's lemma, half-plane version) *Let F be a function which is analytic on H and has a pole at ∞. If $F(H) \subset H$, then $F(\zeta) = b\zeta + O(1)$ with $b > 0$ as $\zeta \to \infty$ and $\mathrm{Re}\,F(\zeta) \geq b\,\mathrm{Re}\,\zeta$ for $\zeta \in H$. Further, equality holds for some $\zeta \in H$ if and only if $F(\zeta) \equiv b\zeta + ci$ with a real constant c.*

Proof It is easy to verify that $b > 0$. Also, we can deduce that $F(\zeta) = b\zeta + c' + ci + O(1/\zeta)$ with $c' \geq 0$ and real c. Let $G(\zeta) = F(\zeta) - b\zeta$. Then, G is analytic at ∞, $G(\infty) = ci$, and

$$\liminf_{\zeta \to \zeta'} \mathrm{Re}\,G(\zeta) \geq 0$$

for any $\zeta' \in \partial H$. By the maximum principle, for any $\zeta \in H$, we have

$$\mathrm{Re}\,G(\zeta) = \mathrm{Re}\,F(\zeta) - b\,\mathrm{Re}\,\zeta \geq 0.$$

If equality holds for some $\zeta \in H$, i.e., $\mathrm{Re}\,G(\zeta)$ attains its minimum at some point in H, then the maximum principle asserts that $\mathrm{Re}\,G(\zeta) \equiv 0$. Consequently, $G(\zeta) \equiv ci$ with a real constant c. The proof is complete. $\qquad \square$

If L is a linear self-mapping of D such that $L(1) = 1$, then $F = \phi \circ L \circ \phi^{-1}$ is a linear self-mapping of H such that $F(\zeta) = \zeta/L'(1) + c_2 i$ for $\zeta \in H$. For $r > 0$, F maps $\overline{H}_r$ onto $\overline{H}_{L'(1)r}$. Consequently, L maps $\overline{\Delta}_r$ onto $\overline{\Delta}_{L'(1)r}$. Thus, Julia's lemma can be formulated as follows: Let f be a function analytic on $D \cup \{1\}$ such that $f(D) \subset D$ and $f(1) = 1$, and let L be a linear self-mapping L of D such that $L(1) = 1$ and $L'(1) = f'(1)$, then $f(\Delta_r) \subset L(\Delta_r)$ for $r > 0$. We see a similarity between the Schwarz-Pick lemma and Julia's lemma. For Julia's lemma, see [2].

6   The Poincaré metric

The metric $\lambda = \lambda(z)|dz|$ defined by

$$\lambda(z) = \frac{2}{1 - |z|^2}, \quad z \in D,$$

is called the *Poincaré metric* on the unit disk. The Gaussian curvature of a metric $\rho = \rho(z)\,|dz|$ is defined by

$$K_\rho = -\frac{1}{\rho(z)^2}\Delta\rho(z),$$

where Δ is the Laplacian defined by

$$\Delta = \frac{\partial^2}{\partial x^2} + \frac{\partial^2}{\partial y^2}.$$

Note that the Gaussian curvature is conformally invariant. The Poincaré metric λ has constant Gaussian curvature -1.

Let $\rho = \rho(w)\,|dw|$ be a metric defined on a domain G, and $w = f(z)$ be an analytic mapping of a domain G' into G. We use the notation $f^*\rho$ to denote the metric defined by

$$f^*\rho = \rho(f(z))\,|df(z)| = \rho(f(z))|f'(z)|\,|dz|,$$

which is called the pull-back of ρ by f. By (2.1), for any linear mapping $w = \phi(z)$ of D onto itself, we have the identity

$$\frac{|\phi'(z)|\,|dz|}{1 - |\phi(z)|^2} = \frac{|dz|}{1 - |z|^2},$$

i.e. $\phi^*\lambda = \lambda$. This means that the Poincaré metric is conformally invariant under a linear self-mapping of D. On the other hand, (2.1) in the Schwarz-Pick lemma can be formulated in the form $f^*\lambda \leq \lambda$ for each analytic mapping of D into itself, i.e., under such a mapping the Poincaré metric is decreased. For a piece-wise smooth curve $\gamma \in D$, the non-Euclidean length of γ is defined by $\int_\gamma \lambda(z)\,|dz|$. For an analytic mapping of D into itself, we have $\int_{f(\gamma)} \lambda(w)\,|dw| = \int_\gamma \lambda(f(z))|f'(z)|\,|dz| \leq \int_\gamma \lambda(z)\,|dz|$, since $f^*\lambda \leq \lambda$. This means that under such a mapping the non-Euclidian length of a curve is decreased.

7   The Ahlfors lemma — a generalization of the Schwarz-Pick lemma

A twice-differentiable metric with Gaussian curvature ≤ -1 is called a *hyperbolic metric*. Ahlfors [1] introduced the notion of ultrahyperbolic metric and generalized the Schwarz-Pick lemma in terms of such metrics. A metric $\rho = \rho(z)|dz|$ defined on a domain or a Riemann surface G is called an *ultrahyperbolic metric*, if (1) $\rho(z)$ is upper semi-continuous, (2) for each point $z_0 \in G$ with $\rho(z_0) \neq 0$ there exists a hyperbolic supporting metric $\rho_0(z)|dz|$ defined in a neighbourhood U of z_0 such that $\rho(z_0) = \rho(z_0)$ and $\rho_0(z) \leq \rho(z)$ for $z \in U$.

7.1 Theorem (Ahlfors lemma) *If $\rho(z)\,|dz|$ is an ultrahyperbolic metric on the unit disk, then $\rho(z) \leq \lambda(z)$ for $z \in D$.*

Proof First we assume that $\rho(z)$ is continuous on ∂D. Then $\log\rho(z) - \log\lambda(z)$ attains its maximum at some point $z_0 \in D$ and $\rho(z_0) > 0$. Let $\rho_0(z)\,|dz|$ be the supporting metric. $\log\rho_0(z) - \log\lambda(z)$ also attains its maximum at z_0. Thus,

$$\Delta(\log\rho_0(z) - \log\lambda(z))|_{z=z_0} \leq 0.$$

On the other hand,

$$\Delta \log \lambda(z) = \lambda(z)^2, \qquad \Delta \log \rho_0(z) \geq \rho(z)^2.$$

It follows that $\rho(z_0) = \rho_0(z_0) \leq \lambda(z_0)$. Consequently, $\log \rho(z_0) - \log \lambda(z_0) \leq 0$ and $\rho(z) \leq \lambda(z)$ for $z \in D$. Thus the theorem is proved under the assumption that $\rho(z)$ is continuous on ∂D. For the general case, we consider $\rho(rz)\,|dz|$ for $r < 1$ and let $r \to 1$. □

If the metric ρ is ultrahyperbolic and f is analytic, then $f^*\rho$ is also ultrahyperbolic, since the Gaussian curvature is conformally invariant. Let f be an analytic mapping of D into itself. Then

$$f^*\lambda = \frac{|f'(z)|\,|dz|}{1 - |f(z)|^2}$$

is an ultrahyperbolic metric. Using the Ahlfors lemma gives (2.1). This shows that the Schwarz-Pick lemma is a mere consequence of the Ahlfors lemma. Any simply-connected domain $G \subset \mathbb{C}$, except the plane itself, has a Poincaré metric $\lambda_G = \lambda_G(z)|dz|$, which is defined by

$$\lambda_G(z) = \lambda(\phi(z))|\phi'(z)|,$$

where ϕ is a conformal mapping of G onto D. Since the Poincaré metric λ of D is conformally invariant, the metric λ_G defined above is independent of the choice of the mapping ϕ. Moreover, the Ahlfors lemma is valid for any domain conformally equivalent to the unit disk. Bearing this in mind, we note for subsequent use that the Poincaré metric of the right half-plane $H = \{z = x + iy : x > 0\}$ is $|dz|/x$.

8   Julia's lemma for multiple-valued functions

The following theorem, proved by Liu and Minda [27], is an extension of Julia's lemma to certain multiple-valued functions .

8.1 Lemma *Let f be a function analytic on $D \cup \{1\}$ such that $f(1) = 1$. Then, f has no zero of order n in $\Delta_{n/f'(1)}$. Further, f has a zero of order n at $z_0 \in \partial\Delta_{n/f'(1)}$ if and only if $f(z) = (L_{z_0}(z))^n$, where L_{z_0} is the unique linear mapping of D onto itself such that $L_{z_0}(z_0) = 0$ and $L_{z_0}(1) = 1$.*

Proof Assume that $z_0 \in \partial\Delta_r$ is a zero of order n for f. Let $g(z) = f(z)/(L_{z_0}(z))^n$. Note that $L'_{z_0}(1) = 1/r$. By the maximum modulus principle, we have $|g(z)| \leq 1$ for $z \in D$. If $g(z) \equiv 1$, then $f(z) \equiv (L_{z_0}(z))^n$ and $f'(1) = nL'_{z_0}(1) = n/r$, i.e., $r = n/f'(1)$. In the case that g is not a constant, we have $g(1) = 1$ and $g(D) \subset D$. Julia's lemma asserts $g'(1) > 0$. Consequently, $f'(1) > n/r$, i.e., $r > n/f'(1)$. This proves the first statement. If f has a zero of order n at $z_0 \in \partial\Delta_{n/f'(1)}$, then the function g defined above must be the constant 1, i.e., $f = L^n_{z_0}$. Otherwise, we have $g'(1) > 0$, $L'(z_0)(1) = f'(1)/n$, $f'(1) = f'(1) + g'(1) > f'(1)$. This is a contradiction. On the other hand, if $f = L^n_{z_0}$ with $z_0 \in \partial\Delta_r$, then $f'(1) = n/r$. This shows $z_0 \in \partial\Delta_{n/f'(1)}$, completing the proof. □

8.2 Theorem *Let f be a function analytic on $D \cup \{1\}$ such that $f(1) = 1$ and f has only zeros of order at least n. Then the above lemma asserts that f has no zero in $\Delta_{n/f'(1)}$. Let g be the single-valued branch of $f^{1/n}$ in $\Delta_{n/f'(1)}$ such that $g(1) = 1$. Then, $g(\overline{\Delta}_r) \subset \overline{\Delta}_{g'(1)r}$*

for $0 < r \le n/f'(1)$. Further, a point $z' \in \partial\Delta_r$ with $0 < r \le n/f'(1)$ is mapped by g into $\partial\Delta_{g'(1)r}$ if and only if $g = L$ and $f = L^n$, where L is a linear mapping of D onto itself.

Proof We use h to denote the multiple-valued analytic function $f^{1/n}$. Let

$$\rho = \rho(z)\,|dz| = \lambda(h(z))|h'(z)|\,|dz|.$$

ρ is well defined since two different branches of h in a neighbourhood of any point at which f does not vanish differ only by a constant factor with modulus 1. ρ has constant curvature -1 except for the points where f has zeros of order $\ge n+1$, since it is locally the pull-back of λ by a conformal mapping (a local branch of $f^{1/n}$). Now, assume that f has a zero of order $k \ge n+1$ at z_0. In a neighbourhood of z_0 we have

$$h(z) = (z - z_0)^{k/n}h_0(z),$$

where h_0 is single-valued and $h(z_0) \ne 0$. Thus,

$$\rho(z) = \lambda(h(z))|z - z_0|^{k/n-1}|k/n + h_0'(z)| \to 0 \quad \text{as} \quad z \to z_0.$$

This shows that ρ is an ultrahyperbolic metric on the unit disk.

As in Section 5, let

$$\zeta = \phi(z) = \frac{1+z}{1-z}, \qquad G = \phi \circ g \circ \phi^{-1}.$$

Then, G is analytic on $H' \cup \{\infty\}$, where $H' = \{\zeta = \xi + i\eta : \xi > f'(1)/n\}$, and

$$w = G(\zeta) = \frac{n}{f'(1)}\zeta + O(1) \quad \text{as} \quad \zeta \to \infty.$$

We only need to prove that

$$\operatorname{Re} G(\zeta) \ge \frac{n}{f'(1)}\operatorname{Re}\zeta \quad \text{for} \quad \zeta \in H'.$$

Let $\rho_1 = (\phi^{-1})^*\rho$, which is an ultrahyperbolic metric on H. By the Ahlfors lemma $\rho_1(\zeta) \le \lambda_H(\zeta) = 1/\xi$ for $\zeta \in H$, where $\zeta = \xi + i\eta$. Restricted to H', we have

$$\rho_1 = (\phi^{-1})^*\rho = (\phi^{-1})^*g^*\lambda = (\phi^{-1})^*g^*\phi^*\lambda_H = G^*\lambda_H.$$

Fix $\zeta_0 = \xi_0 + i\eta_0 \in H'$. Let $\zeta' = \xi' + i\eta_0$ with large ξ, $w_0 = G(\zeta_0) = u_0 + iv_0$ and $w' = G(\zeta') = u' + iv'$. We have

$$\int_{[\zeta_0,\zeta']} \rho_1(\zeta)\,|d\zeta| = \int_{w_0}^{w'} \lambda_H(w)\,|dw| = \int_{w_0}^{w'} \frac{|dw|}{u} \ge \log\frac{u'}{u_0}.$$

On the other hand,

$$\int_{[\zeta_0,\zeta']} \rho_1(\zeta)\,|d\zeta| \le \int_{[\zeta_0,\zeta']} \frac{|d\zeta|}{\xi} = \log\frac{\xi'}{\xi_0}.$$

Thus,

$$\frac{u'}{u_0} \le \frac{\xi'}{\xi_0}.$$

Letting $\xi' \to +\infty$, we obtain

$$u_0 \geq \frac{n}{f'(1)} \xi_0.$$

This is what we wanted to prove.

Let $G(\zeta) = U(\zeta) + iV(\zeta)$. We have proved that $U(\zeta) \geq n\xi/f'(0)$ for $\zeta \in H'$. Assume that there exists a point $z_0 \in \Delta_r$, $r < \Delta_{n/f'(1)}$, such that $g(z_0) \in \partial\Delta_{rf'(1)/n}$. Let $\zeta_0 = \xi_0 + i\eta_0 = \phi(z_0)$. Then $\zeta_0 \in H'$ and $U(\zeta_0) = n/f'(0)\xi_0$. By the maximum principle, we have $U(\zeta) = \xi$ for $\zeta \in H'$. Consequently, $G(\zeta) = n/f'(0)\zeta + ai$ for $\zeta \in H'$ and g is a linear mapping of D onto itself. This completes the proof. $\square$

9 Distortion theorems for Bloch functions

An analytic function f on the unit disk is called a *Bloch function*, if

$$\sup_{z \in D}(1 - |z|^2)|f'(z)| < \infty.$$

Now, we consider the subclass B of those Bloch functions such that

$$f(0) = 0, \qquad f'(0) = 1, \qquad (1 - |z|^2)|f'(z)| \leq 1 \quad \text{for} \quad z \in D.$$

For any positive integer n, let

$$B_n = \{f \in B : f'(z) = 0 \iff f^k(z) = 0 \text{ for } k = 1, \ldots n\}.$$

Let B_∞ denote the class of locally univalent functions in B. Then,

$$B_\infty = \bigcap_{n=1}^{\infty} B_n.$$

Let $f \in B$ and $f'(z) = 1 + a_1 z + a_2 z^2 + \cdots$. Then it is obvious that $a_1 = 0$ and $|a_2| \leq 1$. Bonk [6] proved that $|a_3| \leq 5$. Chen and Gauthier [10] proved that $|a_3| \leq 4.2$. For $f \in B_\infty$, The author and others proved that $|a_3| \leq 3.914$.

Let $a_n = \sqrt{n/(n+2)}$ and, for $z \in D$, define

$$L_n(z) = \frac{a_n - z}{1 - a_n z},$$

$$f_n(z) = \frac{(n+2)a_n}{2(n+1)}(1 - a_n^{-(n+1)} L_n(z)^{n+1}).$$

We have

$$f_n'(z) = \frac{(a_n - z)^n}{a_n^n(1 - a_n z)^{n+2}},$$

$$(1 - |z|^2)|f_n'(z)| = (1 - |L(z)|^2) \cdot \frac{1}{2}(n+2)a_n^{-n}|L(z)|^n.$$

Thus, $(1 - |z|^2)||f_n'(z)|$ attains its maximum on the circle of diameter $[0, a_n]$, where $|L(z)| = a_n$. It is easy to verify that $f_n(0) = 0$ and $f'(0) = 1$. This shows that $f_n \in B_n$.

Bonk [6] first proved a distortion theorem for functions in B and then, with its help, gave an improvement of the lower bound for Bloch's constant. Bonk's theorem is formulated as follows.

9.1 Theorem *If $f \in \mathcal{B}$, then*

$$\operatorname{Re} f'(z) \geq \frac{1 - \sqrt{3}|z|}{(1 - |z|/\sqrt{3})^3} \quad for \quad |z| \leq \frac{1}{\sqrt{3}}$$

with equality for some $z = re^{i\theta} \neq 0$ if and only if $f(z) = e^{i\theta} f_1(e^{-i\theta}z)$.

By using their generalization of Julia's lemma, Liu and Minda [27] improved Bonk's result for the classes $\mathcal{B}_n$, and obtained the following distortion theorems.

9.2 Theorem *If $f \in \mathcal{B}_n$, then*

(i)

$$|f'(z)| \geq f'_n(|z|) = \frac{(a_n - |z|)^n}{a_n^n(1 - a_n|z|)^{n+2}} \quad for \quad |z| \leq a_n \tag{9.1}$$

   with equality for a $z = re^{i\theta} \neq 0$ if and only if $f(z) = e^{i\theta} f_n(e^{-i\theta}z)$;

(ii)

$$\operatorname{Re} f'(z) \geq f'_n(|z|) \quad for \quad |z| \leq \frac{(n+2)a_n}{2n+1}, \tag{9.2}$$

   with equality for some $z = re^{i\theta} \neq 0$ if and only if $f(z) = e^{i\theta} f_n(e^{-i\theta}z)$;

(iii)

$$\operatorname{Re} f'(z) > 0 \quad for \quad |z| < b_n = \sqrt{\frac{n\sin(\pi/2n)}{2 + n\sin(\pi/2n)}}. \tag{9.3}$$

Proof Assume that $f \in \mathcal{B}_n$. Let

$$\phi(\zeta) = \frac{\zeta - a_n}{1 - a_n\zeta}, \quad h(\zeta) = f(\phi(\zeta)), \quad \zeta \in D.$$

Then,

$$(1 - |\zeta|^2)|h'(\zeta)| = (1 - |\zeta|^2)|f'(\phi(\zeta))||\phi'(\zeta)| = (1 - |\phi(\zeta)|^2)|f'(\phi(\zeta))|,$$

where the identity $|\phi'(\zeta)|/(1 - |\phi(\zeta)|^2) = 1/(1 - |\zeta|^2)$ is used. So, $(1 - a_n^2)h'(a_n) = 1$ and

$$(1 - a_n^2)|h'(\zeta)| < 1 \quad for \quad \zeta \in D_{a_n}.$$

Let $g(\omega) = (1 - a_n^2)h'(a_n\omega)$ for $\omega \in D$; we have $g(1) = 1$, $|g(\omega)| < 1$ for $\omega \in D$ and

$$g'(1) = (1 - a_n^2)a_n h''(a_n) = (1 - a_n^2)a_n\phi''(a_n) = \frac{2a_n^2}{1 - a_n^2} = n,$$

since $f'(0) = 1$ and $f''(0) = 0$. Note that the zeros of g are just the zeros of f' and g has only zeros of order at least n. Applying the extension of Julia's lemma, Theorem 8.2, to the function g, we see that g has no zero in the horodisk Δ_1 and, letting G be the single-valued analytic branch of $g^{1/n}$ with $G(1) = 1$, we have $G(\overline{\Delta}_r) \subset \overline{\Delta}_r$ for $0 < r \leq 1$. In particular,

$|G(\xi)| \geq \xi$ and $|g(\xi)| \geq \xi^n$ for $0 \leq \xi < 1$, i.e., $(1 - a_n^2)|f'(\phi(a_n\xi))\phi'(a_n\xi)| \geq \xi^n$ for $0 \leq \xi < 1$. This is equivalent to

$$|f'(x)| \geq \frac{(x + a_n)^n}{a_n^n(1 + a_n x)^{n+2}} \quad \text{for} \quad -a_n \leq x < 0. \tag{9.4}$$

Thus (9.1) is proved for $z \in [-a_n, 0]$. We can prove (9.1) completely by considering $f(e^{i\theta}z)$ with $0 \leq \theta \leq 2\pi$. If for some $x \in [-a_n, 0)$ equality in (9.4) holds, i.e., $|G(\xi)| = \xi$ for some $\xi \in [0, 1)$, then, by the second part of Theorem 8.2, $G(\xi) = \xi$ and G is a linear mapping of D onto itself, which fixes ξ and 1. Obviously, G is the identity. Thus,

$$f'(z) = \frac{(z + a_n)^n}{a_n^n(1 + a_n z)^{n+2}} = -\frac{1}{a_n^n(1 - a_n^2)}L_n(-z)^n L_n'(-z),$$

$$f(z) = \frac{\sqrt{n(n + 2)}}{2(n + 1)}\left[\left(\frac{n + 2}{n}\right)^{(n+1)/2} L_n(-z)^{n+1} - 1\right] = -f_n(-z).$$

If the equality in (9.1) holds for some $z = re^{i\theta}$ with $0 < r \leq a_n$, considering the function $-e^{i\theta}f(-e^{-i\theta}z)$, we can show that $f(z) = e^{i\theta}f(e^{-i\theta}z)$.
    (9.2) and (9.3) are proved in the same way. $\qquad\square$

For $z \in D$, let $L(z) = (1 + z)/(1 - z)$ and

$$f_\infty(z) = -\frac{e}{2}\exp\{-L(z)\} + \frac{1}{2}.$$

Note that L is a linear mapping of D onto the right half-plane such that $L(0) = 1$ and $L(1) = \infty$. We have

$$f_\infty'(z) = \frac{1}{(1 - z)^2}\exp\left\{-\frac{2z}{1 - z}\right\} = 1 - z^2 - \frac{4}{3}z^3 - \cdots,$$

$$(1 - |z|^2)|f_\infty'(z)| = e\,\mathrm{Re}\,L(z)\exp[-\,\mathrm{Re}\,L(z)].$$

Thus, $(1 - |z|^2)|f_\infty'(z)|$ attains its maximum 1 on the circle of diameter $[0, 1]$, on which $\mathrm{Re}\,L(z) = 1$. This shows that $f_\infty \in B_\infty$.

9.3 Theorem *If $f \in B_\infty$, then*

(i) $$|f'(z)| \geq f_\infty'(|z|) = \frac{1}{(1 - |z|)^2}\exp\left\{-\frac{2|z|}{1 - |z|}\right\} \quad \text{for} \quad z \in D \tag{9.5}$$

with equality for a $z = re^{i\theta} \neq 0$ if and only if $f(z) = e^{i\theta}f_\infty(e^{-i\theta}z)$;

(ii) $$\mathrm{Re}\,f'(z) \geq f_\infty'(|z|) \quad \text{for} \quad |z| \leq \frac{1}{2} \tag{9.6}$$

with equality for some $z = re^{i\theta} \neq 0$ if and only if $f(z) = e^{i\theta}f_n(e^{-i\theta}z)$;

(iii) $$\mathrm{Re}\,f'(z) > 0 \quad \text{for} \quad |z| < b_\infty = \sqrt{\frac{\pi}{4 + \pi}}. \tag{9.7}$$

Proof Let $f \in \mathcal{B}_\infty$. Since

$$\frac{(1+z)^2}{1-|z|^2} = 1 \quad \text{for} \quad z \in \partial D_{1/2}(-1/2),$$

$$\frac{(1+z)^2}{1-|z|^2} < 1 \quad \text{for} \quad z \in D_{1/2}(-1/2),$$

letting $h(z) = (1+z)^2 f'(z)$, we have $h(1) = 1$ and $|h(z)| < 1$ for $z \in D_{1/2}(-1/2)$. Let $g(\omega) = h(-1/2 + \omega/2)$. Then, $g(D) \subset D$ and $g(1) = 1$. Let $G(\omega)$ be a single-valued analytic branch of $\text{Log}\, g(\omega)$ such that $G(1) = 0$, and let

$$H(\omega) = \frac{1 + G(\omega)}{1 - G(\omega)}.$$

We have $H(1) = 1$, $H(D) \subset D$ and $H'(1) = 2$. Applying Julia's lemma to H, we see that for $\xi \in (-1,1)$, $G(\xi)$ lies in the closed disk with diameter $[2(\xi - 1)/(\xi + 1)]$. Thus,

$$|g(\xi)| \geq \exp\left\{\frac{2(\xi - 1)}{\xi + 1}\right\}.$$

This is equivalent to

$$|f'(x)| \geq \frac{1}{(1+x)^2} \exp\left\{\frac{2x}{1+x}\right\} \quad \text{for} \quad x \in (-1,0).$$

(9.5) is proved for $z \in (-1,0)$. By considering the function $-e^{-i\theta} f(-e^{i\theta} z)$ with real θ, we can prove (9.5) for any $z = re^{i\theta} \in D$. The proof of the theorem can be completed in a similar fashion to the proof of the previous theorem. $\quad\square$

10 Improvements of (9.3) and (9.7)

The numbers b_n and b_∞ in the preceding section are not optimal. Now, we give some improvements.

10.1 Lemma *Suppose that g is a function holomorphic in $D \cup \{1\}$, with $g(1) = 1$ and $\beta^{-1} < |g(z)| < 1\}$ for $z \in D$, where $\beta > 1$. Let*

$$k = \frac{\log \beta}{\pi g'(1)}, \qquad a = \exp\left\{\frac{\pi^2}{2\log\beta}\right\}, \qquad r = \frac{1}{2}\left(a - a^{-1}\right).$$

Then $\text{Re}\{g(z)\} > 0$ for $z \in \Delta_{kr}$.

Proof Let

$$\phi(z) = \exp\left\{-\frac{\pi i}{\log \beta} \log g(z)\right\},$$

$$h(z) = i \cdot \frac{\phi(z) - i}{\phi(z) + i}.$$

Then, $h(D) \subset D$, $h(1) = 0$ and $h'(1) = k^{-1}$. Let $\theta = \arctan r$. If $z \in \Delta_{kr}$ $h(z) \in \Delta_r$, then, by Julia's lemma, $\phi(z)$ lies in the disk with diameter $[a^{-1}e^{i\theta}, ae^{i\theta}]$ and, consequently, $\text{Re}\{g(z)\} > 0$. This shows the lemma. $\quad\square$

10.2 Theorem *If $f \in \mathcal{B}_\infty$, then*

$$\operatorname{Re} f'(z) > 0 \quad \text{for} \quad |z| < \rho_\infty \approx 0.6654.$$

Proof Let $0 < \alpha < 1$, $\Delta = \{z \in \mathbb{C} : |z + \alpha/(1+\alpha)| < \alpha/(1+\alpha^2)\}$, $\phi(z) = (z+\alpha)/(1+\alpha z)$ and $h(z) = (1+az)^2 f'(z)$. Since

$$1 - |z|^2 = \frac{1 - |\phi(z)|^2}{|\phi'(z)|} \quad \text{for} \quad z \in D,$$

$|\phi(z)| = a$ for $z \in \partial\Delta$ and $|\phi(z)| < a$ for $z \in \Delta$, we have

$$|h(z)| < (1 - |z|^2)|f'(z)| \leq 1 \quad \text{for} \quad z \in \Delta,$$

and

$$|h(z)| = (1 - |z|^2)|f'(z)| \quad \text{for} \quad z \in \partial\Delta.$$

Note that $h(0) = 1$. For $z \in \partial\Delta$, by (9.5),

$$|h(z)| = (1 - |z|^2)|f'(z)| \geq \left(1 - \frac{4\alpha^2}{(1+\alpha^2)^2}\right) f'_\infty\left(\frac{2\alpha}{1+\alpha^2}\right)$$

$$= \left(\frac{1+\alpha}{1-\alpha}\right)^2 \exp\left\{-\frac{4\alpha}{(1-\alpha)^2}\right\} = \beta^{-1}.$$

Consequently, from the maximum principle,

$$1 > |h(z)| > \left(\frac{1+\alpha}{1-\alpha}\right)^2 \exp\left\{-\frac{4\alpha}{(1-\alpha)^2}\right\} = \beta^{-1} \quad \text{for} \quad z \in \Delta,$$

since $f'(z) \neq 0$ for $z \in D$.

Let

$$g(z) = h\left(\frac{\alpha}{1+\alpha^2}(z-1)\right).$$

Then, $g(1) = 1$, $\beta^{-1} < |g(z)| < 1\}$ for $z \in D$ and $g'(1) = 2\alpha^2/(1+\alpha^2)$. Applying Lemma 10.1 to the function $g(z)$, we have $\operatorname{Re} g(z) > 0$ for $z \in \Delta_{r'}$, where

$$r' = \frac{r(1+\alpha^2)\log\beta}{2\alpha^2\pi},$$

$$r = \frac{1}{2}\left(\exp\left\{\frac{\pi^2}{2\log\beta}\right\} - \exp\left\{-\frac{\pi^2}{2\log\beta}\right\}\right).$$

This is equivalent to $\operatorname{Re} h(z) > 0$ for $|z + \rho/2| < \rho/2$, where

$$\rho = \frac{2\alpha r\log\beta}{2\alpha^2\pi + r(1+\alpha^2)\log\beta}.$$

Thus $\operatorname{Re} f'(x) > 0$ for $-\rho < x \leq 0$. Considering the functions $f'(e^{i\theta}z)$ for $0 \leq \theta \leq 2\pi$, we obtain $\operatorname{Re} f'(z) > 0$ for $|z| < \rho$. A numerical calculation shows that ρ attains its maximum $\rho_\infty \approx 0.6654$ at $\alpha = \alpha_0 \approx 0.6412$. This proves the theorem. □

The above method also provides constants bigger than b_n in (9.3). Now, let $n \geq 10$ and $f \in \mathcal{B}_n$. Take $\alpha = \alpha_0$; then $\sqrt{n/(n+2)} > 2\alpha/(1+\alpha^2)$ for $n \geq 10$. Let Δ, h, g be the same as in the above proof. Then, $|h(z)| < 1$ for $z \in \Delta$ and, by Theorem 9.2,

$$|h(z)| = (1-|z|^2)|f'(z)| \geq (1-|z|^2)f_n'(|z|) \geq \left(1 - \frac{4\alpha^2}{(1+\alpha^2)^2}\right) f_n'\left(\frac{2\alpha}{1+\alpha^2}\right)$$

$$= \left(1+\frac{2}{n}\right)^{n/2} \left(\frac{1-\alpha^2}{1+\alpha^2}\right)^2 \frac{\left(\sqrt{n/(n+2)} - 2\alpha/(1+\alpha^2)\right)^n}{\left(1 - \sqrt{n/(n+2)}(2\alpha/(1+\alpha^2))\right)^{n+2}} = \beta_n^{-1}.$$

Consequently, $1 > |h(z)| > \beta_n^{-1}$ for $z \in \Delta$, since $f'(z) \neq 0$ for $z \in \Delta$.

Let
$$r_n = \frac{1}{2}\left(\exp\left\{\frac{\pi^2}{2\log\beta_n}\right\} - \exp\left\{-\frac{\pi^2}{2\log\beta_n}\right\}\right).$$

Then the same reasoning gives

$$\mathrm{Re}\, f'(z) > 0 \quad \text{for} \quad |z| < \rho_n = \frac{2\alpha r_n \log\beta_n}{2\alpha^2\pi + r_n(1+\alpha^2)\log\beta_n}.$$

It is obvious that $\rho_n \to \rho_\infty \approx 0.6654$ since $f_n'(2\alpha/(1+\alpha^2)) \to f_\infty'(2\alpha/(1+\alpha^2))$.

A numerical calculation gives $\rho_{10} \approx 0.6635$, $\rho_{15} \approx 0.6646$, $\rho_{20} \approx 0.6650$, $\rho_{50} \approx 0.6653$, $\rho_{100} \approx 0.6654$. Note that

$$b_n < b_\infty, \qquad b_n \to b_\infty = \sqrt{\pi/(4+\pi)} \approx 0.6633.$$

11 Bloch constants and their upper bounds

We have defined Bloch's constant $\mathbf{B}$ in Section 4. Now we define different Bloch constants for different classes of functions analytic in the unit disk. For a positive integer n, define

$$\mathbf{B}_n = \inf\{B_f : f'(0) = 1,\ f'(z) = 0 \iff f^k(z) = 0 \text{ for } k = 1,\ldots,n\}.$$

Also, define
$$\mathbf{B}_\infty = \inf\{B_f : f'(0) = 1,\ f'(z) \neq 0 \text{ for } z \in D\}.$$

Note that $\mathbf{B}_1 = \mathbf{B}$.

In 1938, Ahlfors and Grunsky [3] established the following upper bound for $\mathbf{B}$:

$$\mathbf{B} \leq \frac{1}{\sqrt{1+\sqrt{3}}} \frac{\Gamma(1/3)\Gamma(11/12)}{\Gamma(1/4)} \approx 0.4719.$$

By a natural extension of Goluzin's method, Minda [22] obtained upper bounds for all $\mathbf{B}_n$ and $\mathbf{B}_\infty$. For $q \in [0, 1/3)$, let Ω_q be the circular triangle, which has all interior angles of size πq and vertices at the points r_q, $r_q e^{2\pi i/3}$ and $r_q e^{4\pi i/3}$. Let $\Omega_1/3$ be the Euclidian triangle with vertices 1, $e^{2\pi i/3}$ and $e^{4\pi i/3}$. A calculation gives

$$r_q = \left[\frac{\sin\pi(5/6 + q/2)}{\sin\pi(1/6 + q/2)}\right]^{1/2}.$$

Let f_q be the unique conformal mapping of D onto Ω_q/r_q, the circular triangle with all interior angles of size πq and vertices at the points 1, $e^{2\pi i/3}$ and $e^{4\pi i/3}$, such that $f_q(0) = 0$ and $f(e^{2k\pi i/3}) = e^{2k\pi i/3}$ for $k = 0, 1, 2$. Then

$$f_q'(0) = \frac{\Gamma(5/6 + q/2)\Gamma(2/3)}{\Gamma(1/6 + q/2)\Gamma(4/3)}.$$

For $n = 1, 2, \ldots$ and $n = \infty$, let $q_n = 1/(3(n+1))$

$$g_n(z) = f_{1/3}(f_{q_n}^{-1}(z/r_{q_n})).$$

This is the conformal mapping of Ω_{q_n} onto $\Omega_{1/3}$ which sends 0 to 0 and makes the vertices correspond. We have

$$g_n'(0) = \frac{\Gamma(1/6 + 1/(6(n+1)))}{r_{q_n}\Gamma(1/3)\Gamma(5/6 + 1/(6(n+1)))}.$$

By using the Schwarz reflection principle in conjunction with the fact that Ω_{q_n} leads to a triangulation of D via reflection, we extend g_n to be an analytic function defined on D, which is locally univalent if $n = \infty$, and is univalent except at each vertex of the triangulation, where it assumes a value with exact multiplicity $n + 1$. This shows that

$$\mathbf{B}_n \leq 1/g_n'(0).$$

In particular, we have the upper bound of Ahlfors and Grunsky for $\mathbf{B}$, and an upper bound for $\mathbf{B}_\infty$:

$$\mathbf{B}_\infty \leq \frac{\Gamma(1/3)\Gamma(5/6)}{\Gamma(1/6)} \approx 0.5433$$

12   The lower bounds for Bloch constants

In 1938, Ahlfors [1] proved that $\mathbf{B} \geq \sqrt{3}/4$ by using his generalization of the Schwarz-Pick lemma, the Ahlfors lemma in Section 7, and then, in 1962, Heins [18] showed that the strict inequality holds. Ahlfors's method also provided lower bounds for every $\mathbf{B}_n$:

$$\mathbf{B}_\infty \geq \frac{1}{2}, \qquad \mathbf{B}_n \geq \frac{\sqrt{n(n+2)}}{2(n+1)}, \qquad n = 1, 2, \ldots.$$

This approach to Bloch constants will be explained in the next section. Here, we formulate Bonk's proof and Liu-Minda's proof for these lower bounds.

The following lemma is due to Landau [20]. It shows that, when studying Bloch constants, we need only consider functions in $\mathcal{B}$.

12.1 Lemma *For $n = 1, 2, \ldots, \infty$, we have $\mathbf{B}_n = \inf\{B_f : f \in \mathcal{B}_n\}$.*

Now, Ahlfors' lower bound for Bloch's constant $\mathbf{B}$ is just a direct consequence of Bonk's distortion theorem for functions in $\mathcal{B}$ (Theorem 9.1), see [6]. Let $f \in \mathcal{B}$. First we see that f

is univalent on $D_{1/\sqrt{3}}$, since $\operatorname{Re} f'(z) > 0$ for $z \in D_{1/\sqrt{3}}$. Then, for any $z' = e^{i\theta}/\sqrt{3} \in D_{1/\sqrt{3}}$, by Theorem 9.1, we have

$$|f(z')| = \left| \int_0^{1/\sqrt{3}} f'(re^{i\theta})\,dr \right| \geq \int_0^{1/\sqrt{3}} \operatorname{Re} f'(re^{i\theta})\,dr$$

$$\geq \int_0^{1/\sqrt{3}} \frac{1 - \sqrt{3}r}{(1 - r/\sqrt{3})^3}\,dr = \frac{\sqrt{3}}{4}.$$

This shows that $f(D_{1/\sqrt{3}})$ contains the disk $D_{\sqrt{3}/4}$. Thus, $B_f \geq \sqrt{3}/4$. Using Landau's lemma, we obtain a new Ahlfors lower bound for $\mathbf{B}$.

Ahlfors' lower bounds for $\mathbf{B}_n$, $n > 1$, can also be obtained from Liu-Minda's distortion theorems in a similar way. In this procedure, the following lemma is needed.

12.2 Lemma *Let f be analytic and locally univalent on G. If $D_r(f(z'))$ is the largest univalent disk in $f(G)$ with center $f(z')$ and f maps a domain $G' \subset G$, $z' \in G'$, onto $D_r(f(z'))$ injectively, then there is a point $w' \in \partial D_r(f(z'))$ such that $f_{G'}^{-1}(w) \to \partial G$ as $w \to w'$ and $w \in D_r(f(z'))$, where $f_{G'}$ is the restriction of f on G'.*

Let n be a positive integer, $f \in \mathcal{B}_n$, and $D_{r'}(f(0))$ be the largest univalent disk contained in $f(D)$ with center $f(0)$. By Lemma 12.2, there is a point $w' \in \partial D_{r'}(f(0))$ such that $f_{D_{a_n}}^{-1}(w) \to \partial D^0$ as $w \to w'$ and $w \in D_{r'}(f(0))$, where D^0 is the domain obtained by removing the zeros of f' from D. Let $\tilde{\gamma} = f^{-1}([0, w'))$. Then there is a sequence $z_j \in \tilde{\gamma}$ such that $w_j = f(z_j) \to w'$ and z_j converges to a zero of f' or to a point in ∂D. Thus, we have

$$r(0, f) = |w'| = \int_{\tilde{\gamma}} |f'(z)|\,|dz| \geq \int_0^{a_n} \frac{(a_n - r)^n}{a_n^n(1 - a_n r)^{n+2}}\,dr$$

$$= \frac{a_n}{(n+1)(1 - a_n^2)} = \frac{\sqrt{n(n+2)}}{2(n+1)}.$$

This gives the lower bound for $\mathbf{B}_n$. In the same way, we can prove that $r(0, f) \geq 1/2$ for $f \in \mathcal{B}_\infty$ and $\mathbf{B}_\infty \geq 1/2$.

By a normal family argument, we can find a function $f \in \mathcal{B}_\infty$ such that $B_f = \mathbf{B}_\infty$. If $r(0, f) > 1/2$, then $\mathbf{B}_\infty \geq r(0, f) > 1/2$. If $r(0, f) = 1/2$, then we can see from the above proof that $f(z) = e^{-i\theta} f_\infty(e^{i\theta} z)$. Thus, $\mathbf{B}_\infty = B_f = e/4 > 1/2$. In the same way we can prove that the strict inequality holds also for $n = 1, 2, \ldots$.

13   Improvements of the lower bounds

In 1990, Bonk [6] first proved a distortion theorem for functions in $\mathcal{B}$, and then reobtained Ahlfors' lower bound for $\mathbf{B}$ as a direct consequence of his distortion theorem. This was explained in the preceding section. Furthermore, he proved the following lemma with the help of which he improved this lower bound to

$$\mathbf{B} > \sqrt{3}/4 + 10^{-14}.$$

13.1 Lemma *Let* $f \in B$, $\varphi_0 = 2\arcsin(1/20)$ *and* $r_0 = 1/1000$. *Then, for* $\varphi_0 \leq \varphi \leq 2\pi - \varphi_0$ *and* $|z| \leq r_0$, *we have*

$$\frac{1}{2}\operatorname{Re}[f'(z) + f'(e^{i\varphi}z)] \geq \frac{1 - \sqrt{3}\,|z|}{(1 - |z|/\sqrt{3})^3} + \frac{1}{2}|z|^3.$$

We emphasize that the above estimate is local since it holds only if $|z|$ is very small. This is the reason why only a quite small improvement is obtained from this estimate. However, this was the first improvement since Ahlfors had established his lower bound $\sqrt{3}/4$ a half century earlier.

Let $f \in B$ and $f'(z) = 1 + a_2z^2 + a_3z^3 + \cdots$. Chen and Gauthier [10] discovered that the arguments $\arg a_2$ and $\arg a_3$ have evident global influences on the lower bound of $\operatorname{Re} f'(z)$. They proved the following estimates.

13.2 Lemma *Let* $f \in B$ *and* $f'(z) = 1 + |a_2|e^{i\theta_0}z^2 + a_3z^3 + \cdots$. *Then, for* $\theta \in [0, 2\pi]$, *we have*

$$\int_0^{1/\sqrt{3}} \operatorname{Re} f'(re^{i\theta})\,dr \geq \frac{\sqrt{3}}{4} + 0.0109(1 + |a_2|\cos(2\theta + \theta_0)).$$

13.3 Lemma *Let* $f \in B$ *and* $f'(z) = 1 + a_2z^2 + |a_3|e^{i\theta_0}z^3 + \cdots$. *Then, for* θ *with* $\cos(3\theta + \theta_0) \geq 0$, *we have*

$$\int_0^{1/\sqrt{3}} \operatorname{Re} f'(re^{i\theta})\,dr \geq \frac{\sqrt{3}}{4} + 0.005.$$

It follows from these two lemmas that when $f \in B$ is given, the integral $\int_0^{1/\sqrt{3}} \operatorname{Re} f'(re^{i\theta})\,dr$ has a lower bound bigger than $\sqrt{3}/4$ for every $\theta \in [\alpha, \beta]$ with $\beta - \alpha > \pi$. Using this fact, they improved the lower bound by showing

$$\mathbf{B} > \sqrt{3}/4 + 0.0002.$$

Chen and others have also found that the distortion theorem for functions in B_∞ can be improved in a small neighbourhood of the origin except for a small sector with its vertex at the origin. They proved the following estimates:

13.4 Lemma *Let* $f \in B_\infty$ *and let*

$$f'(z) = 1 + a_2z^2 + a_3z^3 + \cdots,$$

where $0.9 \leq a_2 \leq 1$ *and* $\operatorname{Im} a_3 \geq 0$. *Then, for* $|z| \leq 0.03$ *and* $|\arg z - \pi/2| \leq \eta = \arcsin(0.2)$, *we have*

$$|f'(z)| \geq 1 - |z|^2 - 1.13|z|^3.$$

13.5 Lemma *Let* $f \in B_\infty$ *and let*

$$f'(z) = 1 + a_2z^2 + a_3z^3 + \cdots,$$

where $0.9 \leq a_2 \leq 1$. *Then, for* $|z| \leq 0.03$ *and* $|\arg z| \leq \pi/2 - \arcsin(0.19)$ *or* $|\arg z - \pi| \leq \pi/2 - \arcsin(0.19)$, *we have*

$$|f'(z)| > 1 - 0.93|z|^2 - 4.1|z|^3.$$

13.6 Lemma *Let $f \in \mathcal{B}_\infty$ and let*

$$f'(z) = 1 + a_2 z^2 + a_3 z^3 + \cdots,$$

where $|a_2| \leq 0.9$. Then, for $|z| \leq 0.03$, we have

$$|f'(z)| > 1 - 0.9|z|^2 - 4.3|z|^3.$$

13.7 Lemma *Let $f \in \mathcal{B}_\infty$ and*

$$f(z) = z + az^3 + \cdots$$

Then, for $|z| \leq 0.3$, we have

$$\left| \arg \frac{f(z)}{z} \right| \leq 0.73|z|^2.$$

On the basis of these lemmas and the original distortion theorem for functions in $\mathcal{B}_\infty$, they established the following lower bound (unpublished):

$$\mathbf{B}_\infty > \frac{1}{2} + 2 \times 10^{-8}.$$

By using a different method, Yanagihara [39] had proved that

$$\mathbf{B}_\infty > \frac{1}{2} + 10^{-335}.$$

14   More about the Ahlfors lemma

In 1938, Ahlfors [1] introduced the notion of ultrahyperbolic metric and obtained an extension of the classical Schwarz lemma (see Section 7). As an application, he obtained a remarkable lower bound for Bloch's constant: $\mathbf{B} \geq \sqrt{3}/4$. Later, Heins [18] introduced the class of SK metrics, which includes ultrahyperbolic metrics, and verified that the Ahlfors lemma remains valid for SK metrics. In addition, he showed that equality $\rho(z') = \lambda(z')$ at a single point z' implies $\rho(z) \equiv \lambda(z)$. However, his proof of the equality statement is not as elementary as the proof of the Ahlfors lemma since it relies on an integral representation for a solution of the nonlinear partial differential equation $\Delta u = \exp\{2u\}$. At the same time, Heins proved that $\mathbf{B} > \sqrt{3}/4$. Minda [23] and Royden [32] gave elementary proofs of the equality statement.

For an upper semicontinuous function u, the *generalized lower Laplacian* of u at a point z_0 where $u(z_0) > -\infty$ is defined by

$$\underline{\Delta} u(z_0) = \liminf_{r \to 0} \left(\frac{1}{2\pi} \int_0^{2\pi} u(z_0 + re^{i\theta}) \, d\theta - u(z_0) \right).$$

If u is actually of class C^2 in a neighbourhood of z_0, then it is straightforward to show that $\underline{\Delta} u(z_0) = \Delta u(z_0)$, the usual Laplacian of u at z_0. Note that the generalized lower Laplacian is conformally invariant. A metric $\rho = \rho(z) \, |dz|$, defined on a domain (or, generally, a Riemann surface), is called an SK metric, if $\rho(z)$ is upper semicontinuous and $\underline{\Delta} \log \rho(z) \geq \rho(z)^2$ holds at each point z such that $\rho(z) > 0$. It is obvious that an ultrahyperbolic metric is an SK metric. Also, almost the same proof shows that the Ahlfors lemma remains true for SK metrics.

As indicated above, Heins, Minda, Royden proved the equality statement of the Ahlfors lemma for an SK metric. However, their proofs are much more complicated than that of the Ahlfors lemma itself. The author discovered that in fact, the same reasoning that is used in the proof of the Ahlfors lemma also provides a result which is stronger than the equality statement.

14.1 Lemma *Let $0 < r_0 < 1$ and ρ be an SK metric on the annulus $\{z : r_0 < |z| < 1\}$. If*

$$\limsup_{z \to \partial D_{r_0}} = \rho_0 < \lambda(r_0) = \frac{2}{1 - r_0^2},$$

then

$$\rho(z) \le \frac{2\alpha |z|^{\alpha-1}}{1 - |z|^{2\alpha}}, \qquad \text{for} \quad r_0 < |z| < 1,$$

where $\alpha > 1$ is the number such that $\rho_0 = 2\alpha r_0^{\alpha-1}/(1 - r_0^{2\alpha})$.

Proof It easy to verify that $2\alpha r_0^{\alpha-1}/(1 - r_0^{2\alpha})$ decreases from $\lambda(r_0)$ to 0 as α increases from 1 to ∞. So, there is a unique number $\alpha > 1$ such that $\rho_0 = 2\alpha r_0^{\alpha-1}/(1 - r_0^{2\alpha})$. Consider the metric

$$\rho_\alpha(z)\,|dz| = \frac{2\alpha |z|^{\alpha-1}}{1 - |z|^{2\alpha}},$$

which is the pull-back of the Poincaré metric $\lambda(w)|dw|$ by the mapping $w = z^\alpha$ and, consequently, has constant Gaussian curvature -1. For a fixed $r \in (r_0, 1)$, $r\rho(rz)\,|dz|$ is also an SK metric defined on $A = \{z : r_0/r < |z| < 1\}$. Let

$$\phi(z) = \log \rho_\alpha(z) - \log(r\rho(rz)), \qquad a = \inf\{\phi(z) : z \in A\}.$$

Assume that $a < 0$. Then there is a point $z' \in A$ such that $\phi(z') = a$, since $\liminf_{r \to \partial A} > 0$. Thus,

$$0 \le \underline{\Delta}\phi(z') = \underline{\Delta} \log \rho_\alpha(z)|_{z=z'} - \underline{\Delta} \log(r\rho(rz))|_{z=z'}$$
$$\le \rho_\alpha(z')^2 - r^2\rho(rz')^2.$$

It follows that $\phi(z') \ge 0$, a contradiction. Thus we have proved that $a \ge 0$, i.e., $\rho_\alpha(z) \ge r\rho(rz)$ for $z \in A$. Letting $r \to 1$ completes the proof. $\qquad\square$

15 Ahlfors' method

We now introduce Ahlfors' proof for $\mathbf{B} \ge \sqrt{3}/4$, see [1, 2]. Let f be an analytic function on D such that $f'(0) = 1$. We regard $f(D)$ as a covering surface and, for $z \in D$, regard $w = f(z)$ as both a point in the covering surface and a complex number (a point in the plane). We define a metric on $f(D)$ by

$$\tilde{\rho} = \frac{A\,|dw|}{r_w^{1/2}(A^2 - r_w)},$$

where $r_w = r(z, f)$ for $w = f(z)$, $A^2 > 3B_f$. Note that $\tilde{\rho}$ is not defined for the branch points at this moment.

Let w_0 be a branch point of multiplicity 2. We have $r_w = |w - w_0|$ in a small neighbourhood of w_0. If we take the local parameter $\zeta = \sqrt{w - w_0}$, then

$$\tilde{\rho} = \frac{A |dw|}{|w - w_0|^{1/2}(A^2 - |w - w_0|)} = \frac{2A |d\zeta|}{A^2 - |\zeta^2|},$$

which is the pull-back of the Poincaré metric $\lambda(\omega) |d\omega| = 2 |d\omega|/(1 - |\omega|^2)$ by the mapping $\omega = \zeta/A$. So, $\tilde{\rho}$ is actually regular at w_0 and has constant curvature -1 in a neighbourhood of w_0. For a branch point w_0 of multiplicity $n \geq 3$, with the local parameter $\zeta = \sqrt[n]{w - w_0}$, we have

$$\tilde{\rho} = \frac{nA|\zeta|^{n/2-1/2} |d\zeta|}{A^2 - |\zeta^2|}.$$

This shows that $\tilde{\rho} \to 0$ as $w \to w_0$.

Now, assume that $w_0 \in f(D)$ is a regular point. There is a boundary point or branch point w' on $\partial D_{w_0}(r_w)$. We have $r_{w_0} = |w_0 - w'|$ and $r_w \leq |w - w'|$ for $w \in D_{w_0}(r_w)$. If w is very close to w_0, then

$$\frac{A |dw|}{r_w^{1/2}(A^2 - r_w)} \geq \frac{A |dw|}{|w - w'|^{1/2}(A^2 - |w - w'|)},$$

since the function $t^{1/2}(A^2 - t)$ is increasing for $0 < t < A^2/3$. The above equality holds at $w = w_0$. The same reasoning as above shows that the right metric has constant curvature -1. This shows that the right metric is a supporting metric of $\tilde{\rho}$ at w_0. This proves that $\tilde{\rho}$ is an ultrahyperbolic metric on $f(D)$.

Let

$$\rho = \tilde{\rho}(f(z))|f'(z)| |dz|,$$

which is an ultrahyperbolic metric on D. Applying the Ahlfors lemma for ρ at $z = 0$, we have

$$\frac{A}{B_f^{1/2}(A^2 - B_f)} \leq \frac{A}{r_{f(0)}^{1/2}(A^2 - r_{f(0)})} \leq 2.$$

Letting $A \to \sqrt{3B_f}$ gives $B_f \geq \sqrt{3}/4$. This is what we want to prove.

The same method also provides lower bounds for other Bloch constants. For $\mathbf{B}_n$, $n = 1, 2, \ldots$, we consider metrics

$$\tilde{\rho} = \frac{2A |dw|}{nr_w^{1-1/n}(A^2 - r_w^{2/n})},$$

where $A^2 > B_f^{2/n}(n + 1)/(n - 1)$. For $\mathbf{B}_\infty$, we consider the metric

$$\tilde{\rho} = \frac{|dw|}{r_w(\log A - \log r_w)},$$

where $A > eB_f$.

16   Spherical Bloch constants

Let f be a non-constant meromorphic function on a domain G. Now, we identify the extended complex plane $\overline{\mathbb{C}}$ with the Riemann sphere, and regard $f(G)$ as a covering surface of the Riemann sphere, and regard $w = f(z)$ as both a point in the covering surface and a complex number or ∞ (a point in the extended plane $\overline{\mathbb{C}}$). For every $z_0 \in G$, we define $b(z_0, f)$ to be the angular radius of the largest univalent spherical disk of center $f(z_0)$ contained in $f(G)$. If z_0 is a critical point, i.e., f has a multiplicity bigger than 1 at z_0, then $b(z_0, f) = 0$. We define

$$B_f^s = \sup\{b(z, f) : z_0 \in G\}.$$

The following estimate was proved by Pommerenke [29].

16.1 Theorem *Let f be a meromorphic function on D. If $B_f^s \leq \pi/3 - \eta$, $0 < \eta < \pi/3$, then*

$$(1 - |z|^2)f^\#(z) \leq C_\eta, \quad z \in D,$$

where $f^\# = |f'|/(1 + |f|^2)$ is the spherical derivative of f and C_e is a constant depending on η only.

Proof For a regular point $w = f(z) \in f(D)$, let $r_w = \tan(b(z, f)/2)$. We define a metric $\tilde{\rho}$ on $f(D)$ by

$$\tilde{\rho} = \frac{A(1 + r_w^2)}{r_w^{1/2}(A^2 - r_w)} \cdot \frac{|dw|}{1 + |w|^2}.$$

It follows from $B_f^s \leq \pi/3 - \eta$ that $r_w \leq 1/\sqrt{3} - \delta$, where $0 < \delta < 1/\sqrt{3}$ depends on η only. The constant A is to be determined.

For w in a neighbourhood of a branch point w_0 with multiplicity $n > 1$, letting

$$\zeta = \frac{w - w_0}{1 + \overline{w}_0 w},$$

we have

$$r_w = |\zeta|, \qquad \frac{|d\zeta|}{1 + |\zeta|^2} = \frac{|dw|}{1 + |w|^2},$$

and

$$\rho = \frac{A|d\zeta|}{|\zeta|^{1/2}(A^2 - |\zeta|)}.$$

Using the local parameter $t = \sqrt[n]{\zeta}$ gives

$$\rho = \frac{nAt^{n/2-1}|dt|}{(A^2 - |t|^n)}.$$

Thus, ρ is a hyperbolic metric in a neighbourhood of w_0, or is continuous and has a zero at w_0, according as $n = 2$ or $n > 2$.

Now, assume w_0 to be a regular point. Let Δ be the largest univalent spherical disk centered at w_0 and contained in $f(D)$. Then there is a point $w' \in \partial\Delta$ which is a branch point or a boundary point of $f(D)$. Letting

$$\zeta = \frac{w - w'}{1 + \overline{w'}w}, \qquad \zeta_0 = \frac{w_0 - w'}{1 + \overline{w'}w_0},$$

we have

$$r_{w_0} = |\zeta_0|, \qquad r_w \le |\zeta|, \qquad \frac{|d\zeta|}{1 + |\zeta|^2} = \frac{|dw|}{1 + |w|^2},$$

and

$$\rho = \frac{A(1 + r_w^2)}{r_w^{1/2}(A^2 - r_w)} \cdot \frac{|d\zeta|}{1 + |\zeta|^2} = \rho_1(\zeta) \, |d\zeta|.$$

Let

$$\rho_0 = \rho_0(\zeta) \, |d\zeta| = \frac{A \, |d\zeta|}{|\zeta|^{1/2}(A^2 - |\zeta|)}.$$

Then, $\rho_1(\zeta_0) = \rho_0(\zeta_0)$ and, if we take A to be sufficiently large, $\rho_1(\zeta) \ge \rho_0(\zeta)$ in a neighbourhood of ζ_0, since $r_w \le |\zeta| < |\zeta_0| + \delta/2 = t_{w_0} + \delta/2 \le 1/\sqrt{3} - \delta/2$ in a neighbourhood of ζ_0 and the function $t^{1/2}(A^2 - t)/(1 + t^2)$ is increasing for $t \in (0, 1/\sqrt{3} - \delta/2)$ provided that A is sufficiently large. Now, let A be such a number, which is determined by η only. Then ρ_0 is a supporting metric and $\bar{\rho}$ is an ultrahyperbolic metric. Using the Ahlfors lemma gives $f^\#(0) \le 2A$, where $C_\eta = 2A \to \infty$ as $\eta \to 0$ and $C_\eta \to 0$ as $\eta \to \pi/3$. $\qquad \square$

Let $\mathcal{M}$ denote the collection of all non-constant meromorphic functions on the plane $\mathbb{C}$. Define the *spherical Bloch constant* for $\mathcal{M}$ by

$$\mathbf{B}^s = \inf\{B_f^s : f \in \mathcal{M}\}.$$

Then, as a direct consequence of the above theorem, we have $\mathbf{B}^s \ge \pi/3$. Recently, Bonk and Eremenko [7] obtained the correct value of $\mathbf{B}^s$. They proved:

16.2 Theorem $\mathbf{B}^s = \arctan\sqrt{8} \approx 1.231 \approx 70°32'$.

One can define $\mathbf{B}_n^s$ similarly. Bonk and Eremenko also proved that $\mathbf{B}_n^s = \pi/2$ for $n = \infty$ and $n = 2, 3, \ldots$. $\mathbf{B}_\infty^s = \pi/2$ was proved first by Minda [24, 25]. In [7], the authors show that the famous five islands theorem of Ahlfors is a consequence of the above theorem.

16.3 Theorem *Given five Jordan domains on the Riemann sphere $\overline{\mathbb{C}}$ with disjoint closures, every non-constant meromorphic function on the plane $\mathbb{C}$ has a holomorphic branch of the inverse on one of these domains.*

Proof Let

$$e_1 = 0, \qquad e_2 = 1, \qquad e_3 = \infty, \qquad \text{and} \qquad e_{4,5} = \exp(\pm 2\pi i/3),$$

and let ψ be a diffeomorphism of $\overline{\mathbb{C}}$ onto itself, which maps the given Jordan domains G_j onto spherical disks Δ_j of radius η_0 centered at e_j, where η_0 is a sufficiently small positive number. By the uniformization theorem and the fact that there is no quasiconformal mapping of $\mathbb{C}$ onto the unit disk D, we have a quasiconformal mapping ϕ of $\mathbb{C}$ onto itself such that $g = \psi \circ f \circ \phi$ is a meromorphic function on $\mathbb{C}$. By Theorem 16.2, there is a spherical disk Δ of radius $\arctan\sqrt{8}$ on which a holomorphic branch of g^{-1} is defined. Such a Δ must contain one of Δ_j, since η_0 is sufficiently small. Consequently, a holomorphic branch of f^{-1} is defined on the corresponding G_j. The theorem is proved. $\qquad \square$

17 Hyperbolic Bloch constants

For a non-constant function f analytic on the unit disk D such that $f(D) \subset D$, we can define $r(z_0, f)$ for any $z_0 \in D$ to be the pseudo-radius of the largest univalent hyperbolic disk centered at $f(z_0)$ and contained in $f(D)$, and define

$$B_f^p = \sup\{r(z, f) : z \in D\}.$$

17.1 Theorem *Let f be a non-constant analytic function on the unit disk D such that $f(D) \subset D$. Let n be a positive integer. If f' has only zeros of order at least n, then, for $z \in D$, we have*

$$\frac{(1 - |z|^2)|f'(z)|}{1 - |f(z)|^2} \leq \frac{2(n+1)B_f^p}{\sqrt{n(n+2)(B_f^p)^4 + 2(n^2 + 2n + 2)(B_f^p)^2 + n(n+2)}}.$$

If f is locally univalent, then, for $z \in D$, we have

$$(1 - |z|^2) \cdot \frac{|f'(z)|}{1 - |f(z)|^2} \leq \frac{2B_f^p}{1 + (B_f^p)^2}.$$

Proof Assume that f' has zeros of order at least n. A reasoning similar to Pommerenke's shows that the metric

$$\tilde{\rho} = \frac{2A(1 - r_w^2)}{m r_w^{1-1/m}(A^2 - r_w^{2/m})} \cdot \frac{|dw|}{1 - |w|^2}$$

is ultrahyperbolic, where $m = n + 1$, $r_w = r(z, f)$, and

$$A^2 > \frac{(B_f^p)^{2/m}(m + 1 + (m-1)(B_f^p)^2)}{m - 1 + (m+1)(B_f^p)^2}.$$

Using the Ahlfors lemma, replacing r_w by B_f^p, and letting

$$A^2 \to \frac{(B_f^p)^{2/m}(m + 1 + (m-1)(B_f^p)^2)}{m - 1 + (m+1)(B_f^p)^2},$$

we obtain the first conclusion of the theorem. Letting $n \to \infty$, we obtain the second one. □

For a positive integer n and $\alpha \in (0, 1)$, let $\mathcal{H}_{n,\alpha}$ denote the collection of analytic functions on D such that $f(D) \subset D$, $|f'(0)|/(1 - |f(0)|^2) = \alpha$, and f' has only zeros of order at least n. Define the Bloch constants:

$$\mathbf{B}_{n,\alpha}^p = \inf\{B_f^p : f \in \mathcal{H}_{n,\alpha}\}.$$

Let $\mathcal{H}_{\infty,\alpha}$ denote the collection of locally univalent analytic functions on D such that $f(D) \subset D$ and $|f'(0)|/(1 - |f(0)|^2) = \alpha$, and

$$\mathbf{B}_{\infty,\alpha}^p = \inf\{B_f^p : f \in \mathcal{H}_{\infty,\alpha}\}.$$

Then, as consequences of the above theorem, we have

$$\mathbf{B}^p_{n,\alpha} \geq \phi_n^{-1}(\alpha), \qquad \mathbf{B}^p_{\infty,\alpha} \geq \frac{\alpha}{1 + \sqrt{1 - \alpha^2}},$$

where

$$\phi_n(t) = \frac{2(n+1)t}{\sqrt{n(n+2)t^4 + 2(n^2 + 2n + 2)t^2 + n(n+2)}}, \qquad 0 < t < 1.$$

These lower bounds were obtained by Minda [22].

Actually, the above method also provides an improvement of Landau's theorem, Theorem 3.1, in the case that f' has only zeros of order at least $n \geq 2$ or is locally univalent.

17.2 Theorem *Let f be an analytic function on D such that $f(D) \subset D$, $f(0) = 0$ and $0 < f'(0) = \alpha < 1$. Let n be a positive integer. If f' has only zeros of order at least n, then $r(0, f) \geq \beta_n$, where β_n is the unique solution of the equation*

$$\frac{mt^{1-1/m}(1 - t^{2/m})}{1 - t^2} = \alpha$$

for $0 < t < 1$, where $m = n + 1$. If f is locally univalent, then $\rho(0, f) \geq \beta_\infty = \lim_{n \to \infty} \beta_n$, and β_∞ is the unique solution of the equation

$$\frac{2t \log(1/t)}{1 - t^2} = \alpha$$

for $0 < t < 1$. The lower bounds β_n and β_∞ are best possible.

Proof If f' has only zeros of order at least $n \geq 1$, as in the above proof, we use the ultrahyperbolic metric

$$\tilde{\rho} = \frac{2A(1 - r_w^2)}{mr_w^{1-1/m}(A^2 - r_w^{2/m})} \cdot \frac{|dw|}{1 - |w|^2},$$

where $m = n + 1$, and

$$A^2 > \frac{B_f^{2/m}(m + 1 + (m - 1)B_f^2)}{m - 1 + (m + 1)B_f^2}.$$

One can verify that the function $t^{2/m}(m + 1 + (m - 1)t)/(m - 1 + (m + 1)t^2)$ is increasing on $[0, 1]$ and assumes the value 1 at $t = 1$. So, we can take A to be any number bigger then 1. Applying the Ahlfors lemma at $z = 0$ and letting $A \to 1$ gives

$$\frac{mr^{1-1/m}(1 - r^{2/m})}{1 - r^2} \geq \alpha,$$

where $r = r(0, f)$. Let

$$\omega_m(t) = \frac{mt^{1-1/m}(1 - t^{2/m})}{1 - t^2}.$$

Then, $\omega_m(t)$ is increasing for $0 \leq t \leq 1$, $\omega_m(0) = 0$ and $\omega_m(1) = 1$. This shows that $r(0, f) \geq \beta_n$.

If f is locally univalent, then $r(0, f) \geq \beta_n$ for $n = 1, 2, \ldots$. Thus, $b_f(0) \geq \beta_\infty$. Since the function $mt^{1-1/m}(1 - t^{2/m})/(1 - t^2)$ is increasing with respect to $t \in [0, 1]$ and tends to

$2t \log(1/t)/(1 - t^2)$ decreasingly as $m \to \infty$, we see that β_∞ is the unique solution of the second equation in the theorem.

To show the sharpness, for a given $\alpha \in (0,1)$ and each integer $n \geq 1$, let $\beta = \beta_n^{1/(n+1)}$,

$$f_n(z) = \frac{((z+\beta)/(1+\beta z))^{n+1} - \beta^{n+1}}{1 - \beta^{n+1}((z+\beta)/(1+\beta z))^{n+1}}.$$

Then, $f_n(0) = 0$, $f_n'(0) = \alpha$ and $f_n(D)$ is a $(n+1)$-sheeted complete covering of D with a branch point of multiplicity $n+1$ at $w = -\beta^{n+1} = -\beta_n$. Thus, $r(0, f_n) = \beta_n$. This shows that the numbers β_n for $n = 1, 2, \cdots$ are best possible. For a given $0 < \alpha < 1$, the sharpness of β_∞ is shown by the function

$$f_\infty(z) = \frac{\exp\{\log \beta_\infty (1-z)/(1+z)\} - \beta_\infty}{1 - \beta_\infty \exp\{\log \beta_\infty (1-z)/(1+z)\}}.$$

It is easy to verify that $f_\infty(0) = 0$, $f_\infty'(0) = \alpha$. $f_\infty(D)$ is the universal covering of $D \setminus \{-\beta_\infty\}$, so $r(0, f_\infty) = \beta_\infty$. This proves the theorem. □

18 Harmonic mappings

Let G be a domain in $\mathbb{R}^n$. A mapping $f = (f_1, \cdots, f_n)$ of G into $\mathbb{R}^n$ is called a harmonic mapping, if every component $u_j = f_j(x_1, \cdots, x_n)$ is a harmonic function of n real variables, i.e.,

$$\Delta f_j = \frac{\partial^2 f_j}{\partial x_1^2} + \cdots + \frac{\partial^2 f_j}{\partial x_n^2} = 0, \quad j = 1, \ldots, n.$$

Denote the matrix $(\partial f_j/\partial x_k)_{k,j=1,\ldots,n}$ by $f'(x)$ and regard a point $x \in \mathbb{R}^n$ as a column vector.

For a harmonic mapping f of dimension n, the definition of $r(x, f)$ and B_f are the same as in Section 4 with the only modification that "univalent disk" is replaced by "univalent ball" (of dimention n).

More than fifty years ago, Bochner [5] first extended Bloch's theorem to quasiregular harmonic mappings.

18.1 Theorem (Bochner) *Corresponding to any integer $n \geq 2$ and any positive constant $K \geq 1$, there exists a radius $R_0 = R_0(n, k)$ having the following property: If f is a harmonic mapping of the unit ball in $\mathbb{R}^n$ into $\mathbb{R}^n$ such that*

$$\sum_{k,j=1}^n \left(\frac{\partial f_j}{\partial x_k} \right)^2 \leq K|J(x)|^{1/n}$$

holds at every point of the ball and $J(0) = 1$, where $J(x) = \partial(f_1, \ldots, f_n)/\partial(x_1, \ldots, x_n)$ is the Jacobian of f, then we have $B_f \geq R_0(n, k)$.

J. Clunie and T. Sheil-Small [13] showed the very striking result that the Koebe theorem for univalent analytic function of the unit disk remains true for univalent planar harmonic mappings, and it is natural to ask whether Bloch's theorem is true for planar harmonic mappings without the assumption of quasiregularity. In a very recent paper of the author with Gauthier and Hengartner [11], it was shown that Bloch's theorem is not true for general harmonic mappings f with the normalization that $f'(0)$ is the identity, nor even for univalent harmonic mappings in dimension ≥ 3. However, they proved a Bloch theorem for open planar harmonic mappings with the same normalization.

18.2 Theorem *Let f be an open harmonic mapping of the unit disk D such that $f_z(0) = 1$ and $f_{\bar{z}}(0) = 0$. Then $f(D)$ contains a schlicht disk of radius at least*

$$R_1 = \frac{\pi\sqrt{2}}{16}(7 - 4\sqrt{3}) \approx 0.02.$$

They also extended Landau's theorem for bounded analytic functions of the unit disk, Theorem 3.1, to bounded planar harmonic mappings.

18.3 Theorem *Let f be a harmonic mapping of the unit disk D such that $f(0) = 0$, $f_{\bar{z}}(0) = 0$, $f_z(0) = 1$, and $|f(z)| < M$ for $z \in D$. Then, f is univalent on a disk D_{ρ_0} with*

$$\rho_0 = \frac{\pi^2}{16mM},$$

and $f(D_{\rho_0})$ contains a disk D_{R_0} with

$$R_0 = \rho_0/2 = \frac{\pi^2}{32mM},$$

where $m \approx 6.85$ is an absolute constant.

It is interesting that the power of M is the same as in Theorem 3.1.

19   Bloch constants in several complex variables

We now consider holomorphic mappings of the unit ball $B_n \subset \mathbb{C}^n$ into $\mathbb{C}^n$. For such a mapping $f = (f_1, \ldots, f_n)$, every compoment $f_j(z) = f_j(z_1, \ldots, z_n)$ is a holomorphic function of n complex variables. We denote the Jacobian matrix of f by

$$f'(z) = \left(\frac{\partial f_i}{\partial x_j}\right)_{i,j=1,\ldots,n}.$$

In dimension $n \geq 2$, there is no Bloch theorem for holomorphic mappings f, even with the normalization that $f'(0)$ be the identity I. This can be shown by the following example of Duren and Rudin. Let $f_\delta(z_1, z_2) = (z_1, z_2 + (z_1/\delta)^2)$ for $\delta > 0$. Then $f(B_n)$ does not contain any ball of radius δ, so $B_f \leq \delta$. However, $f'_d(0) = I$ for every $\delta > 0$. So, in order to obtain Bloch theorems in several variables, one must impose extra restrictions on the mappings besides the normalization at the origin.

In 1951, Takahashi [35] estimated the Bloch constant β for the class of mappings f, normalized by $\det f'(0) = 1$, satisfying the weaker condition

$$\max_{|z| \leq r} \|f'(z)\| \leq K \max_{|z| \leq r} |\det f'(z)|^{1/n} \quad \text{for each} \quad 0 \leq r < 1,$$

where $|f'(z)|$ denotes the operator norm of the matrix $f'(z)$, which is defined as

$$|f'(z)| = \max\{f'(z)\theta : \theta \in \mathbb{C}^n, \ |\theta| = 1\}.$$

Let us call such mappings Takahashi K-mappings. For such normalized Takahashi K-mappings, Sakaguchi [33], in 1956, improved Takahashi's estimate to

$$\beta \geq \frac{(n-1)^{n-2}}{8K^{2n-1}}.$$

Let λ^2 and Λ^2 (with $0 \leq \lambda \leq \Lambda$) denote the smallest and largest characteristic values of the Hermitian matrix A^*A, where $A = f'$ and A^* is the conjugate transpose of A. Let us say that the holomorphic mapping f is a Hahn K-mapping if

$$\max_{|z|=r} \Lambda(z) \leq K \max_{|z|=r} \lambda(z) \quad \text{for each} \quad 0 \leq r < 1.$$

In 1973, Hahn [16, Corollary 1 of Theorem 4] obtained the estimate

$$\beta \geq \frac{K^{1/n}}{4K(2K+1)},$$

for the Bloch constant for normalized Hahn K-mappings.

These classes of mappings are related to the important class of so-called quasiregular mappings. A holomorphic mapping f from the unit ball B^n of $\mathbb{C}^n$ into $\mathbb{C}^n$ is said to be K-quasiregular, if

$$|f'| \leq K |\det f'|^{1/n},$$

at each point $z \in B^n$. Wu [38] proved the Bloch theorem for normalized K-quasiregular mappings. In fact, as indicated by Wu, the Bloch theorem for K-quasiregular holomorphic mappings follows already from the result of Bochner mentioned in the above section.

We denote by $\mathcal{F}_{K,n}$ the family of all K-quasiregular mappings of B^n into $\mathbb{C}^n$. For $n > 1$ and $K \geq 1$, we define the Bloch constant for n-dimensional K-quasiregular mappings as

$$\beta(K,n) = \inf\{B_f : f \in \mathcal{F}_{K,n}, \ \det f'(0) = 1\}.$$

Since a K-quasiregular mapping is a Takahashi K'-mapping for $K' = \sqrt{n}K$, the above result of Sakaguchi yields the following lower estimate:

$$\beta(K,n) \geq \frac{1}{8K^{2n-1}n^{3/2}} \left(1 - \frac{1}{n}\right)^{n-2} > \frac{1}{8eK^{2n-1}n^{3/2}}. \tag{19.1}$$

Also, since a Wu K-mapping is a Hahn K^n-mapping, the above result of Hahn yields the following lower estimate:

$$\beta(K,n) \geq \frac{1}{8K^{2n-1}(1 + 1/(2K^n))}. \tag{19.2}$$

Harris [17] gave the following lower bound:

$$\beta(K,n) \geq \frac{1}{8K^{2n}}. \tag{19.3}$$

In a very recent paper, Chen and Gauthier [9] proved two lower estimates for $\beta(K,n)$:

$$\beta(K,n) \geq \frac{1}{10K^{2n-1}}. \tag{19.4}$$

$$\beta(K,n) \geq \frac{1}{12K^{n-1}}. \tag{19.5}$$

The following example shows that (19.5) is most reasonable and is best possible in terms of powers of K. Let $f(z) = Az$, where A is a $n \times n$ diagonal matrix with positive elements $K, \cdots, K, 1/K^{n-1}$, $K > 1$. Then $f \in \mathcal{F}_{K,n}$ and $B_f = 1/K^{n-1}$. However, (19.5) is not always best. Comparing the inequalities (19.1)–(19.5) we see first that (19.4) is always better than (19.1), and then (19.2) is worse than (19.4) if $1 \leq K^n < 2$ and is worse than (19.5) if $K^n \geq 2$. Finally, among (19.3), (19.4) and (19.5), none can be covered by the other two.

It is known [21] that for $n > 1$, a quasiregular holomorphic mapping is locally biholomorphic. It was proved by Titov [36] and also by Marden and Rickman [21] that entire quasiregular holomorphic mappings are affine. In fact, Poletsky [29] showed that quasiregular holomorphic mappings (in any bounded domain) are rather rigid. On the other hand, there are still quite a few quasiregular holomorphic mappings, since, for example, Poletsky showed that the holomorphic mapping

$$(z_1, z_2) \longmapsto (z_1 + 3^{-1}(z_1 - 1)^{3/2}, z_2)$$

is quasiregular in the unit ball.

It was indicated in Section 9 that Minda and Liu [27] generalized Bonk's distortion theorem by using Julia's lemma. Their method also provides distortion theorems for holomorphic functions F on the unit disk D, which satisfy $F(0) = 1$ and

$$(1 - |z|^2)^\alpha |F(z)| \leq 1, \quad z \in D,$$

where α may be an arbitrary positive number. By applying such distortion theorems to the Jacobian of a holomorphic mapping, Liu estimated the Bloch constant for the class of Bloch mappings.

A holomorphic mapping f from the unit ball B^n of $\mathbb{C}^n$ into $\mathbb{C}^n$ is called a Bloch mapping, if

$$||f|| = \sup\{|(f \circ \phi)'(0)| : \phi \in \mathrm{Aut}(B^n)\} < +\infty,$$

where $\mathrm{Aut}(B^n)$ is the class of self-biholomorphic mappings of B^n. There are several equivalent definitions for Bloch mappings. Liu proved the following results:

19.1 Theorem *If f is a Bloch mapping such that $||f|| \leq K$ and $\det f'(0) = 1$, then*

$$B_f \geq \frac{1}{K^{n-1}} \frac{1}{e} \frac{\sqrt{n+2}}{n} \left(\left(1 + \frac{1}{n+1}\right)^{n+1} - 2 \right).$$

19.2 Theorem *If f is a univalent Bloch mapping such that $||f|| \leq K$ and $\det f'(0) = 1$, then*

$$B_f \geq \frac{1}{K^{n-1}} \int_0^1 \frac{(1 - t^2)^{n-1}}{(1 - t)^{n+1}} \exp\left\{ -\frac{(n+1)t}{1-t} \right\} dt.$$

Following Liu, Gong and others [15] established Bloch theorems for Bloch mappings on other classical domains.

References

[1] L. V. Ahlfors, An extension of Schwarz's lemma, *Trans. Amer. Math. Soc.* **43** (1938), 359–364.

[2] L. V. Ahlfors, *Conformal Invariants: Topics in Geometric Function Theory*, McGraw-Hill, New York, 1973.

[3] L. V. Ahlfors and H. Grunsky, Über die Blochsche Konstante, *Math. Z.* **42** (1937), 671–673.

[4] A. Bloch, Les théorèmes de M. Valiron sur les fonctions entières et la théorie de l'uniformisation, *Ann. Fac. Sci. Univ. Toulouse III* **17** (1926), 1–12.

[5] S. Bochner, Bloch's theorem for real variables, *Bull. Amer. Math. Soc.* **52** (1946), 715–719.

[6] M. Bonk, On Bloch's constant, *Proc. Amer. Math. Soc.* **110** (1990), 889–894.

[7] M. Bonk and A. Eremenko, Covering properties of meromorphic functions, negative curvature and spherical geometry, *Ann. of Math. (2)* **152** (2000), 551–592.

[8] S. Bshouty and W. Hengartner, Univalent harmonic mappings in the plane, *Ann. Univ. Mariae Curie-Sklodowska Sect. A* **48** (1994), 12–42.

[9] H. Chen and P. M. Gauthier, Bloch constants in several variables, *Trans. Amer. Math. Soc.* **353** (2001), 1371–1386.

[10] H. Chen and P. M. Gauthier, On Bloch's constant, *J. Anal. Math.* **69** (1996), 275–291.

[11] H. Chen, P. M. Gauthier, and W. Hengartner, Bloch constants for planar harmonic mappings, *Proc. Amer. Math. Soc.* **128** (2000), 3231–3240.

[12] C. Carathéodory, Über die Winkelderivierten von beschränkten analytischen Funktionen, *Sitzungsber. Preuss. Akad. Wiss. Berlin Phys.-Math. Kl.* 1929, 1–18.

[13] J. Clunie and T. Sheil-Small, Harmonic univalent functions, *Ann. Acad. Sci. Fenn. Ser. A I Math.* **9** (1984), 3–25.

[14] P. Duren and W. Rudin, Distortion in several variables, *Complex Variables Theory Appl.* **5** (1986), 323–326.

[15] Carl H. FitzGerald and S. Gong, The Bloch theorem in several complex variables, *J. Geom. Anal.* **4** (1994), 35–58.

[16] K. T. Hahn, Higher dimensional generalisations of the Bloch constant and their lower bounds, *Trans. Amer. Math. Soc.* **179** (1973), 263–274.

[17] L. A. Harris, On the size of balls covered by analytic transformations, *Monatsh. Math.* **83** (1977), 9–23.

[18] M. Heins, On a class of conformal metrics, *Nagoya Math. J.* **21** (1962), 1–60.

[19] E. Landau, Der Picard-Schottkysche Satz und die Blochsche Konstanten, *Sitzungsber. Preuss. Akad. Wiss. Berlin Phys.-Math. Kl.* 1926, 467–474.

[20] E. Landau, Über die Blochsche Konstante und zwei verwandte Weltkonstanten, *Math. Z.* **30** (1929), 608–643.

[21] A. Marden and S. Rickman, Holomorphic mappings of bounded distortion, *Proc. Amer. Math. Soc.* **46** (1974), 225–228.

[22] C. D. Minda, Bloch constants, *J. Anal. Math.* **41** (1982), 54–84.

[23] C. D. Minda, The strong form of Ahlfors' lemma, *Rocky Mountain J. Math.* **17** (1987), 54–84.

[24] C. D. Minda, Euclidian, hyperbolic and spherical Bloch constants, *Bull. Amer. Math. Soc.* **6** (1982), 441–444.

[25] C. D. Minda, Bloch constants for meromorphic functions, *Math. Z.* **181** (1982), 83–92.

[26] X. Liu, Bloch functions of several variables, *Pacific J. Math.* **152** (1992), 347–363.

[27] X. Liu and C. D. Minda, Distortion theorems for Bloch functions, *Trans. Amer. Math. Soc.* **333** (1992), 325–338.

[28] E. A. Poletsky, Holomorphic quasiregular mappings, *Proc. Amer. Math. Soc.* **92** (1985), 235–241.

[29] Ch. Pommerenke, Estimates for normal meromorphic functions, *Ann. Acad. Sci. Fenn. Ser. A I Math.* **476** (1970).

[30] G. Pick, Über eine Eigenschaft der konformen Abbildung kreisförmiger Bereiche, *Math. Ann.* **77** (1915), 1–6.

[31] G. Pick, Über die Beschränkungen analytischer Funktionen, welche durch vorgeschriebene Werte bewirkt werden, *Math. Ann.* **77** (1915), 7–23.

[32] H. L. Royden, The Ahlfors-Schwarz lemma: the case of equality, *J. Anal. Math.* **46** (1986), 261–270.

[33] K. Sakaguchi, On Bloch's theorem for several complex variables, *Sci. Rep. Tokyo Kyoiku Daigaku. Sect. A* **5** (1956), 149–154.

[34] H. A. Schwarz, *Gesammelte Abhandlungen*, vol. 2, Springer, Berlin, 1890.

[35] S. Takahashi, Univalent mappings in several complex variables, *Ann. Math.* **53** (1951), 464–471.

[36] O. V. Titov, Quasiconformal harmonic mappings of Euclidean space, *Dokl. Akad. Nauk SSSR* **194** (1970), 521–523.

[37] G. Valiron, *Lectures on the General Theory of Integral Functions*, Toulouse, 1923; reprinted by Chelsea, New York, 1949.

[38] H. Wu, Normal families of holomorphic mappings, *Acta Math.* **119** (1967), 193–233.

[39] H. Yanagihara, On the locally univalent Bloch constant, *J. Anal. Math.* **65** (1995), 1–17.

Approximation of subharmonic functions with applications

David DRASIN

Department of Mathematics
Purdue University
West Lafayette, IN 47907
USA

Abstract

If $f(z)$ is analytic in a domain $G \subset \mathbb{C}$, the function $v(z) = \log|f(z)|$ is subharmonic in G.

We discuss the extent to which the converse is true, and show that approximation of general subharmonic functions $u(z)$ by those of the special form $v(z) = \log|f(z)|$ provides a powerful tool to create analytic and meromorphic functions.

1 Introduction

Let G be a domain in the plane. The function u is *subharmonic* in G if

(1) $u \not\equiv -\infty$;

(2) u is upper semi-continuous;

(3) if $z \in G$ and $r_0 > 0$ is sufficiently small, then

$$u(z_0) \le \frac{1}{2\pi} \int_0^{2\pi} u(z_0 + re^{i\theta})\, d\theta \quad (r < r_0).$$

It is standard that this means that in the sense of distributions Δu is a (positive) Borel measure μ on G such that

$$(\Delta u, v) = (u, \Delta v) \quad (v \in C_0^\infty(G)),$$

where $(u, v) = \int_G u\bar{v}\, dx$. A function u is δ-subharmonic if $u = u_1 - u_2$ with u_1, u_2 subharmonic, a representation which will never be unique. A subharmonic function may be $-\infty$ on a set of zero capacity, and hence a δ-subharmonic function may well be undefined on such a set.

The theory properly begins with papers of F. Riesz in the 1920s, although Riesz credits Osgood and Poincaré with earlier key insights.

The characteristic property of subharmonic functions is: u is subharmonic in G if whenever $D \subset \bar{D} \subset G$ (i.e., $D \subset\subset G$) and h is harmonic in D with continuous boundary values such that if $u \le h$ on ∂D, then $u \le h$ in D. Thus subharmonic functions can be characterized by

N. Arakelian and P.M. Gauthier (eds.), Approximation, Complex Analysis, and Potential Theory, 163–189.
© *2001 Kluwer Academic Publishers. Printed in the Netherlands.*

their relation to harmonic functions. For example, the standard method to solve the *Dirichlet problem* is the Perron method. Given a continuous function $\phi(\zeta)$ on ∂D we consider the (very large) class $\mathcal{P}$ of subharmonic funcions in D whose 'boundary values' are less than ϕ. The harmonic function h in G with 'boundary values' ϕ is

$$h(z) = \sup_{\mathcal{P}} u(z),$$

in the sense that $h = \phi$ at regular points of ∂G and $h - \phi$ is bounded near other points.

According to Riesz [25], if u is subharmonic in G and $D \subset\subset G$, we may write

$$u = h - P \quad (z \in D), \tag{1.1}$$

where h is harmonic and

$$P = P^\mu(z) = \int_D \log \frac{1}{|1 - z/\zeta|} \, d\mu(\zeta) \tag{1.2}$$

is the *potential* of the measure μ. In (1.2), μ is the measure generated by the density $(1/2\pi)\Delta u$, either in the classical sense when u is C^2, or in the sense of distributions otherwise. Sometimes

$$p^\mu = \int_D \log \frac{1}{|z - \zeta|} \, d\mu(\zeta)$$

is used in place of P^μ, since this usually differs from (1.2) by a constant.

1.1 Lemma *Let u be subharmonic in a simply-connected domain G and suppose its laplacian consists of unit masses (Dirac measures) concentrated at a discrete set of points $\{\zeta_j\}$ in G. Then there is a function $f(z)$ analytic in G such that*

$$u(z) = \log|f(z)| \quad (z \in G).$$

Proof We follow the argument of W. Al-Katifi, cf. [16]. There exists an entire function $g(z)$ whose zeros coincide with the $\{\zeta_j\}$ [17]. Then $h = u - \log|g|$ is harmonic in G, and (since G is simply-connected) has a harmonic conjugate k. If we set $F = e^{h+ik} \neq 0$, then $u - \log|g| = \log|F|$, and so $u = \log|Fg| = \log|f|$ as claimed

1.2 Remark The same analysis shows that if u is δ-subharmonic in a simply-connected domain G, then $u = \log|f|$, now with f meromorphic in G.

Note that although our concern is with functions that are entire or meromorphic in the plane, Lemma 1.1 and methods used here can also be used to construct functions in proper domains $G \subset \mathbb{C}$.

This plan of the next several sections follows from these remarks. If u is a subharmonic (or, perhaps, δ-subharmonic) function, we will approximate the Riesz measure $\mu = \Delta u$ by a system of unit (Dirac) masses, and Lemma 1.1 will show that these masses correspond to zeros of some analytic function. Hence the error in approximation will be controlled by how well μ can be approximated by these masses. Note, however, that whenever μ is a discrete measure, the potential in (1.2) tends to $-\infty$ as z tends to the support of μ; thus our theorems can be expected to apply only outside certain exceptional sets. If u is δ-subharmonic, we call the signed measure $\Delta u = \mu$ a *charge*, and usually approximate separately the positive and negative constituent measures; an exception to this is in Section 6.

Using approximation of suitable subharmonic functions to construct specific entire or holomorphic functions goes back at least to N. Levinson (cf. [21, p. 165]) and S. N. Mergelyan (cf. [24, p. 29]); in their survey, A. A. Goldberg, B. Ja. Levin and I. V. Ostrovskii [14] cite M. Keldysh (1945) and N. U. Arakelyan (1966). However, modern attention follows from two antecedents. First, the thesis of B. Kjellberg [19], with further progress from W. K. Hayman, P. B. Kennedy and W. Al-Katifi, an account of which is in [16]. These authors consider the situation that u is subharmonic in the plane, and harmonic off a curve or a small family of curves. It was V. Azarin [2] who introduced the general problem of approximation with no restriction on the support of μ. The most significant penetration is due to R. Yulmukhametov [30], but we also profit from the manuscript of Y. Lyubarski and E. Malinnikova [22]. I am especially grateful for conversations at this Séminaire with Prof. M. Hirnyk (Gyrnyk), and consultations with his joint work with Goldberg. Lemma 2.4, which allows a simple direct approach, is due to helpful suggestions from A. Stray and H. Donnelly.

Our emphasis is more on methods and applications, rather than attempting to produce an 'ultimate' form of approximation; by now, I doubt there is such, since different situations invite exploitation of various flexibilities in the method. The material in Section 2 is the key to all applications, and its formulation is new. Since the precise form of Theorem 2.1 has been in doubt, we give a scrupulous proof.

In many approaches to approximation, the goal is to approximate well on part of the domain, and then accept that in some other parts one loses control. What is presented here often applies in the most rigid of situations. That is, our 'model' is usually globally defined, and global approximation is sought.

The following notations will be used: $B(a, r) = \{z; |z - a| = r\}$ (occasionally we will explicitly include some of $\partial B(a, r)$ with $B(a, r)$); $B(r) = B(0, r)$, $S(a, r) = \partial B(a, r)$, $S(r) = S(0, r)$. Lebesgue measure (in the relevant dimension) is dm, $a \sim b$ means that $|\log(a/b)| = O(1)$, while $a \cong b$ means that $a/b \to 1$.

Only a small number of applications of this principle appear here, and the cornucopia seems far from exhausted. We refer to work of K. Seip [27] and Yulmukhametov, as well as [16, Section 10.5].

This article continues a tradition of having complex approximation theory as one of the themes of the Séminaire de Mathématiques Supérieures. My first international conference was in 1967 at this Séminaire, where the impact of W. H. J. Fuchs's presentation of Arakelyan's achievements had an important impact on me and many others of my generation.

I thank A. Cantón, A. Granados and Y. Lyubarski for helpful comments, and my colleague P. Cook for the illustration in Section 6.

2 Decomposition of measures

We follow the elegant argument of R. Yulmukhametov [30].

2.1 Theorem [1] *Let μ be a measure with compact support in the plane, such that*

$$\mu(\mathbb{R}^2) = N \quad (N > 1, \ N \in \mathbb{Z}).$$

---

[1] Another proof, valid in dimension 2, is by A. F. Grishin and S. V. Makarenko, *Math. Notes* **67** (2000).

Then there exists a system of pairs

$$\mathcal{S} = \{(R_k, \mu_k)\}_1^N,$$

where each R_k is a closed rectangle with sides parallel to the coordinate axes, and μ_k is a measure with support a convex subset of R_k such that

(1) $\mu = \sum_1^N \mu_k$; $\mu_k(\mathbb{R}^2) = 1$;

(2) *the interiors of the convex hulls of the supports of the $\{\mu_k\}$ are disjoint;*

(3) *the ratio of sides of each R_k lies in the interval $[3^{-1}, 3]$;*

(4) *each point of the plane lies in the interior of at most four rectangles R_k.*

Remarks. We call a rectangle with sides parallel to the coordinate axes, with ratio of sides in the interval $[3^{-1}, 3]$ an *almost-square*. In general, the boundaries of the various rectangles carry nontrivial mass.

In [30], Yulmukhametov assumed that μ was absolutely continuous, so the form of Theorem 2.1 here is new.

The key step is the following lemma. In it (and in Lemmas 2.4 and 2.5) an additional property of the measure is required: a measure ν is *prepared* if $\nu(p) < 1$ for each point p. If μ is the measure which appears in the statement of Theorem 2.1, we subtract the integer portions $[\mu(p)]$ of each point mass so that the measure which remains, μ', is prepared. After the proof of Lemma 2.4 (which applies to μ') we discuss the routine transition from μ' to μ, and so deduce Theorem 2.1.

2.2 Lemma *Let ν be a prepared measure with compact support and*

$$\nu(\mathbb{R}^2) = N \quad (N > 1, \ N \in \mathbb{Z}).$$

Assume that for any line L parallel to either coordinate axis there is at most one point $p \in L$ such that $0 < \nu(p)(< 1)$, while $\nu(L \setminus p) = 0$. Let R be an almost-square which contains the support of ν. Then there exist almost-squares R_0 and R_1 and measures ν_0 and ν_1 such that for $i = 1, 2$,

(1) $\nu = \nu_0 + \nu_1$; $\nu_i(\mathbb{R}^2) \in \mathbb{Z}$;

(2) *the interiors of the convex hulls of the supports of ν_0, ν_1 do not intersect;*

(3) *the support of ν_i is contained in R_i.*

This procedure always refines at least one of ν or R. In particular, if either ν_0 or ν_1 is the zero measure, then the Euclidean areas $|\cdot|$ of R_0 and R_1 satisfy

$$\frac{1}{3}|R| \leq |R_i| \leq \frac{2}{3}|R| \quad (i = 0, 1).$$

Proof Choose coordinates so that $R = \{0 \le x \le 1, a \le y \le b\}$, and suppose that $b - a \ge 1$. Partition R into congruent subrectangles R', R'', R''', where $R' = R \cap \{a \le y \le a' = a + (b - a)/3\}$ and $R''' = R \cap \{b' = b - (b - a)/3 \le y \le b\}$. (We use the word 'partition' somewhat informally since some boundaries of these subrectangles may intersect). This is the *initial partition* of R, and we adhere to the convention that R''' is the top subrectangle and R' the bottom rectangle of the initial partition.

We first dispose of the trivial case, that $\nu(R')\nu(R''') = 0$. In this situation, choose $R_1 \in \{R', R'''\}$ such that $\nu(R_1) = 0$ and take $\nu_1 = \nu_{|R_1}$ (restriction to R_1). This yields $R_0 = R \setminus R_1$ and $\nu_0 = \nu_{|R_0}$. This is always applied when possible to a given rectangle R, so that we ignore the resulting R_1 and identify R_0 with the original R.

In all other cases, both measures ν_i $(i = 0, 1)$ will be nontrivial. Let

$$R_t = R \cap \{a \le y < t\},$$

and for $1 \le j \le N - 1$ let $t_j^- = \sup\{t; \nu(R_t) < j\}$, $t_j^+ = \inf\{t; \nu(R_t) > j\}$, and $I_j = [t_j^-, t_j^+]$. We assert that if

$$\bigcup_{j=1}^{N-1} I_j \cap [a', b'] \ne \emptyset, \tag{2.1}$$

the lemma may be satisfied with rectangles R_0 and R_1 whose interiors have disjoint supports. To see this, choose $J \in [1, N-1]$ and $t_J \in I_J$ with $t_J \in [a', b']$. If $\nu(R_{t_J}) = J$, take $R_1 = R_{|t_J}$, $\nu_1 = \nu_{R_{t_J}}$ and $R_0 = R \setminus R_1$, $\nu_0 = \nu - \nu_1$. If $\nu(R_{t_J}) \ne J$, our hypothesis on ν implies that

$$0 \le \nu(\bigcap_{t > t_J} R_t) - \nu(R_{t_J}) < 1,$$

and that $\nu(\{R \cap \{y = t_J\})$ is supported at a point $p_J = (x_J, t_J)$, with $0 < \nu(p_J) < 1$. Take $R_0 = R_{t_J} \cup ([0, x_J] \times \{t_J\})$, with $\nu(p_J)$ apportioned so that $\nu(R_{t_J}) = J$. When $\nu(p_J) > 0$, this may mean assigning only some of this mass to R_1. We then set $\nu_0 = \nu_{|R_0}$, $R_1 = R \setminus R_0$ and $\nu_1 = \nu - \nu_0$. It is routine to check that the $\{R_i\}$ are almost-squares, and that (1)–(3) hold.

Otherwise, $\nu(R') > 0$, $\nu(R''') > 0$ and (2.1) fails, so that $\bigcup_1^{N-1} I_j \subset [a, a'] \cup [b', b]$. In this situation the interiors of R_0 and R_1 will likely intersect. If $\bigcup_1^{N-1} I_j$ meets $[a, a']$, let J be the largest $j \in [1, N - 1]$ with $I_j \cap [a, a'] \ne \emptyset$. Choose $R_1 = R'$ and $R_0 = R \setminus R_{t_J}$, again with the understanding that a point mass on $\{y = t_J\}$ may be assigned to either or divided among both $\{R_i\}$. We also take $\nu_1 = \nu_{|R_{t_J}}$ and $\nu_0 = \nu - \nu_1$. Finally, if this system of lines only meets R''', we take $R_1 = R'''$, $\nu_1 = \nu_{|\{y \ge t_1\}}$, $R_0 = R \cap \{t \le t_1\}$, $\nu_0 = \nu - \nu_1$, again apportioning any point mass on $\partial R_0 \cap R_1$ as warranted. Conditions (1)–(3) of the lemma can be verified by inspection, and although $R_0^\circ \cap R_1^\circ \ne \emptyset$, both R_0, R_1 are almost-squares (X° is the interior of X).

2.3 Corollary *When $\nu(R')\nu(R'') \ne 0$ and (2.1) fails, then*

(1) *both $\nu(R') > 0$ and $\nu(R''') > 0$;*

(2) *R_1 is one of R', R''';*

(3) *there are positive integers k and ℓ such that $\nu_0(\mathbb{R}^2) = k$, $\nu_1(\mathbb{R}^2) = \ell$;*

(4) $\nu_1(R_0{}^\circ \cap R_1) = 0$, $\nu_1(R_0 \cap R_1) < 1$;

(5) let $R^\natural \in \{R', R'''\}$ so that $\{R', R'''\} = \{R_1, R^\natural\}$. Then

$$\nu_0(R_0 \setminus R^\natural) < 1.$$

Note that (1)–(4) also hold when (2.1) is satisfied.

The assumption in Lemma 2.2 that each line parallel to the axes contains at most one point mass is realistic, if we choose a favorable rotation.

2.4 Lemma *Let ν be a prepared Borel measure whose support is contained in a compact subset R of $\mathbb{R}^2$, with $\nu(R) = N, N > 1$, $N \in \mathbb{Z}$. Then there is a rotation to a system of orthogonal coordinates such that if L is any line parallel to either of the coordinate axes, there is at most one point $p \in L$ with $\nu(p) > 0$, while always $\nu(L \setminus p) = 0$.*

Proof Start with a fixed coordinate system; all slopes introduced will be with respect to this system. Let $\mathcal{D}$ consist of points p with $\nu(p) > 0$. The set of points $(p, p') \in \mathcal{D} \times \mathcal{D}$ is countable, so eliminating slopes parallel or perpendicular to the slopes determined by such pairs eliminates an at most countable set of rotations.

Next we remove all point masses, so that ν has no atoms. For each $0 < \alpha < \infty$, let B_α be the lines of slope α or $-(\alpha)^{-1}$. Assuming no suitable rotation exists so that the lemma holds for ν, we may choose for each α a line $L_\alpha \in B_\alpha$ with $\nu(L_\alpha) > 0$. If this were possible for each α, it would follow that for some $k > 0$ there exists a countably infinite subset S of $\mathbb{R}^+$ with $\nu(L_\alpha) > k$ for each $\alpha \in S$. However, if $\alpha \neq \alpha'$, $L_\alpha \cap L_{\alpha'}$ consists of one point, which we have already arranged to carry no mass. Thus, since S is countable

$$\nu(\bigcup_S L_\alpha) = \infty,$$

contradicting the hypothesis that the total mass is finite.

This suggests a natural way to prove Theorem 2.1. By subtracting a measure $\mu_1 = \{[\mu(p)]; \ \mu(p) > 0\}$, we arrive at the measure μ' which satisfies Lemmas 2.2 and 2.4. After we consider μ', we re-adjoin μ_1, which we consider as supported in degenerate rectangles, each of integer mass; these new rectangles play no role in assertions (2) and (4) of Theorem 2.1. Hence in proving Theorem 2.1 we may assume μ is already prepared.

Choose an almost-square R which contains the support of μ. Let $I = i_1, \ldots, i_n$ be a finite word using the digits 0 and 1. By repeated application of Lemma 2.2 we obtain a (finite) sequence of almost-squares $\{R_I\}$, whose sides are parallel to the coordinate axes, and measures $\{\mu_I\}$, each supported in R_I, such that if $\mu_I(\mathbb{R}^2) = n > 1$, $n \in \mathbb{Z}$, then R_I may be divided into almost-squares $R_{I,0}$ and $R_{I,1}$, carrying nontrivial measures $\mu_{I,0}$ and $\mu_{I,1}$ such that

$$\mu_I = \mu_{I,0} + \mu_{I,1}.$$

We arrange indexing so that if $I = i_1, \ldots, i_n$ and $J = j_1, \ldots, j_k$, then R_I contains R_J if and only if $n \leq k$ and $i_s = j_s$ for all $s \leq n$. This means that $\mu_I = \mu_J + \tau$, where τ is a measure, and we say that $I > J$.

Almost-squares R_I and R_J are of the *same type* if the lines used in their initial partition are parallel. Thus rectangles corresponding to pairs (R_I, μ_I) with $\mu_I(R_I) > 1$ are partitioned into at most two classes.

2.5 Lemma *Let ν satisfy the hypotheses of Lemma 2.2, and let $R_I \supset R_J$ be of the same type. Then*

$$E = \{R_{I,0} \cap R_{I,1}\} \cap \{R_{J,0} \cap R_{J,1}\} = \emptyset.$$

Proof Write $R = R_I$ and $R^* = R_J$. We assume that the initial partitions of R and R^* are made by lines parallel to the x-axis, so that $R_1 (= R''')$ is on the top of R. Since R and R^* are of the same type and $R^* \subset R$, we identify them by their projections on the y-axis; the Lemma can only fail if Corollary 2.3 applies in each division, so we assume that is the case; chose coordinates so that $R \leftrightarrow [0,3]$, $R_1 \leftrightarrow [2,3]$, $R_0 \leftrightarrow [0,t]$, where $t > 2$, and $R^* \leftrightarrow [a,b]$, and $a' = a + (b-a)/3$, $b' = b - (b-a)/3$, with $a < \tau < b$ playing the role for R^* that t plays for R.

First, suppose that (2.1) fails for R itself. Then $R_0 \cap R_1$ is a line segment L of mass < 1, and L is the upper boundary of one R_i and the lower boundary of the other. Thus, if Lemma 2.5 were false, we may assume that L contains the top of R^*. Suppose also $L \cap R_0^* \neq \emptyset$. Then since $\nu(R^*) = M > 1$ ($M \in \mathbb{Z}$) and $\nu(L \cap R_0^*) < 1$, it follows that the bottom of R_0^* lies strictly below L. Thus, Corollary 2.3(3) yields that $R_1^* \cap L = \emptyset$, and Lemma 2.5 holds.

Next suppose that (2.1) fails only for $R = R^*$. Then $R^* \subset R_0$ or $R^* \subset R_1$, and separate arguments are required for each possibility.

First let $R^* \subset R_0$. According to Corollary 2.3(4), we must have $b \leq t < 2$. Thus, if the lemma were false, and if $L^* = R_0^* \cap R_1^* \subset \{y = \ell\}$, we have $2 \leq \ell \leq t$. If L^* lies in the middle third of R^*, then (2.1) shows that

$$\ell \leq b - \frac{b-a}{3} < \frac{4}{3} + \frac{a}{3}.$$

Hence, if $a \leq 2$, we have $\ell < 2$, a contradiction, while if $a > 2$, property (5) of Corollary 2.3 shows that

$$\nu^*(R^*) \leq \nu_0(R_0 \setminus R') < 1,$$

again a contradiction.

Suppose next that $R^* \subset R_1$. Let us assume that $L^* = R_0^* \cap R_1^*$ is the top of R_0^*. Then by (4) of the corollary,

$$\nu_0^*(R_0^*) \leq \nu_1(R_0^* \cap R_0) < 1,$$

another contradiction.

Hence we suppose the hypotheses of Corollary 2.3 apply to both R and R^*. By Corollary 2.3(2), we know that one of R', R''' will be R_1, and for concreteness suppose that $R_1 = R'''$.

There are two cases: (I) $R^* \subset R_1$; and (II) $R^* \subset R_0$.

Case (I): $R^* \subset R_1$. Now R_1^* is at the top of R^* or the bottom, and first suppose that R_1^* is the top third of R^*. Recall, in the notation of Lemma 2.2, that $(R^*)'$ is the bottom third of R^*. Since $E \neq \emptyset$, it follows that $b' < t$; hence $a' < t$. Since $R^* \subset R_1$, this means that

$$(R^*)' \subset R_0{}^\circ \cap R_1.$$

Corollary 2.3(4) implies that the carrier of μ_1 is in $R_1 \setminus R_0{}^\circ$, and the measure ν_1 associated to R_1 dominates the measure ν^*, which is concentrated in R^*. Thus, by (4) of the corollary, $\nu^*((R^*)') \leq \nu_1(R_0{}^\circ \cap R_1) = 0$, and this contradicts Corollary 2.3(1).

If R_1^* is on the bottom of R^*, so that $R_0^* \leftrightarrow [\tau, b]$, it is routine to see that $E \neq \emptyset \iff \tau \leq t$. That $R_1^* \subset R_0 \cap R_1$ is a consequence of the readily checked inclusion $[a, \tau] \subset [2, t]$. Thus Corollary 2.3(4) yields

$$\nu_1^*(R_1^*) \leq \nu_1(R_0 \cap R_1) < 1,$$

contradicting Corollary 2.3(3).

Case (II): $R^* \subset R_0$. Now we first suppose that R_1^* is on the top of R^*. The hypothesis $E \neq \emptyset$ yields that $\tau \geq 2$, and since $R^* \subset R_0$, we have that $b \leq t$: $[\tau, b] \subset [2, t]$, and so $R^* \setminus R_0^* \subset R_0 \cap R_1$. But this time ν_1^* is dominated by ν_0, and hence on applying (4) and (5) from Corollary 2.3, we have that

$$\nu_1^*(R_1^*) = \nu_1^*(R_1^* \setminus (R_0^*)^\circ) + \nu_1^*((R_0^*)^\circ \cap R_1^*)$$
$$= \nu_1^*(R_1^* \setminus (R_0^*)^\circ) \leq \nu_0(R_0 \cap R_1) \leq \nu_0(R_0 \setminus R') < 1,$$

a contradiction to conclusion (3) of the corollary.

Finally, suppose R_1^* is on the bottom of R^*. We claim now that $R^* \subset R_0 \setminus R'$, where $R' \leftrightarrow [0, 1]$ is the bottom third of R. Indeed, $R^* \leftrightarrow [a, b]$ and since $E \neq \emptyset$, $a' > 2$. Obviously, $b < 3$ so we conclude that $a > 3/2 > 1$, and this implies that $R^* \subset R_0 \setminus R'$. One more appeal to Corollary 2.3 (this time (5)) yields that $\nu^*(R^*) \leq \nu_0(R_0 \setminus R') < 1$, and so (3) cannot hold either. This completes the proof.

Proof of Theorem 2.1 As we observed following the proof of Lemma 2.4, we may assume that μ satisfies the hypotheses of Lemma 2.2, so that none of the rectangles $\{R_I\}$ obtained are degenerate. We claim that the collection of pairs $\{R_I, \mu_I\}$ with $\mu_I(\mathbb{R}^2) = 1$ satisfies the theorem; these rectangles are called *minimal*. It is clear from Lemma 2.2 that all rectangles are almost-squares and satisfy assertions (1)–(3). It remains to check (4), and by assuming the contrary suppose there is a point z_0 common to the interior of five minimal rectangles, $R_{I(j)}$, $1 \leq j \leq 5$.

We consider three 'stopping times' relative to the five words $I(j)$, corresponding to the associated $R_{I(j)}$. Choose the minimal p, $0 \leq p$, so that all five words $I(j)$ agree in the first p entries, but not in the next. Let I be the word consisting of these p digits, so that I is common to all five $I(j)$; of course p may be zero and I the empty word. It follows that the words $I, 0$ and $I, 1$ appear as initial elements of the five $I(j)$, and thus three of them must begin with, say, $I, 0 \equiv I'$. We call these three indices $I'(j)$. Note that $|I'| = p + 1$.

We repeat this argument with these three words $I'(j)$, and select the first $q \geq p+1$ so that all words corresponding to the indices $I'(j)$ coincide for the first q digits, but not the next. It follows that at least two indices $I'(j)$ must begin with $I', J, 0$ or $I', J, 1$, where $|J| = q - (p+1)$, and we assume it is the latter. Set $J' = J, 1$ and denote these two indices which begin with $I', J, 1$ as $I''(j)$. Finally we choose $\ell \geq q + 1$ and a word K with $|K| = \ell - (q+1)$ so that the two words $I''(j)$ agree for the first ℓ indices, but not the next.

Thus z_0 belongs to the interior of the rectangles $R_I, R_{I',J}, R_{I',J',K}$, and of course two of these must have the same type. This contradicts Lemma 2.5, and proves Theorem 2.1.

Remark. Yulmukhametov also formulates an n-dimensional version of Theorem 2.1.

In Section 4 we discuss the recent paper [22], but one feature should be mentioned here. Let Q be a cube in $\mathbb{R}^n$, $n \geq 3$, ν a probability measure on Q, and consider the Newtonian potential

$$V^\nu(x) = \int_Q \frac{d\nu_y}{|x - y|^{n-2}}.$$

A classical way to obtain an equilibrium measure μ on Q is to take the weak limit of the measures generated by the $(n + 1)N$ Fekete points as $N \to \infty$ [4, Section 4.4]. Theorem 2.1 provides another way to approximate μ using Dirac masses. For each $n \geq 1$, choose $(n+1)N$ points using the procedure of Theorem 2.1, each now with mass $[(n + 1)N]^{-1}$. Then in [22] it is observed that the approximating potential

$$V^N(x) = \frac{1}{(n+1)N} \sum_k \frac{1}{|x - y|^{n-2}}$$

satisfies

$$\int_{\mathbb{R}^n} |V^\mu(x) - V^N(x)| \, dm < C \frac{1}{N^{2/n}},$$

which is better than the best estimates using the Fekete points.

3 Approximation of entire functions of finite order

In this section, we describe Yulmukhametov's application of Theorem 2.1 to approximate entire functions of finite order ρ, where if $g(r)$ ($r \geq 0$) is an increasing function we define the order

$$\rho = \limsup_{r \to \infty} \frac{\log g(r)}{\log r}$$

(when f is entire, we take $g(r) = \log M(r, f)$, where $M(r) = \max_{|z|=r} |f(z)|$, and if f is meromorphic, $g(r) = T(r, f)$, the Nevanlinna characteristic).

In Section 4 we consider more recent advances, but we feel that this special case already contains the heart of the method. Note that the hypothesis of finite order (cf. (3.1)) is used only in the elementary Lemma 3.5.

3.1 Theorem *Let $u(z)$ be subharmonic in the plane and of finite order ρ (mean type):*

$$u(re^{i\theta}) \leq Cr^\rho, \tag{3.1}$$

and suppose that $\mu(\mathbb{C}) = \infty$. Then there is an entire function $f(z)$ such that for any $\alpha > \rho$

$$|u(z) - \log |f(z)|| \leq C_\alpha \log |z| \quad (z \to \infty, \ z \notin E_\alpha),$$

where the exceptional set E_α may be covered by a family of balls $B_i = B_i(z_i, r_i)$ such that $\sum_{|z_i|>R} r_i^{-\alpha} \leq C_\alpha R^{-\alpha}$.

The bound given is qualitatively best possible, since the function

$$u_0(z) = (1/2) \log |z|$$

is subharmonic in the plane, but any approximation by $v(z) = \log |f(z)|$ cannot be $o(\log |z|)$ or within $(1/2) \log |z|$.

Remark. When $\mu(\mathbb{C}) < \infty$ the estimate of Theorem 3.1 also holds, but the arguments are somewhat different, and we ignore this possibility here.

We use the scheme of Theorem 2.1 to approximate $\mu = \Delta u$ by a system of point masses. Note that the hypotheses in Lemmas 2.2–2.5 that μ be prepared are natural here, since the integer portions of the points masses which have been removed transform naturally to be zeros of f.

3.2 Lemma *Let $\{R_k\}$ be a union of almost-squares with the property that each point lies in at most p almost-squares. Then the number of squares of diameter greater than bt which intersect any disk $B = B(\zeta, t)$ $(b, t > 0)$ does not exceed $K(p, b)$.*

Proof If $d = \mathrm{diam}(R)$, then R contains a disk of radius $d/10$. In particular, if $d > 10t$, then each side of R has length at least $2t$. Hence if $d > 10t$ and $R \cap B \neq \emptyset$, then R must meet one of the vertices of the square which circumscribes B: by assumption there can be at most $4p$ such rectangles.

Consider next those rectangles R with $d \leq 10t$ as above, and K_1 the number of such. If $R \cap B \neq \emptyset$, then $R \subset B(\zeta, 11t)$. Property (4) of Theorem 2.1 shows that

$$K_1 \pi (bt/2)^2 \leq p\pi(11t)^2,$$

and hence $K(p, b) \leq p(4 + (22)^2 b^{-2})$.

Yulmukhametov introduces a measure of the local intensity of a measure μ near a point z. Let $\beta(z)$, $s(z)$ be positive functions on $\mathbb{C}$, and μ a measure which is finite on compact subsets. A point z is called (β, s)-*normal* for μ if

$$\mu(B(z,t)) \equiv \mu(z,t) \leq \beta t \quad (\beta = \beta(z),\ 0 < t < s(z)). \tag{3.2}$$

3.3 Lemma *Let ν be a measure in $\mathbb{C}$ with $\nu(1) \equiv \nu(B(1)) = 0$ and $\nu(\mathbb{C}) < \infty$. If z is (β, s)-normal for ν and P^ν is as in (1.2), then*

$$|P^\nu(z)| \leq \beta s(|\log s| + 1) + \nu(\mathbb{C})(\log|z| + 2) \quad (|z| \geq 2).$$

Proof If $|w| \geq 2|z|$, then $|\log|1 - z/w|| \leq 1$, and so

$$\left| \int_{|w| \geq 2|z|} \log|1 - z/w|\, d\nu(w) \right| \leq \nu(\mathbb{C}).$$

We consider next the integral over $B(2|z|)$. Since $\nu(1) = 0$,

$$\int_{B(2|z|)} \log|1 - z/w|\, d\nu(w) \leq \nu(\mathbb{C})(\log|z| + 1)$$

(no absolute value!). On the other hand, since z is (β, s)-normal,

$$\left| \int_{B(z,s)} \log|z - w|\, d\nu(w) \right| = \left| \log s\nu(z,s) - \int_0^s \nu(z,t)t^{-1}\, dt \right|$$

$$\leq \beta s(|\log s| + 1),$$

from which we deduce that

$$\int_{B(2|z|)} \log|1 - z/w|\, d\nu(w) \geq \int_{B(z,1)} \log|z - w|\, d\nu - \int_{B(2|z|)} \log|w|\, d\nu$$

$$\geq -\beta s(|\log s| + 1) - \nu(\mathbb{C})(\log|z| + 1).$$

This proves the lemma.

We postpone the proof of the next theorem until we see how it captures the main issues.

3.4 Theorem *Let μ be a measure in the plane with compact support such that $\mu(p) < 1$ for each point-mass. Suppose that in addition*

$$\mu(2) = 0, \qquad \mu(\mathbb{C}) = N \qquad (N \geq 3, \; N \in \mathbb{Z}).$$

Then there is a measure η consisting P of N Dirac masses such that if $2|\beta(z)| < |z|$ with z (β, s)-normal for μ and η, then for sufficiently large z we have

$$|P^\mu(z) - P^\eta(z)| \equiv \left| \int \log \frac{1}{|1 - z/\zeta|} d(\mu - \eta)(\zeta) \right| \tag{3.3}$$
$$\leq A \log |z| + B\beta s(|\log s| + 1) + C,$$

the constants independent of μ and the functions β and s.

Theorem 3.1 follows from Theorem 3.4 As usual, we assume that μ is prepared. Choose once and for all $a > 2$ so that $\mu(a) = N_1$, where in computing $\mu(a)$ we may need to include some mass supported on $S(a)$.

With $\alpha > \rho$ given from Theorem 3.1, choose $\beta(z) = |z|^\alpha$, $s(z) = |z|^{-\alpha}$. Let μ_1 be the restriction of μ to $B(a)$. Let η_1 consist of N_1 point masses in $B(a)$, chosen in accord with Theorem 2.1, although that is not essential here. Then

$$|P^{\mu - \eta}(z)| = o(1) \quad (|z| \to \infty). \tag{3.4}$$

Now consider $\mu^* = \mu - \mu_1$, and for simplicity of notation replace μ^* by μ in what follows (with a similar gloss for the corresponding η^*). It is clear that the modified μ satisfies the hypothesis of Theorem 3.4.

We use a simple lemma:

3.5 Lemma *Let μ be a Borel measure on the plane with*

$$\mu(r) \equiv \mu(0, r) \leq Cr^\rho \quad (r > 1). \tag{3.5}$$

Let $\alpha > \rho$ be given, and set $\beta(z) = \beta(|z|) = |z|^\alpha$, $s(z) = s(|z|) = |z|^{-\alpha}$. Then the set $E = E_\alpha$ of non-(β, s)-normal points may be covered by a network of disks $B_i = \{|z - z_i| < r_i\}$ with

$$\sum_{|z_i| > R} r_i < KR^{\rho - \alpha}.$$

Proof Corresponding to each $z \in E$ is $t_z \in (0, |z|^{-\alpha})$ so that if $B_z = \{|\zeta - z| < t_z\}$, then $\mu(B_z) \geq |z|^\alpha (t_z)^{-\alpha}$. By a standard covering lemma of Besicovitch type (cf. [23, p. 30],) there is a countable subfamily $B_i = B_{z_i}$ which covers E, and such that no point of E is in more than K of the $\{B_i\}$, K a constant depending only on the dimension. Let I_n be those B_i with $|z_i| \in [2^n, 2^{n+1})$. Then, by (3.5) and the defining property of the $\{B_i\}$,

$$K2^{n\rho} \geq \sum_{I_n} \mu(B_i) \geq \sum |z_i|^\alpha t_i \geq 2^{n\alpha} \sum t_i$$

we have

$$\sum_{I_n} t_i \leq C2^{n(\rho - \alpha)}. \tag{3.6}$$

If u is the function of Theorem 3.1, Jensen's formula (cf. [14, 15]) and (3.1) ensure that the Riesz mass of μ satisfies (3.5) for all $r > 1$. For each $n(\geq N_1)$, choose $\Gamma_n = \{|z| = r_n\}$ so that $\mu(B(r_n)) = n$. Let Ω_n be the bounded component of $\mathbb{C} \setminus \Gamma_n$, and μ_n the restriction of μ to Ω_n, so that μ_n may assign positive measure to $\partial\Omega_n$. For $r > r_0$, each μ_n satisfies (3.5) with C independent of n; this and Lemma 3.2 ensure that each approximating measure η_n, consisting of unit masses, also satisfies (3.5) for $r > 1$ with a fixed C. Thus Theorem 3.4 and in particular (3.3) hold for each of the measures μ_n and η_n, with $\mu_n(\mathbb{C}) = \eta_n(\mathbb{C}) = n - N_1$, $N_1 = \mu(a)$.

Thus $K_n(z) \equiv u(z) + [P^{\eta_n}(z) - P^{\mu_n}(z)]$ is of the form $K_n(z) = \log|f_n(z)|$, where f_n is analytic in Ω_n. However, (3.6) holds with constants independent of n, and so if $[\mu(\bigcap_{r<r_n} B(r))] = M_n$, then $M_n \to \infty$, and for $1 \leq m < M_n$ we may choose r'_m with $r'_m - r_m = o(1)$ such that the circle $\{|z| = r'_m\}$ avoids the exceptional set of Lemma 3.5. Thus with our choice of β and s, (3.3) and (3.4) give that

$$|\log|f_n(z)| - u(z)| \leq A \log|z| + B \quad (z \in \Gamma_m, \ m \geq m_0), \tag{3.7}$$

so the theory of normal families ensures that a subsequence of the $\{f_n\}$ converges on compact sets of the plane to an entire function f, which satisfies Theorem 3.1.

Proof of Theorem 3.4 The procedure of Theorem 2.1 yields N almost-squares R_k, so that η is a union of point-masses $\{w_k\}$, with $w_k \in R_k$. We set $\kappa = \mu - \eta$, $|\kappa| = \mu + \eta$, with similar arrangements for restrictions of μ and η, a convention which will be used for the rest of this section.

Yulmukhametov's insight was to place the mass of each R_k at $w_k = u_k + iv_k \in R_k$ so that

$$\int (w - w_k) \, d\mu_k(w) = 0. \tag{3.8}$$

For $|z| > 2$, let $\mathcal{A}$ be the ring domain $\mathcal{A} = \{w; |w - z| > s/2, \ |w| < 2|z|\}$ (since $s < |z|/2$, this is doubly-connected). Since $\mu(2) = 0$ there are at most $K(4, 1)$ R_k which meet $B(1)$, and their contribution to the error in approximation in Theorem 3.1 can be absorbed in (3.4). We divide the R_k with $R_k \cap B(1) = \emptyset$ into three groups: (I) those R_k which meet the bounded component of the complement of $\mathcal{A}$; (II) those R_k which meet the unbounded component and (III) those R_k strictly contained in $\mathcal{A}$ (of course, a given R_k may fit more than one of these conditions), Let $\delta_k = \text{diam}(R_k)$ and $d(P, Q)$ be the distance between sets P and Q.

Analysis of (I). Lemma 3.2 asserts that the number of R_k in group (I) with $\delta_k > s/2$ is at most $K(4, 1)$, and so the contribution of these terms to the bound (3.3) is controlled by Lemma 3.3.

Now suppose $\delta_k < s/2$. It follows that each R_k is contained in $B(z, s)$. Since z is (β, s)-normal for the measure μ, we have

$$\left| \int_{B(z,s)} \log|z - w| \, d\mu(z, t) \right| = \left| \int_0^s \log \tau \, d\mu(z, \tau) \right| \leq \beta s(|\log s| + 1);$$

the same estimate applies to the corresponding contribution to η, and thus to $|\kappa|$. This bound is almost what is needed in (3.3). Suppose that η concentrates the mass of R_k at w_k. Since

$\delta_k < s/2 \leq |z|/4$, we have that if $w \in R_k$ and z is normal, then

$$\sum_I \int_{R_k} \log\left(\left|\frac{w}{w_k}\right|\right) d\mu_k(w) = \int_0^s \log t \, d\kappa(I)(t) < \beta s(|\log s| + 1).$$

It follows, using the notation of (1.2), that

$$|P^{\kappa(I)}(z)| \leq 3\beta s(|\log s| + 1) + 2K(4,1)(\log|z| + 2). \tag{3.9}$$

Analysis of (II). First suppose that $3\delta_k < d(R_k, 0)$. Then $R_k \subset \{|w| \geq (3/2)|z|\}$. With z fixed, let

$$\mathcal{L}(w) = \log|1 - z/w|.$$

The choice of w_k in (3.8) and Taylor's formula yield that

$$\left|\int [\log|1 - z/w| \, d\mu_k(w) - \log|1 - z/w_k|] - [(\mathcal{L})_u(w_k)(u - u_k) + (\mathcal{L})_v(w_k)(v - v_k)]\right|$$

$$\leq 27|z| \max_{R_k} \frac{\delta_k^2}{|w|^3} \equiv 27|z| \frac{\delta_k^2}{|w_k'|^3}.$$

For $n \geq 0$ let $J(n)$ consist of R_k in class (II) with $2^n|z| \leq |w_k'| < 2^{n+1}|z|$. Since each R_k is an almost-square,

$$27|z| \sum_n \sum_{J(n)} \frac{\delta_k^2}{|w_k'|^3} \leq 27|z| \sum_n \frac{A(J(n))}{2^{3n}|z|^3},$$

where $A(J(n))$ is the sum of areas of all R_k with $w_k' \in J(n)$. Each point belongs to at most four R_k and so if $w_k' \in J_n$, the hypothesis that $\delta_k < (1/3)|z|2^n$ implies that $R_k \subset B(2^{n+2}|z|)$, hence

$$27|z| \sum_n \sum_{J(n)} \frac{d_k^2}{|w_k'|^3} \leq 216 \cdot 8\pi \sum_{n \geq 0} 2^{-n}.$$

We next consider $R_k \in$ (II) with $3\delta_k \geq d(R_k, 0)$. Note that the number of $R_k \in$ (II) which penetrate $B(z, 3|z|/2)$ is at most $K(4, 1/2)$, and hence the contribution of these R_k to the estimate (3.3) is controlled by Lemma 3.3.

If μ' is the measure the portion of μ assigned to those $R_k \in$ (II) with $3\delta_k \geq d(R_k, 0)$ and $P_k \cap B(3|z|/2) = \emptyset$, we claim that

$$\mu'(10t/9) - \mu'(t) \leq K(4, 3) \quad (t > |z|), \qquad \mu'(t) = 0 \quad (t < 3|z|/2). \tag{3.10}$$

The second assertion is immediate from the defining property of the R_k which contribute to μ'. In addition, if R_k is included in the left side of (3.10) and meets a ring $\{t < |z| < (10t/9)\}$, the hypothesis on δ_k ensures that R_k must meet one of the ring's boundary components. This shows that μ' satisfies (3.10), and so $\mu'(t) \leq CK(4,3)\log(t/|z|)$ with $C = (\log(10/9))^{-1}$. On the other hand, in the support of μ'

$$|\log|1 - z/w|| \leq 2\frac{|z|}{|w|},$$

and so

$$|P^{\mu'}(z)| \le 2 \int_{\mathbb{C}\backslash B(3|z|/2)} \frac{|z|}{|w|} \, d\mu'(w) \le \frac{K(4,3)}{\log(10/9)} \int_{3/2}^{\infty} \frac{1}{t^2} \, dt$$
$$\le 20K(4,3)(\log|z| + 1) \quad (|z| > 1),$$

and the same holds for η'. We have shown that

$$|P^{\kappa(II)}(z)| \le 216 \cdot 8\pi + 2(K(4,1/2) + \beta s)(|\log s| + 1)$$
$$+ 2K(4,1/2)(\log|z| + 3) + 20K(4,3)(\log|z| + 1). \tag{3.11}$$

Analysis of (III). This involves two sub-cases: (IIIa): $d(z, P_k) \le \delta_k$; (IIIb): $d(z, P_k) > \delta_k$. Let us consider the first. Let P'_k be similar to P_k with the same barycenter but side-lengths multiplied by 3. Then $z \in P'_k$, and so if $P_k \cap B(z, t) \ne \emptyset$ for some $t > (s/2)$, we must have $\text{diam}(P'_k) \ge t$: $\delta_k = \text{diam}(P_k) \ge t/3$. According to Lemma 3.2, the number of such P_k for a given t is at most $K(4, 1/3)$, and since $w_k \in P_k$,

$$|\kappa(\text{IIIa})(z, t)| = |\mu(\text{IIIa})(z, t) - \eta(\text{IIIa})(z, t)| \le 2K(4, 1/3).$$

It is obvious that $\kappa(\text{IIIa})(z, s/2) = \kappa(\text{IIIa})(z, 3|z|) = 0$, and so

$$P^{\kappa(\text{IIIa})}(z) = \left| \int_{s/2}^{3|z|} \frac{\kappa(\text{IIIa})(z, \tau)}{\tau} \, d\tau \right| \le 2K(4, 1/3) \log \frac{6|z|}{s}. \tag{3.12}$$

Finally, there is $\kappa(\text{IIIb})$. Taylor's formula and (3.8) now show that for some $w_k \in P_k$,

$$\left| \int \log|z - w| \, d\mu_k - \log|z - w_k| \right| \le 2 \left(\frac{\delta_k}{\inf_{P_k} |w - z|} \right)^2 \equiv 2 \left(\frac{\delta_k}{|w'_k - z|} \right)^2.$$

The $\{P_k\}$ are almost-squares and $2|z - w'_k| > |z - w|$ for all $w \in P_k$. Hence, for $P_k \in (\text{IIIb})$,

$$\left(\frac{\delta_k}{|w'_k - z|} \right)^2 \le \frac{40}{3} \int_{P_k} \frac{du dv}{|z - w|^2},$$

and property (4) of Theorem 2.1 shows that

$$\left(\sum_{(\text{IIIb})} \frac{\delta_k}{|z - w'_k|} \right)^2 \le \frac{160}{3} \int_A \frac{du dv}{|z - w|^2} \le \frac{320\pi}{3} \int_{s/2}^{3|z|} \frac{dt}{t} \le \frac{320\pi}{3} \log \left(\frac{6|z|}{s} \right).$$

We also must consider $\sum_{(\text{IIIb})} \log(|w_k/w|) \, d\mu_k$. Since $\log(|w'_k|/|w|) = \int_{|w|}^{|w'_k|} t^{-1} \, dt$ and the $\{R_k\}$ are almost squares, we have

$$\sum_{(\text{IIIb})} \left| \log \left| \frac{w'_k}{w} \right| \right| \le 2 \cdot 4\pi \int_{\frac{s}{2}/\frac{6|z|}{s}}^{3|z|} \frac{dt}{t} \le 8\pi \log \left(\frac{6|z|}{s} \right),$$

and hence

$$P^{\kappa(\text{IIIb})}(z) \le \frac{328\pi}{3} \log \left(\frac{6|z|}{s} \right). \tag{3.13}$$

Theorem 3.4 follows from (3.9), (3.11), and (3.12).

4   Variations on a theme

So far we have presented the situation as of 1985, except that the formulation and some of the steps in the proof of Theorem 2.1 are new. Yulmukhametov himself as well as Girnyk [10] produced a variant for functions of infinite order, where instead of $|z|^{\rho-\alpha}$, the error term is measured in terms of a general measure of growth $V(r)$ introduced by O. Blumenthal.

It is natural to consider mean approximation, and, for example, Girnyk and Goldberg [12] obtained a result in which there is no exceptional r-set for functions of finite order. If v is a function in the plane, set

$$\|v\|_{q,r} = \left(\int_0^{2\pi} |v(re^{i\theta})|^q\, d\theta \right)^{1/q}.$$

4.1 Theorem *Let u be subharmonic in the plane and $\mu = \Delta u$. Then there is an entire function $f(z)$ such that*

$$\| \log|f| - u \|_{q,r} = Q(r,u).y \tag{4.1}$$

In (4.1), $Q(r,u)$ is a function which is $O(\log r)$ when u has finite order, and

$$Q(r,u) = O(\log r + \log \mu(B(R))) \quad (r \to \infty)$$

when the order of u is unrestricted. Perhaps some words should be said concerning the role of expressions such as $Q(r,u)$. Error terms of this nature occur naturally in Nevanlinna's theory of meromorphic functions, where they appear in the 'error term' of the Second Fundamental Theorem or the growth of the logarithmic derivative. The precise form for Q in this context has attracted special interest in recent years, because a formal analogy discovered between Nevanlinna theory and classical number theory [18, 26, 29], and this has led to progress in both directions. Girnyk and Gol'dberg produce specific choices for the $Q(r,u)$ which arises in Theorem 4.1.

In [11], Girnyk develops a sharp form of approximation for functions subharmonic in the disk, a theme also treated in [30].

The recent manuscript [22] by Lyubarsky and Malinnikova (already mentioned at the end of Section 2) provides a refined analysis and re-examination of this procedure. For an arbitrary subharmonic function, the authors show:

4.2 Theorem *Let u be subharmonic in $\mathbb{C}$. Then for each $q > 1/2$, there is $R_0 > 0$ and an entire function f such that*

$$\frac{1}{\pi R^2} \int_{B(R)} |u(z) - \log|f(z)||\, dm_z < q \log R \quad (R > R_0).$$

Note that no assumption is made about the growth of u.

It is not surprising that one cannot in general expect a smaller error term, as the function $u_0(z) = (1/2) \log|z|$, which was mentioned after the statement of Theorem 3.1 shows. Both [12] and [22] present more complicated functions u (based on u_0) that cannot be approximated within $q \log|z|$ with $q < 1/2$, even though u has infinite mass or large finite order. It is open whether Theorem 4.2 holds with error $(1/2) \log|z| + O(1)$.

It seems difficult to go beyond Theorem 4.2 without additional information on the function u. The example u_0 suggests, and [22] confirms, that many of the complications in good approximation are related to gaps in the support of μ, or measures of very small density. For example, let $\mathcal{A}_n = \{r_n \leq |z| \leq r_{n+1}\}$, with the $\{r_n\}$ chosen so that $\mu(\mathcal{A}_n) \equiv 1$. If μ has small Radon-Nikodým derivative with respect to Lebesgue measure, then r_{n+1}/r_n can be arbitrarily large, and approximating μ in $\mathcal{A}_n$ by a single point-mass is not likely to be effective.

When there is additional information about u (or Δu) to rule this out, sometimes one can do better. The authors of [22, Theorem 3'] present an example of this form, which is too complicated to state here, but depends on specific properties of Δu. In Section 6 we present another example.

5   Applications

We now present a modern treatment of the classical family of *Levin-Pfluger* entire functions, introduced independently at the end of the 1930s by B. Ja. Levin and A. Pfluger: they are also known as *functions of completely regular growth*. This theory occupies dozens of pages in Levin's classical monograph [20], but the ideas discussed here and other notions introduced more recently make a rapid presentation possible. This class of functions includes most of the entire functions used in applications. Let h be a $W^{1,1}$ function with a 2π-periodic representative: this means that h and h' have representatives which are in L^1, derivatives taken in the sense of distributions. Then h is called *subtrigonometric* (of order $\rho > 0$) or *trigonometrically convex of order ρ* if

$$L_\rho(h) \equiv h'' - \rho^2 h = d\mu \geq 0, \qquad (5.1)$$

where μ is a (positive) measure. (The official definition is that

$$\int_0^{2\pi} (h'\phi' + \rho^2 h\phi)\, d\theta \geq 0 \quad (\phi \in \mathcal{D})$$

($\mathcal{D}$ are the test functions).) For any $\rho > 0$, the functions $\sin \rho\theta$, $\cos \rho\theta$ satisfy (5.1) with equality, but when $\rho \notin \mathbb{Z}$ these functions are not periodic. Exactly as convex functions arise as the upper envelope of linear functions, we create trigonometrically convex functions of any order ρ by considering some sinusoid functions of order ρ, and then taking their upper envelope, so long as the resulting function is periodic. In [16] these are called harmonic splines. The next proposition is either elementary (for C^2 functions), or a nice exercise in distribution theory (where we are considering distributions on $\mathbb{R}^+ \times S^1$).

5.1 Proposition *The function $u(z) = u(re^{i\theta}) = r^\rho h(\theta)$ is subharmonic in $\mathbb{C} \setminus \{0\}$ if and only if h is trigonometrically convex. More generally, u is δ-subharmonic if and only if $L_\rho(h)$ is a charge.*

Remark. The computation does not apply at $z = 0$ where the polar form of Laplace's equation is not valid. However, when $\rho > 0$, we usually can show that the putative singularity at the origin will be removable. It is also not hard to see that $W^{1,1}$ functions on a line are continuous, and have representatives whose derivative increases with θ; exactly as with convex functions, these may be considered as left or right derivatives. Functions h which satisfy (5.1) occur prominently in the theory of entire functions. If f is entire of order $0 < \rho < \infty$, mean type, its Phragmén-Lindelöf *indicator* is defined as

$$h(\theta) = h_f(\theta) = \limsup_{r \to \infty} \frac{\log |f(re^{i\theta})|}{r^\rho}, \tag{5.2}$$

and a standard Phragmén-Lindelöf argument shows that if h is finite for some θ_0, then h satisfies (5.1); in addition to [20], a standard reference is [6]. If f is entire but perhaps not mean type, it is possible to replace the factor r^ρ in (5.2) by a growth function $V(r)$ (proximate order) such that the modified expression is finite for all θ and satisfies (5.2); cf. [6, 8].

The class of Levin-Pfluger functions has the property that 'lim sup' is replaced by 'lim' in (5.2) (with r avoiding a set of linear density zero), and plays a key role in the study of entire functions: functions in this class often exhibit extremal behavior. A basic result in the theory is:

5.2 Proposition *To every (2π-periodic) ρ-subtrigonometric function h corresponds an entire function f such that*

$$\log |f(re^{i\theta})| = r^\rho h(\theta) + O(\log r) \quad (z = re^{i\theta} \to \infty, \ z \notin E), \tag{5.3}$$

where E is a small exceptional set.

Proof Let $u(z) = r^\rho h(\theta)$. Then u is subharmonic in the plane and of finite order ρ. Theorem 3.1 applies, with both a very good error term and control of the exceptional set.

There is a substantial literature on generalizations of Proposition 5.2, which describes the exceptional set or supplements (5.3) with detailed estimates of the remainder.

The zeros of the function f of (5.3) are concentrated near rays corresponding to the support of the measure L_ρ of (5.1), and so if h is the upper envelope of some sinusoid functions, the mass will be concentrated on the rays where h' has a jump. For example, if $h'_+(\theta_0) - h'_-(\theta_0) = \Delta$, it follows that if $n(r, \theta_0, \epsilon)$ is the number of zeros z_n with $|z_n| < r$, $|\arg z_n - \theta_0| < \epsilon$, then

$$n(r, \theta_0, \epsilon) = (1 + o(1))\Delta \pi \rho r^\rho. \tag{5.4}$$

This orientation is philosophically somewhat different than that of the founders of the theory, which is displayed even in the title of Levin's famous book [20]. While both approaches are equivalent, traditionally the focus was directed to how the asymptotics of $\log |f(z)|$ (or, in sectors where h is sinusoidal, those of $\arg f(z)$) is influenced by the location of the zeros. This had one serious annoyance: when f has integral order $\rho = p$, then f and $f \exp(P(z))$ (with $\deg P = p$) have the same zeros but different asymptotics. For this reason, the theory had to be developed in a parallel manner depending on whether the order ρ was an integer or not.

In contrast, our starting point is the graph of a given subtrigonometric function h. If h is subtrigonometric of order ρ, we can always construct a function f which satisfies (5.3) and then determine the distribution of its zeros. Our discussion (cf. (5.4)) shows that we can capture this distribution at least to the extent of the terms which have positive density with respect to the growth function r^ρ. The following two examples may make these comments more concrete.

(1) Consider the range $0 < \rho \leq 1$. If $\rho < 1$, the Lindelöf functions of order ρ are defined as those with zeros on the negative axis such that for some $\Delta > 0$, $n(t) = [\Delta t^\rho]$ ([ ] is the greatest integer function). The corresponding Lindelöf function, f_ρ, has the asymptotic representation

$$\log |f_\rho(re^{i\theta})| = (1 + o(1))r^\rho \frac{\pi \Delta}{\sin \pi \rho} \cos \rho\theta, \tag{5.5}$$

at least away from the zeros of f.

As $\rho \to 1$, (5.5) degenerates because of the factor $\csc \pi\rho$. On the other hand, if instead of relying on (5.5) we use the entire function $g_\rho(z)$ associated to the subtrigonometric family $\{h(\theta) = \cos \rho\theta; \rho < 1, (|\theta - \pi| < \pi)\}$ (thus relying on Proposition 5.2), we obtain a collection of functions which formally converges to the limit function $(f_1(z) = e^z)$ with no zeros. Of course,

$$\log |g_\rho(z)| \cong \pi^{-1} \sin \pi\rho \log |f_\rho(z)|$$

away from the zeros of the two functions.

(2) The function $g(z) = 1/\Gamma(z)$ has zeros at the nonpositive integers, and is of order one, but maximal type. Thus $M(r, g) = (1 + o(1))r \log r$, while the density with respect to $\log M(r, g)$ of the zero-counting function $n(r) = (1 + o(1))r$ degenerates. This phenomenon, which occurs only when $\rho \in \mathbb{Z}$, is difficult to capture in the theory of the indicator; when the zeros are balanced (for example when $f(z) = \sin z$) these can be captured from the graph of h.

In order to include functions such as (2) in this theory, full treatments replace the comparison function r^ρ with the more general *proximate order*, which we do not discuss here.

The principles of Proposition 5.2 are an essential ingredient for constructing many examples, and there is a rich literature on this theme. However, in the last few years, it has been used to create *meromorphic* functions. While a general subtrigonometric function h is the upper envelope of a family of translates of the sinusoids of order ρ, the functions $h(\theta)$ associated to meromorphic functions are periodic $W^{1,1}$ functions, so that L_ρ in (5.1) is a signed measure (charge). Any charge μ can be written as $\mu = \mu^+ - \mu^-$ in many ways, but there is a unique way to minimize $\|\mu\| \equiv \|\mu^+\| + \|\mu^-\|$. The 'atomization' procedure of Section 2 is usually applied independently to μ^+ and μ^-, unless circumstances warrant closer scrutiny, and by inspecting the *graph* of h, we can read off both the asymptotic behavior of f (especially where $|\log |f||$ is large) and the distribution of its zeros and poles. Thus, although (5.2) is difficult to adapt to meromorphic functions, the theory of the extremal functions survives. In the next section we outline a recent application of this technique, one with a surprisingly sharp estimate on the error.

6 Quasiconformal modifications and meromorphic functions with precise set of asymptotic values

A nonconstant meromorphic function f has the *asymptotic value* $a \in \mathbb{C} \cup \infty$ if there is a curve γ tending to ∞ such that $f(z) \to a$ as $z \to \infty$, $z \in \gamma$. A. Denjoy conjectured that if f

is entire of order ρ with k distinct finite asymptotic values, then

$$\rho \geq \frac{1}{2}k, \qquad (6.1)$$

which L. Alhfors proved in his thesis. In particular, if $\rho < \infty$, f can have only a finite number of asymptotic values. As Nevanlinna wrote in 1967, "Denjoy's theorem might appear as a rather particular question. Its general significance depends on the influence it has exerted on the development of important new methods in the theory of functions." If the asymptotic curves are radial segments, (6.1) follows from the Phragmén-Lindelöf principle, and it is also elementary that $|f|$ must be unbounded in unbounded regions between two asymptotic curves. The hypothesis that $\rho < \infty$ forces the curves corresponding the different asymptotic values to be separated. Ahlfors captured this effect in the sharp form (6.1).

The situation is quite different for meromorphic functions, and to describe it requires a bit of Nevanlinna theory. Let f be meromorphic in the plane, and for convenience suppose that $f(0) \neq 0, \infty$. If $n(r) = n(r, \infty)$ is the number of poles in $B(r)$, and $N(r) = N(r, f) = \int_0^r n(t)t^{-1}\,dt$, then Nevanlinna's characteristic $T(r, f)$ $(r > 0)$ is

$$T(r, f) = N(r) + \frac{1}{2\pi}\int_0^{2\pi} \log^+ |f(re^{i\theta})|\,d\theta \equiv N(r, f) + m(r, f). \qquad (6.2)$$

When f is entire, it is elementary that $T(r) \leq \log M(r) \leq 3T(2r)$, $(M(r) = \max_{|z|=r}|f(z)|)$ so the order of an entire function can be computed using either $T(r)$ or $\log M(r)$. Alternative definitions of the characteristic are due to Ahlfors and H. Cartan.

When f is rational, then $T(r) = O(\log r)$, and the converse also holds. The Levin-Pfluger functions of Proposition 5.2 all have finite order, and in fact $T(r) \sim cr^\rho$.

Of course a rational function can have only one asymptotic value, and Valiron showed that this persists as long as $T(r) = O(\log^2 r)$. He also showed that if $\psi(r) \to \infty$, no matter how slowly, there was a meromorphic function with uncountably many asymptotic values, and yet

$$T(r) = o(\psi(r)\log^2(r)) \quad (r \to \infty). \qquad (6.3)$$

Thus the functions considered here are all of order zero.

Let $\text{Asym}(f)$ be the set of asymptotic values of f. Eremenko [9] improved Valiron's example to produce functions $f(z)$ with $\text{Asym}(f) = \hat{\mathbb{C}}$ with growth controlled by (6.3). It is classical that $\text{Asym}(f)$ is an *analytic set* in the sense of set theory [28], and so the next result is best possible [5].

6.1 Theorem *Let A be an analytic set in the plane and $\psi \uparrow \infty$ as above. Then there is a meromorphic function f whose growth is controlled by (6.3) with*

$$\text{Asym}(f) \equiv A.$$

Details will appear elsewhere, but we note some themes which resonate with Sections 2 and 3. The method can also be adapted to show that there are entire functions f with $\text{Asym}(f) = A$ and $\log(\log M(r))/(\log r) \to \infty$ as slowly as desired.

If $A = \emptyset$ or a single point, the construction is elementary, and here we assume that $0 \in A$, $\infty \in A$ and that $A \setminus \{\infty\} \subset \{|w| < 1\}$. The strategy of the construction is explained in terms of of *asymptotic tracts* or, more simply, tracts. Recall that if f is a function in a domain D, then a *tract* for $w = a$ is a network of sets $U(a, \epsilon)$, $\epsilon > 0$, such that

(1) $U(a, \epsilon)$ is a component of $\{|f(z) - a| < \epsilon\}$;

(2) if $\epsilon_1 < \epsilon_2$, then $U(a, \epsilon_1) \subset U(a, \epsilon_2)$;

(3) $\bigcap_{\epsilon > 0} U(a, \epsilon) = \emptyset$

(obvious modifications when $a = \infty$).

The construction consists of two steps. The first, which displays the techniques already introduced, yields a meromorphic function F whose growth satisfies (6.3), but with $A(F) = \{0\} \cup \{\infty\}$. However, there is concrete information about the tract(s) of F for $w = 0$ (and $w = \infty$). For a fixed $\epsilon > 0$, $\mathcal{U}_F(0, \epsilon) \equiv \cup U(0, \epsilon)$ will eventually meet $S(r)$ in $p(r) \uparrow \infty$ intervals $U(r, \epsilon)$, and if z is in one of the complementary intervals, then z may be connected to ∞ by a path $\gamma \subset \mathbb{C} \setminus \mathcal{U}(0, \epsilon)$ on which $F(z) \to \infty$. Thus, for any r_0, $|F|$ is unbounded in the domains which meet $\mathbb{C} \setminus B(r_0)$ outside $\mathcal{U}(0, \epsilon)$. We will see that the angular measure between two tracts $U(0, \epsilon)$ can tend to zero, which explains the distinction between entire and meromorphic functions.

The second stage uses *quasiconformal modification*. This is the theme of Chapter 7 of [13], and now is a standard way of constructing functions; [7] gives a modern outline as well as a far shorter proof of the sharpness of Nevanlinna's defect relation than earlier accounts. Here are the basic definitions. Let $g : \Omega \to \hat{\mathbb{C}}$ be a homeomorphism, and suppose $|g(z_0)| \neq \infty$ at z_0. Then g is $K(= (1 + k)/(1 - k))$-quasiconformal near z_0 if there is a neighborhood $\Omega_0 \ni z_0$ with $g \in W^{1,2}(\Omega_0)$ and

$$\|\mu_g\|_{\infty, \Omega_0} \equiv \left\| \frac{g_{\bar{z}}}{g_z} \right\|_{\infty, \Omega_0} \leq k.$$

(When $g(z_0) = \infty$, take Ω_0 to be a neighborhood of z on which $|g| > 1$, and use the same definition with $1/g$ in place of g.) A homeomorphism g is quasiconformal in Ω if it is K-quasiconformal at each point with some fixed K. A K-quasiregular mapping is defined in exactly the same manner, without the requirement that g be a homeomorphism.

Analytic (or meromorphic) functions have $K \equiv 1$ and give the simplest examples of quasiregular mappings. In a certain sense, these are the only examples in the plane (cf. (6.6) below).

We need two lemmas.

6.2 Lemma *Let $\eta > 0$ be given, and consider the (δ, η, R) problem: for given a ($|a| < \delta$), find a quasiconformal self-map Φ of $\mathbb{C}$ such that*

(1) $\Phi(w) = w$ ($|w| \geq R$, $R < \infty$);

(2) $\Phi(w) = a + w$ ($|w| < \delta$, $\delta > 0$);

(3) $\|\mu_\Phi\|_\infty < \eta$.

Then given η and either R or δ there are choices of $R(\delta, \eta)$, $\delta(R, \eta)$ which solve the problem.

6.3 Lemma *Let $h(z)$ be defined in $\mathbb{C}$ with $\|h\|_\infty < 1$. Then the (Beltrami) equation*

$$\phi_{\bar{z}}(z) = h(z)\phi_z(z)$$

admits a homeomorphic solution $w = \phi(z)$. If in addition

$$L(r) = \frac{1}{2\pi} \int_0^{2\pi} h(re^{i\theta}) \, d\theta \to 0 \quad (r \to \infty) \tag{6.4}$$

then $\lim_{z \to \infty} w/z = A \neq 0$.

The first assertion is in the classical text [1]; the second is Belinskii's refinement of Te-ichmüller's theorem [3].

Quasiconformal modifications arise in a natural way. We will have almost complete information about the function F of the first step, and in particular the tract structure corresponding to the unique finite asymptotic value $w = 0$. Let γ be one of the curves on which $F \to 0$. We wish to convert γ so that f tends to a preassigned value a on a curve modelled after γ. This is achieved by a (usually infinite) family of compositions defined near γ of the nature

$$F_{n+1} = \Phi_n \circ F_n \qquad (F_0 = F) \tag{6.5}$$

(the Φ_n from Lemma 6.2), in such a manner that the limiting function G is continuous in the plane. If the dilatations of the $\{\Phi_n\}$ are controlled using Lemma 6.2(3), G will be a quasiregular formal solution to Theorem 6.1, in that G has A as its precise set of asymptotic values. (This will be outlined in the discussion after the Main Lemma 6.5.) Although G is not meromorphic, $\|\mu_G\|_\infty$ can be taken as close to zero as desired, and (6.4) holds with $h = \mu_G$. In addition, if we formally compute $T(r, G)$ using (6.2) (or other variants) it follows that G satisfies (6.3). If ϕ is the solution to the Beltrami equation of Lemma 6.3, it is then routine to verify

$$f \equiv G \circ \phi^{-1} \tag{6.6}$$

is meromorphic in the plane and that (6.4) holds so that $T(r, f)$ satisfies (6.3).

6.4 Tracts and the structure of A Let us consider A as the limit of a general Cantor-type procedure: $A = \lim A(m)$ $(m \to \infty)$ where each $A(m)$ consists of 2^m closed sets $B_{j,m}$, $1 \leq j \leq 2^m$. Define the (immediate) successors of $B_{j,m}$ as $B_{2j-1,m+1}$ and $B_{2j,m+1}$ (we amalgamate these as $B_{j',m+1}$). According to [28], $A = h([0,1])$, where h is a left-continuous function. Define

$$B_{j,m} = \overline{h(I_{j,m})},$$

where $I_{1,1} = [0,1]$, $I_{2,1} = [\frac{1}{2}, 1]$, and in general

$$I_{j',m+1} = \begin{cases} I_{j,m}, & j' = 2j - 1 \\ \left[\dfrac{j'-1}{2^{m+1}}, \dfrac{j'}{2^{m-1}}\right], & j' = 2j. \end{cases}$$

A chain C is a collection of sets $\{B_{j(m),m}\}_{m=1}^{\infty}$ with $B_{j(m+1),m+1}$ the successor of $B_{j,m}$ such that

$$\bigcap_{B \in C} B = \bigcap_{m \to \infty} B_{j(m),m} = \{a_C\} \in \mathbb{C}, \tag{6.7}$$

a unique point. If we denote the system of chains by $\mathcal{C}$, then the analytic set A is

$$A = \bigcup_{C \in \mathcal{C}} a_C. \tag{6.8}$$

Note there is no assumption about uniformity of the limits in (6.7).

Corresponding to the set A of (6.8) is the structure of the asymptotic paths in $\mathbb{C}$ on which the asymptotic values $a \in A$ are to be attained. It has the combinatorial structure of a dyadic tree T which is modelled on our system of chains $\mathcal{C}$. Thus at generation m there are 2^m branches, and we associate to each branch a point $b = b_{j,m}$ in the associated $B = B_{j,m}$. The uniformity of this combinatorial pattern dissolves in its realization as system of curves in the plane, in the sense that the rate at which each chain is exhausted is highly non-uniform and depends on

(1) the growth allowed in (5.2), and

(2) the structure of the set A, in particular the nature of convergence in (6.7).

We view T as corresponding to a continuum $\Gamma_0 = \bigcup \gamma_{j,m}$ in $\mathbb{C}$, where the goal is to have $f(z) - b_{j,m}$ small on $\gamma_{j,m}$ (where $b_{j,m} \in B_{j,m}$) in a manner that when each $\gamma_{j,m}$ bifurcates to a $\gamma_{j',m+1}$, then f is close to b with b the appropriate $b_{j',m+1}$. This is achieved by using the $\{\Phi_n\}$ in (6.5), since as $z \to \infty$ in the curve γ of Γ_0 which corresponds to a given chain C, we will have

$$\Phi_n \circ \Phi_{n-1} \cdots \circ \Phi_1(0) \to a \in A, \quad (n \to \infty)$$

with $a = a_C$ in (6.7) (the $\{\Phi_n\}$ thus depend on C). The convergence to a_C is extremely slow.

The bifurcations of Γ, which correspond to exhausting the combinatorial tree T, are controlled by a sequence $r_{j,m}$ with

$$\min(r_{2j-1,m+1}, r_{2j,m+1}) > 2r_{j,m}$$

so that $\gamma_{j,m} \subset A(r_{j,m}, r_{j',m+1}) \equiv \{r_{j,m} < |z| < r_{j',m+1}\}$. Let $\nu(r)$ be the number of components of $\Gamma \cap S(r)$.

With this model of Γ_0, we describe the δ-subharmonic function $U(z)$ which is to be approximated by $\log|F|$. It is based on a slowly-increasing function $L(r)$ with

$$\begin{aligned}
L(r) \to \infty, \quad rL'(r) \to 0, \quad r^2 L''(r) \to 0 \quad (r \to \infty), \\
\nu(r) \to \infty, \quad \nu(r) = o(L(r)) \quad (r \to \infty),
\end{aligned} \tag{6.9}$$

the specific rates depending on the *a priori* data: (6.3) and (6.7). For fixed r, the graph of U will be what we call an $L(r)$-*sawtooth function*: it is continuous and built by at most $4\nu(r)$ functions of the form

$$u(re^{i\theta}) = u_{\phi_\ell}(re^{i\theta}) = \pm L(r)(\theta - \phi_\ell(r)) \quad (1 \le \ell \le \ell(r) \le 4\nu(r)), \tag{6.10}$$

where each of the functions $\{\phi_\ell(r)\}$ also satisfies top line of (6.9). The function U is defined using (6.10) in various portions of the plane, the domains chosen so that U is continuous. It is convenient to add one extra condition:

$$U(r,0) = U(r,\pi) = 0; \quad U(r,\theta+\pi) = -U(r,\theta) \quad (0 \le \theta \le \pi). \tag{6.11}$$

Here is a portion of a typical graph of $U(r, \theta)$ for a fixed r; it shows how bifurcations increase $\nu(r)$:

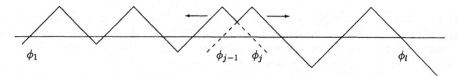

Note that there are an equal number of intervals on which the graph of U is concave up and concave down. Let Γ_0 consist (as r varies) of the local minima of U in those intervals $I_j(r)$ in which the graph is concave up. These intervals will contain the tract $\mathcal{U}_F(0, \epsilon)$, and the asymptotic values of A for the formal solution G will be attained on $\Gamma = \Gamma_0 \cap \{0 \le \arg z < \pi\}$. The figure also indicates how the I_j bifurcate in accord with the combinatorial pattern of Γ_0, and the same remarks apply to Γ', the points of local maxima.

It is obvious that

$$m(r, U) = \frac{1}{2\pi} \int_0^{2\pi} U^+(re^{i\theta}) \, d\theta \le L(r), \tag{6.12}$$

but the computation of $N(r, U)$ is more delicate. It is clear that for u in (6.10) ($\phi = \phi_\ell$),

$$\Delta u = u_{rr} + r^{-1} u_r + r^{-2} u_{\theta\theta}$$
$$= \pm [L''(\theta - \phi) - 2L'\phi' + L\phi'' + r^{-1}(L'(\theta - \phi) - L\phi')],$$

and $U(re^{i\theta})$ has in addition a mass $\pm 2r^{-2} L(r) \delta_\theta$ at points $re^{i\theta}$ which form the continua Γ, Γ', composed of the local extrema of U as θ varies.

Write

$$\Delta U = 2r^{-2} L(r) \Big[\sum_{\theta_j} \delta_{\theta_j} - \sum_{\theta'_j} \delta_{\theta'_j} \Big] + H(r, \theta)$$
$$= D(r, \theta) + H(r, \theta),$$

where the points $re^{i\theta_j}$ and $re^{i\theta'_j}$ are in Γ, Γ' respectively, and H accounts for the contributions from the terms (6.10). Thus if ϕ is a test function,

$$(\Delta u, \phi) = 2 \int_\Gamma |z|^{-2} L(|z|) \phi(z) \, |dz| - 2 \int_{\Gamma'} |z|^{-2} L(|z|) \phi(z) \, |dz|$$
$$+ \iint \phi(z) H(r, \theta) \, dx dy.$$

It is clear that

$$\int_0^{2\pi} D(r, \theta) \, d\theta \equiv 0,$$

and (6.11) then yields that

$$\int_0^{2\pi} \Delta U(re^{i\theta}) \, d\theta = \int_0^{2\pi} H(r, \theta) \, d\theta \equiv 0, \tag{6.13}$$

while

$$N(r, U) = O(\nu(r)L(r)\log^2 r). \tag{6.14}$$

In particular, (6.14) and (6.9) show that the growth of U can be controlled by (6.3). For this to occur, $L(r)$ must grow arbitarily slowly, and so the usual error terms of magnitude $O(\log r)$ in approximation are unacceptable. On the other hand, the rate at which $F \to 0$ on each curve $\gamma_a \subset \Gamma$ (where γ_a corresponds to the value $a \in A$) is of no issue.

The techniques of the first three sections, together with the supplemental information (6.11) lead to:

6.5 Main Lemma *There exists a meromorphic function F such that*

$$|\log|F(z)| - U(z)| = O(1) \quad (z \notin E),$$

where $E = \bigcup B_j$, $B_j = B(z_j, h_j)$, is a union of disjoint disks, such that

$$\sum_{2^n \leq |z_j| \leq 2^{n+1}} h_j = o(2^n),$$

and in each B_j, F has a zero or pole, but not both.

The significance of the Main Lemma is that the exceptional set E effectively vanishes. For example, consider a $B \subset E$ inside of which F has a zero. On ∂B, the lemma ensures that $|U - \log|F||$ is bounded, and so the one-sided bound $\log|F| < U + O(1)$ persists in all of B; this inequality cannot be reversed since F has a zero inside B. In particular, on Γ we have $\log|F(z)| \leq U(z) + O(1)$, and so if K and r_0 are sufficiently large, the points in

$$\mathcal{V}(r_0, K) = \{z; \ |z| \geq r_0, \ \log|F(z)| < U(z) - K\} \tag{6.15}$$

form distinct simply-connected islands, each of which meets Γ. (Similar comments apply to the sets where $\log|F|$ is much greater than U, with respect to the continuum Γ' corresponding to $w = \infty$.) This allows the Φ_n of Lemma 6.5 to be exploited in (6.5) as needed to modify F to a quasiregular function with equivalent growth and precise asymptotic set A.

The proof of the Main Lemma, although long, uses themes that have been already introduced, but at one step, the additional property (6.11) is essential.

Let $\mu_2 = \mu_2^+ - \mu_2^-$ be the measure corresponding to $H(r, \theta)$ of (6.13), and $\mu_1 = \Delta U - \mu_2$, with similar notations for $\mu_1^{\pm}$. The atomization of the measures $\mu_1^{\pm}$ uses that their supports are concentrated very near the curves Γ, Γ' where the various functions u of (6.10) are fused. However, μ_2 is uniformly small, since it is controlled by terms which tend to zero in (6.9), and the problems we raised in Section 4 arise. Thus (6.18) shows that $L(r)$ in (6.9) will almost always satisfy $L(r) = o(\log r)$, and so the effect of any point mass used in the approximation of μ_2^+ or μ_2^- would normally have magnitude comparable to $\log|z|$. In turn, this would destroy the approximation to the asymptotic values. To avoid this is the role of (6.11): it guarantees that for each r, $\mu_2^+(B(r)) = \mu_2^-(B(r))$, and so the point-masses corresponding to annuli in which $\mu_2^{\pm}(B(r))$ increase by one, can be placed at the same point. Hence if we choose $z_j' = z_j$, the corresponding factors

$$\frac{(1 - z/z_j)}{(1 - z/z_j')}$$

cancel, and there is no contribution to the error term.

The remaining support of ΔU occurs from $\mu_1^{\pm}$, supported on points at which $U(re^{i\theta})$ has local extrema in θ. This is simpler to control; we are effectively back into the situation which was studied before [2] appeared, with masses concentrated on curves. We compute the positive and negative contributions independently and obtain the exceptional set described in Lemma 6.5.

We also have

6.6 Corollary *The asymptotic values of F are 0 and ∞.*

Proof There are continua Γ, Γ' such that $U \to -\infty$ as $z \to \infty$, $z \in \Gamma$ and $U \to \infty$ as $z \to \infty$, $z \in \Gamma'$, so the Main Lemma implies that $\mathrm{Asym}(F) \supset \{0, \infty\}$. Suppose there were an asymptotic value $w = a$, with $0 < |a| < \infty$, and γ_a the corresponding asymptotic curve. Lemma 6.5 imples that $\gamma_a \cap \{|z| > R_a\}$ must avoid the exceptional set E, since F is close only to 0 and ∞ in E. For $z \in \gamma_a$ choose an analytic branch of $\log F(z)$ in $\{\Omega(z) \equiv |\log(z'/z)| < h\}$, where h is sufficiently small so that the Cauchy-Riemann equations may be applied to $\log F$ in $\Omega(z)$. The Cauchy-Riemann equations and the lemma then show that

$$\frac{\partial \arg F}{\partial \log r} \equiv -\frac{\partial \log |f|}{\partial \theta} \cong \pm L(r),$$

for a fixed sign. Hence $\arg F$ cannot have a limit on γ_a (if $a \neq 0, \infty$).

References

[1] L. V. Ahlfors, *Lectures on Quasiconformal Mappings*, Van Nostrand, Princeton, NJ, 1966.

[2] V. Azarin, On rays of completely regular growth of an entire function, *Math. USSR-Sb.* **8** (1969).

[3] P. P. Belinskii, *General Properties of Quasiconformal Mappings*, Nauka, Novosibirsk, 1974 (Russian).

[4] L. Carleson, *Selected Problems on Exceptional Sets*, Van Nostrand Math. Stud. **13**, Van Nostrand, Princeton, NJ, 1967.

[5] A. Cantón, D. Drasin, and A. Granados, Sets of asymptotic values for entire and meromorphic functions, to appear.

[6] M. L. Cartwright, *Integral Functions*, Cambridge Tracts in Math. and Math. Phys. **44**, Cambridge Univ. Press, 1956.

[7] D. Drasin, On Nevanlinna's inverse problem, *Complex Variables Theory Appl.* **37** (1998).

[8] A. Edrei, A local form of the Phragmén-Lindelöf indicator, *Mathematika* **17** (1970), 149–172.

[9] A. E. Eremenko, The set of asymptotic values of a finite order meromorphic function, *Mat. Zametki* **24** (6) (1978), 779–783 (Russian); English transl.: *Math. Notes* **24** (1978-9).

[10] M. Girnyk, Approximation of a subharmonic function of infinite order by the logarithm of the modulus of an entire funtion, *Mat. Zametki* **50** (4) (1991), 57–60 (Russian); English transl.: *Math. Notes* **50** (1991), 1025–1027.

[11] M. Girnyk, Approximation of functions subharmonic in a disk by the logarithm of the modulus of an analytic function, English transl.: *Ukrainian Math. J.* **46** (8) (1994), 1188–1192.

[12] M. Girnyk and A. A. Gol'dberg, Approximation of subharmonic functions by logarithms of moduli of entire functions in integral metrics, *Dopov. Nats. Akad. Nauk Ukr. Mat. Prirodozn Tekh. Nauki* (2000).

[13] A. A. Gol'dberg and I. V. Ostrovskii, *The Distribution of Values of Meromorphic Functions* Nauka, Moskow, 1970 (Russian).

[14] A. A. Gol'dberg, B. Ja. Levin, and I. V. Ostrovskii, *Entire and Meromorphic Functions*, Encyclopaedia Math. Sci. **85**, Springer, Berlin, 1997.

[15] W. K. Hayman and P. B. Kennedy, *Subharmonic Functions* vol. 1, Academic Press, London–New York, 1976.

[16] W. K. Hayman, *Subharmonic Functions* vol. 2, Academic Press, London-New York, 1989.

[17] M. Heins, *Selected Topics in the Classical Theory of Functions of a Complex Variable*, Holt, Rinehart and Winston, New York 1962.

[18] A. Hinkkanen, A sharp form of Nevanlinna's second fundamental theorem, *Invent. Math.* **108** (1992), 549–574.

[19] B. Kjellberg, *On Certain Integral and Subharmonic Functions*, Thesis, Univ. of Uppsala, 1948.

[20] B. Ja. Levin, *Distribution of Zeros of Entire Functions*, Transl. Math. Monogr. **5**, Amer. Math. Soc., Providence, RI, 1964.

[21] N. Levinson, *Gap and Density Theorems*, Amer. Math. Soc. Colloq. Publ. **26**, Amer. Math. Soc., Providence, 1940.

[22] Y. Lyubarski and E. Malinnikova, On approximation of subharmonic functions, to appear in *J. Anal. Math.*

[23] P. Mattila, *Geometry of Sets and Measures in Euclidean Space*, Cambridge Univ. Press, 1995.

[24] S. N. Mergelyan, *Uniform Approximations to Functions of a Complex Variable*, Amer. Math. Soc. Transl. **101**, Amer. Math. Soc., Providence, RI, 1954.

[25] F. Riesz, Sur les fonctions subharmoniques et leur rapport à la théorie du potentiel I, *Acta Math.* **48** (1926), 329–343.

[26] M. Ru, Nevanlinna theory and its relation with diophantine approximation, *Bull. Hong Kong Math. Soc.* **1** (1997).

[27] K. Seip, On Korenblum's density condition for the zero sequence for $A^{-\alpha}$, *J. Anal. Math.* **67** (1995), 307–322.

[28] W. Sierpinski, *General Topology*, Univ. of Toronto Press, 1952.

[29] P. Vojta, *Diophantine Approximation and Value Distribution Theory*, Lecture Notes in Math. **1239**, Springer, Berlin–Heidelberg–New York, 1984.

[30] R. Yulmukhametov, Approximation of subharmonic functions, *Anal. Math.* **11** (1985), 257–282.

Harmonic approximation and its applications

Stephen J. GARDINER

Department of Mathematics
University College Dublin
Dublin 4
Ireland

Abstract

These lectures survey some recent developments concerning the theory and applications of harmonic approximation in Euclidean space. We begin with a discussion of the significance of the concept of thinness for harmonic approximation, and present a complete description of the closed (possibly unbounded) sets on which uniform harmonic approximation is possible. Next we demonstrate the power of such results by describing their use to solve an old problem concerning the Dirichlet problem for unbounded regions. The third lecture characterizes the functions on a given set which can be approximated by harmonic functions on a fixed open superset. Finally, we return to applications, and explain how some problems concerning the boundary behaviour of harmonic functions have recently been solved using harmonic approximation.

1 Thinness and approximation

1.1 The basic question

In this opening lecture we will address the following question: on which sets can one approximate by harmonic functions? We are primarily concerned with uniform approximation, but will also consider the possibility of arranging that the error of approximation on a noncompact set decays near "infinity".

In order to be more precise, we introduce some notation. In what follows Ω denotes an open set in $\mathbb{R}^n$ ($n \geq 2$) and E is a relatively closed subset of Ω. Given a set A in $\mathbb{R}^n$, we use $\mathcal{H}(A)$ to denote the collection of all functions which are harmonic on some open set that contains A. Since continuity and harmonicity are both preserved by uniform convergence, any function on E which can be uniformly approximated on E by harmonic functions defined on some prescribed larger open set must belong to $C(E) \cap \mathcal{H}(E^\circ)$. Thus we are concerned with the following question.

Problem *Which pairs of sets* (Ω, E) *have the property that, for any* $u \in C(E) \cap \mathcal{H}(E^\circ)$ *and any* $\varepsilon > 0$, *there exists* $v \in \mathcal{H}(\Omega)$ *such that* $|v - u| < \varepsilon$ *on* E?

(A useful variation is to consider the same problem for u in the smaller class $\mathcal{H}(E)$.)

N. Arakelian and P.M. Gauthier (eds.), Approximation, Complex Analysis, and Potential Theory, 191–219.
© 2001 *Kluwer Academic Publishers. Printed in the Netherlands.*

1.2 Reduced functions

As preparation for describing the solution to the above problem, we recall the notion of a *reduced function* in potential theory. (For a fuller account of this and other potential theoretic concepts in these lecture notes we refer to the book [4].) For the moment we assume that Ω has a Green function (that is, we exclude the case where $n = 2$ and $\mathbb{R}^2 \setminus \Omega$ is a polar set). Let $\mathcal{U}_+(\Omega)$ denote the collection of all positive superharmonic functions on Ω, let $u \in \mathcal{U}_+(\Omega)$ and $A \subseteq \Omega$. The *reduced function* of u relative to A in Ω is defined by

$$R_u^A(x) = \inf\{v(x) : v \in \mathcal{U}_+(\Omega) \text{ and } v \geq u \text{ on } A\} \quad (x \in \Omega);$$

its lower semicontinuous regularization $\widehat{R}_u^A$ is superharmonic on Ω and equals R_u^A outside some polar set. Some basic properties of reduced functions are as follows:

(a) $R_u^A = u$ on A;

(b) R_u^A (and hence also $\widehat{R}_u^A$) is harmonic on $\Omega \setminus \overline{A}$;

(c) if ω is a bounded open set such that $\overline{\omega} \subset \Omega$, then $R_u^{\Omega \setminus \omega}$ is the (generalized) solution to the Dirichlet problem on ω with boundary data u;

(d) if (A_k) is an increasing sequence of sets with union A, then $\widehat{R}_u^{A_k} \uparrow \widehat{R}_u^A$.

Now we will apply the notion of reduced function to address a preliminary, but illuminating, question about harmonic approximation. Let $f \in C(K)$, where K is a compact set in $\mathbb{R}^n$ with empty interior. Can we uniformly approximate f on K by members of $\mathcal{H}(K)$?

To investigate this, let Ω be an open ball containing K. It is an elementary consequence (see Lemma 7.9.1 in [4]) of the Stone-Weierstrass theorem that, given any $\varepsilon > 0$, there exist continuous members u_1 and u_2 of $\mathcal{U}_+(\Omega)$ such that

$$|f - (u_1 - u_2)| < \varepsilon \quad \text{on } K.$$

Thus our question reduces to the following: can we uniformly approximate a continuous member u of $\mathcal{U}_+(\Omega)$ by members of $\mathcal{H}(K)$?

Let

$$U_k = \left\{ x \in \mathbb{R}^n : \operatorname{dist}(x, K) < \frac{1}{k} \right\} \quad (k \in \mathbb{N}). \tag{1}$$

Then the sequence $(\Omega \setminus U_k)$ is increasing, with union equal to $\Omega \setminus K$, so

$$\widehat{R}_u^{\Omega \setminus U_k} \uparrow \widehat{R}_u^{\Omega \setminus K} \quad (k \to \infty). \tag{2}$$

We note that

$$\widehat{R}_u^{\Omega \setminus U_k} \in \mathcal{H}(K)$$

and that

$$\widehat{R}_u^{\Omega \setminus K} = u \quad \text{on the open set } \Omega \setminus K. \tag{3}$$

If we could deduce from (3) that $\widehat{R}_u^{\Omega \setminus K} = u$ on K, then Dini's theorem would ensure that the convergence in (2) is uniform on K, as required. We have assumed that $K^\circ = \emptyset$, but this does not allow us to draw the desired conclusion from (3) since superharmonic functions (in particular, $\widehat{R}_u^{\Omega \setminus K}$) need not be continuous.

1.3   The fine topology

In order to pursue this line of argument, let us now change the underlying topology to achieve continuity of superharmonic functions. The *fine topology* of classical potential theory is the coarsest topology on $\mathbb{R}^n$ which makes every superharmonic function continuous. Further, a set $A \subseteq \mathbb{R}^n$ is said to be *thin* at a point y if y is not a fine limit point of A (that is, there is a fine neighbourhood N of y such that $A \setminus \{y\}$ does not intersect N). Wiener's criterion gives an equivalent formulation of thinness in terms of capacity. For example, when $n \geq 3$, it asserts that a set A is thin at a point y if and only if

$$\sum_k 2^{k(n-2)} C^*(\{x \in A : 2^{-k-1} \leq \|x - y\| < 2^{-k}\}) < +\infty,$$

where $C^*(\cdot)$ denotes (outer) Newtonian capacity.

A classical example of a thin set is the *Lebesgue spine*—the exponentially sharp cusp in $\mathbb{R}^3$ given by

$$A = \{(x_1, x_2, x_3) : x_1 > 0 \text{ and } x_2^2 + x_3^2 < e^{-2/x_1}\}, \tag{4}$$

which is thin at 0. On the other hand, a cone is non-thin at its vertex, and when $n = 2$ a line segment is non-thin at each constituent point.

We remark that the following statements are equivalent:

(a) A is thin at y;

(b) there is a superharmonic function v on a neighbourhood of y such that

$$\liminf_{x \to y,\ x \in A} v(x) > v(y);$$

(c) there is a superharmonic function v on a neighbourhood of y such that

$$\liminf_{x \to y,\ x \in A} \frac{v(x)}{\phi_n(\|x - y\|)} > \mu_v(\{y\}), \tag{5}$$

where

$$\phi_n(t) = \begin{cases} \log(1/t) & \text{if } n = 2 \\ t^{2-n} & \text{if } n \geq 3, \end{cases}$$

and μ_v denotes the Riesz measure associated with v.

Thus (informally), in the case where A is thin at y, the values of v on A may give a misleading impression about $v(y)$ or $\mu_v(\{y\})$.

An account of thinness and the fine topology may be found in Chapter 7 of [4].

1.4   Approximation on compact sets without interior

Returning to our discussion at the end of Section 1.2, we know that the superharmonic function $\widehat{R}_u^{\Omega \setminus K}$ is equal to u on $\Omega \setminus K$ and want to deduce that it is equal to u also on K. This step is a consequence of the fine continuity of the superharmonic function $\widehat{R}_u^{\Omega \setminus K}$ if each point of K is a fine limit point of $\mathbb{R}^n \setminus K$, that is, if

$$\mathbb{R}^n \setminus K \text{ is nowhere thin.}$$

We have now established one implication, namely (b) $\Longrightarrow$ (a), of the following result, which is due to Keldyš [32], Brelot [12], and Deny [16].

1.1 Theorem *Let K be a compact set such that $K^\circ = \emptyset$. The following are equivalent:*

(a) *for any $f \in C(K)$ and $\varepsilon > 0$, there exists $v \in \mathcal{H}(K)$ such that $|v - f| < \varepsilon$ on K;*

(b) $\mathbb{R}^n \setminus K$ *is nowhere thin.*

Below we give an example of a compact set K with empty interior which fails to satisfy condition (b) of Theorem 1.1.

1.2 Example Let $n = 2$, let (y_k) be a dense sequence in the square $[-1,1]^2$, and define

$$u(x) = \sum_k 2^{-k} \log \frac{1}{\|x - y_k\|}$$

and

$$K = \left\{ (x_1, x_2) \in [-1,1]^2 : u(x_1, x_2) \le \log \frac{1}{|x_2|} \right\}.$$

Then K is compact, by the lower semicontinuity of u. Also, $K^\circ = \emptyset$ since $u = +\infty$ on a dense subset of $[-1,1]^2$. However, if y is a point on the line segment $(-1,1) \times \{0\}$, then

$$u(x) > \log \frac{1}{|x_2|} \ge \log \frac{1}{\|x - y\|} \quad (x \in [-1,1]^2 \setminus K),$$

and so $\mathbb{R}^n \setminus K$ is thin at y (compare (5)). Thus approximation is not always possible on this compact set.

As we noted above, we have already established that (b) implies (a) in Theorem 1.1. A useful tool in proving the converse is the following fact (see Theorem 7.3.5 of [4]): on any open set Ω which has a Green function there exists a finite continuous potential $u^\#$ which *determines thinness* in the sense that a set $A \subset \Omega$ is thin at a point $y \in \Omega$ if and only if $\widehat{R}^A_{u^\#}(y) < u^\#(y)$.

Now suppose that condition (a) of Theorem 1.1 holds. We let Ω be an open ball which contains K, choose $u^\#$ to be a potential on Ω as in the previous paragraph, and let $\varepsilon > 0$. By hypothesis, there exists $v \in \mathcal{H}(K)$ such that

$$|v - u^\#| < \varepsilon \quad \text{on } K \tag{6}$$

and thus on U_k for all sufficiently large k, where U_k is defined by (1). By solving the Dirichlet problem on U_k with boundary function $v - u^\#$, we see that

$$-\varepsilon \le v - \widehat{R}^{\Omega \setminus U_k}_{u^\#} \le \varepsilon \quad \text{on } K.$$

Letting $k \to \infty$, we obtain

$$\left| v - \widehat{R}^{\Omega \setminus K}_{u^\#} \right| \le \varepsilon \quad \text{on } K,$$

and we can combine this with (6) to get

$$\left| u^\# - \widehat{R}^{\Omega \setminus K}_{u^\#} \right| < 2\varepsilon \quad \text{on } K.$$

Since ε can be arbitrarily small, we see that $\widehat{R}^{\Omega \setminus K}_{u^\#} = u^\#$ on K. Thus $\Omega \setminus K$ is not thin at any point of K, by the thinness-determining nature of $u^\#$, and so (b) holds.

1.5 Approximation on compact sets with interior

A development of the arguments used above (see Section 7.9 of [4] for details) yields the following more general result in which K° is no longer required to be empty. This theorem is due to Keldyš [32] and Deny [17].

1.3 Theorem *Let K be a compact set. The following are equivalent:*

(a) *for any $u \in C(K) \cap \mathcal{H}(K^\circ)$ and $\varepsilon > 0$ there exists $v \in \mathcal{H}(K)$ such that $|v - u| < \varepsilon$ on K;*

(b) $\mathbb{R}^n \setminus K$ *and* $\mathbb{R}^n \setminus K^\circ$ *are thin at the same points.*

Clearly condition (b) of Theorem 1.3 reduces to condition (b) of Theorem 1.1 when $K^\circ = \emptyset$.

To illustrate condition (b) of Theorem 1.3, let $K = \overline{B} \setminus A$, where B is the open unit ball in $\mathbb{R}^3$ and A is the Lebesgue spine defined by (4). Then the sets $\mathbb{R}^n \setminus K$ and $\mathbb{R}^n \setminus K^\circ$ are both thin at each point of $K^\circ \cup \{0\}$ and both non-thin at all other points. Thus condition (b) holds for this choice of K. On the other hand, condition (b) fails for the set K of Example 1.2. Further, if we want a similar example in which $K^\circ \neq \emptyset$, we can "fill in" the lower half of the set K of Example 1.2 by writing

$$K_1 = K \cup ([-1,1] \times [-1,0]).$$

Condition (b) of Theorem 1.3 fails for K_1 since the set $\mathbb{R}^n \setminus K_1$ is thin at points of the line segment $(-1,1) \times \{0\}$ but $K_1^\circ = (-1,1) \times (-1,0)$ and so $\mathbb{R}^n \setminus K_1^\circ$ is non-thin at these points.

A fact that we will not use, but which is worth remarking in passing, is that condition (b) of Theorem 1.3 is equivalent to:

(b') $C(W \setminus K) = C(W \setminus K^\circ)$ *for every bounded open set W such that $\overline{W} \subset \Omega$,*

where $C(\cdot)$ denotes Newtonian capacity (or Green capacity with respect to some ball containing K if $n = 2$). The implication (b') $\Longrightarrow$ (b) follows from Wiener's criterion, and the converse is also not difficult to prove (see Section 1.3 of [25]).

1.6 Approximation on non-compact sets

So far we have considered approximation only on compact sets. The next result, due to Labrèche [33], extends Theorem 1.3 to non-compact sets.

1.4 Theorem *Let Ω be an open set in $\mathbb{R}^n$ and E be a relatively closed subset of Ω. The following are equivalent:*

(a) *for any $u \in C(E) \cap \mathcal{H}(E^\circ)$ and $\varepsilon > 0$ there exists $v \in \mathcal{H}(E)$ such that $|v - u| < \varepsilon$ on E;*

(b) $\Omega \setminus E$ *and* $\Omega \setminus E^\circ$ *are thin at the same points.*

Roughly speaking, the proof of Theorem 1.4 proceeds as follows (for the details, see Section 3.9 of [25]). We choose a sequence (K_m) of compact sets with nice boundaries such that $K_m \subset K_{m+1}^\circ$ for each m and $\bigcup_m K_m = \Omega$. Define $E_m = E \cap K_m$. Then $\Omega \setminus E_m$ and $\Omega \setminus E_m^\circ$ are thin at the same points, so we can apply Theorem 1.3 to approximate u uniformly on each set E_m by a function $u_m \in \mathcal{H}(E_m)$. Further, as explained in Section 5 of the companion lectures [3] (or see Sections 3.4 and 3.9 of [25]), we can approximate u_m on E_m by harmonic functions with isolated singularities, and then use a fusion lemma to complete the argument.

However, Theorem 1.4 still does not solve the problem posed at the beginning of the lecture, for we want the approximating harmonic function to be defined on the whole of Ω; that is, we are seeking the analogue for harmonic functions of the following landmark result in the theory of holomorphic approximation, due to Arakelyan [1, 2]. Below we use Hol(A), where $A \subseteq \mathbb{C}$, to denote the collection of functions which are holomorphic on some open set containing A, and use Ω^* to denote the Alexandroff (one-point) compactification of the open set Ω. (For a discussion of condition (b) below we refer to Section 8 of [3] or Section 3.2 of [25].)

1.5 Theorem *Let Ω be an open subset of $\mathbb{C}$ and E be a relatively closed subset of Ω. The following are equivalent:*

(a) *for any $f \in C(E) \cap \text{Hol}(E^\circ)$ and $\varepsilon > 0$ there exists $g \in \text{Hol}(\Omega)$ such that $|g - f| < \varepsilon$ on E;*

(b) *$\Omega^* \setminus E$ is connected and locally connected.*

In the case of harmonic approximation we have seen that condition (b) of Theorem 1.4 is necessary and sufficient for the uniform approximation of members of $C(E) \cap \mathcal{H}(E^\circ)$ by members of $\mathcal{H}(E)$. Further, as explained in Section 8 of the lectures by Armitage [3], condition (b) of Theorem 1.5 is sufficient for the uniform approximation of members of $\mathcal{H}(E)$ by members of $\mathcal{H}(\Omega)$. However, condition (b) of Theorem 1.5 is not necessary for this approximation step. Unlike the case of holomorphic approximation, harmonic approximation is possible on certain sets with "holes", as we will now explain.

1.7 Approximation on sets with holes

We continue to use Ω to denote an open set in $\mathbb{R}^n$ and E to denote a relatively closed subset of Ω. If ω is a (connected) component of $\Omega \setminus E$ such that $\overline{\omega}$ is a compact subset of Ω, then we will call ω an Ω-*hole* of E. Further, we define $\widehat{E}$ to be the union of E with the Ω-holes of E. (Thus, in particular, the set $\widehat{E}$ depends on the choice of Ω.) The result below, which is taken from [23] (or see Theorem 3.19 of [25]), gives a complete answer to the problem posed at the beginning of this lecture.

1.6 Theorem *The following are equivalent:*

(a) *for any $u \in C(E) \cap \mathcal{H}(E^\circ)$ and $\varepsilon > 0$ there exists $v \in \mathcal{H}(\Omega)$ such that $|v - u| < \varepsilon$ on E;*

(b) (i) *$\Omega \setminus \widehat{E}$ and $\Omega \setminus E^\circ$ are thin at the same points of E;*

 (ii) *for any compact set $K \subset \Omega$, the Ω-holes of $E \cup K$ whose closure intersects K are all contained in another compact set $L \subset \Omega$. (We refer to this as the K, L-condition.)*

Before discussing the proof of Theorem 1.6, we will briefly explore the meaning of condition (b). Firstly, condition (b)(i) permits the set E to have some Ω-holes. However, if ω is such an Ω-hole, then $\mathbb{R}^n \setminus \omega$ must not be thin at any point of $\partial \omega$; that is, in the terminology of the Dirichlet problem, all boundary points of ω must be regular (see Section 2 for a brief discussion of regularity for the Dirichlet problem). Further, it is not difficult to see that condition (b)(i) implies that $\partial \widehat{E} = \partial E$. In fact, the converse is also true when $n = 2$, for the following reason. Suppose that $\partial \widehat{E} = \partial E$ and that $\Omega \setminus \widehat{E}$ is thin at a point y of E. Then, by a special property of thinness in the plane, there are arbitrarily small circles centred at y which are contained in $\widehat{E}$. By the definition of $\widehat{E}$, it follows that $y \in (\widehat{E})^\circ$. Since $\partial \widehat{E} = \partial E$, we now see that $y \in E^\circ$ and so clearly $\Omega \setminus E^\circ$ is thin at y, as required.

However, condition (b)(i) is more subtle in higher dimensions, as the following example shows.

1.7 Example Let $\Omega = \mathbb{R}^n$, where $n \geq 3$, let $\{y'_k : k \in \mathbb{N}\}$ be a dense subset of $[0,1]^{n-1}$, and define

$$u(x') = \sum_k 2^{-k} \phi_{n-1}(\|x' - y'_k\|) \quad (x' \in \mathbb{R}^{n-1})$$

and

$$K = \partial V \cup \{(x', x_n) \in [0,1]^{n-1} \times (0,1] : u(x') \leq \phi_n(x_n)\},$$

where V is the cube $(0,1)^{n-1} \times (-1,0)$. Then K is compact and $\mathbb{R}^n \setminus K$ has exactly one bounded component, namely V, so $\widehat{K} = K \cup V$. Let $v(x', x_n) = u(x')$, so that v is superharmonic on $\mathbb{R}^n$. If $y \in (0,1)^{n-1} \times \{0\}$, then

$$v(x', x_n) = u(x') > \phi_n(x_n) \geq \phi_n(\|(x', x_n) - y\|)$$

when $(x', x_n) \in (0,1)^n \setminus K$. Since the Riesz measure associated with v does not charge any singleton, it follows that $(0,1)^n \setminus K$, and hence $\mathbb{R}^n \setminus \widehat{K}$, is thin at each point of $(0,1)^{n-1} \times \{0\}$. However, $K^\circ = \emptyset$, so $\mathbb{R}^n \setminus K^\circ$ is certainly non-thin at such points. Thus approximation is not always possible on this compact set.

The K, L-condition (that is, condition (b)(ii) of Theorem 1.6) is a combination of the following two conditions:

(I) $\Omega^* \setminus \widehat{E}$ is locally connected, and

(II) all Ω-holes of E which intersect a given compact set $K \subset \Omega$ are contained in another compact set $L \subset \Omega$ (this is sometimes described by saying that the Ω-holes of E do not form *long islands*).

Thus, if $\Omega = \mathbb{R}^2$, the K, L-condition would exclude the set $E = \partial S$, where

$$S = \bigcup_{k=1}^{\infty} \left\{ (x_1, x_2) : \frac{1}{2k+1} < x_1 < \frac{1}{2k} \text{ and } 0 < x_2 < k \right\}$$

(see Figure 1), but not the set $E = \partial T$, where

$$T = \bigcup_{k=1}^{\infty} \{(x_1, x_2) : 2k < x_1 < 2k+1 \text{ and } 0 < x_2 < k\}$$

(see Figure 2).

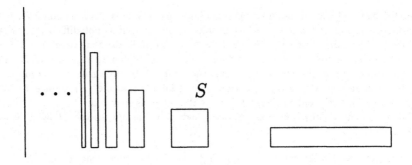

Figure 1

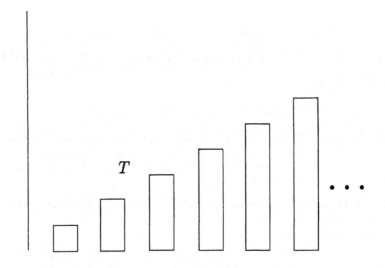

Figure 2

Having explored the meaning of condition (b) of Theorem 1.6, we now outline the proof of the implication (b) $\Longrightarrow$ (a). Suppose that (b) holds and that $u \in C(E) \cap \mathcal{H}(E^\circ)$. The first step is to "fill in" the Ω-holes of E, which we do in the following way. (If E has no Ω-holes, then we simply define $\bar{u} = u$ and move to the next step below.) Let V denote the open set $\widehat{E} \setminus E$. As we noted earlier, condition (b)(i) implies that $\partial \widehat{E} = \partial E$ and that V is regular for the Dirichlet problem. We now obtain an extension $\bar{u}$ of u to $\widehat{E}$ by solving the Dirichlet problem in V with data u on the set ∂V, which is contained in E. We note that, although

u will in general be unbounded, the K, L-condition allows us to work locally, solving the Dirichlet problem on a sequence (U_j) of disjoint bounded open sets such that $\overline{U}_j \cap \overline{U}_k = \emptyset$ whenever $|j - k| > 2$. Further, the resulting function $\overline{u}$ is harmonic on $(\widehat{E})^\circ$, since $\partial \widehat{E} = \partial E$, and is continuous on $\widehat{E}$, since V is regular.

The next step is to approximate $\overline{u}$ by a function harmonic on a neighbourhood of $\widehat{E}$. We note that $E^\circ \subseteq (\widehat{E})^\circ \subseteq \widehat{E}$, so condition (b)(i) implies that $\Omega \setminus \widehat{E}$ and $\Omega \setminus (\widehat{E})^\circ$ are thin at the same points. Given $\varepsilon > 0$, we can thus apply Theorem 1.4 to the function $\overline{u}$ on $\widehat{E}$ to obtain a function h in $\mathcal{H}(\widehat{E})$ such that $|h - \overline{u}| < \varepsilon/2$ on $\widehat{E}$ and so, in particular,

$$|h - u| < \frac{\varepsilon}{2} \quad \text{on } E. \tag{7}$$

The final step is to observe that $\Omega^* \setminus \widehat{E}$ is connected (by the definition of $\widehat{E}$) and locally connected (by the K, L-condition). Thus, as we noted at the end of Section 1.6, there is a harmonic function v on Ω such that $|v - h| < \varepsilon/2$ on $\widehat{E}$. Combining this with (7), we see that $|v - u| < \varepsilon$ on E, as required.

A useful variant of Theorem 1.6 is as follows: if, in condition (a) of that result, we consider functions u in $\mathcal{H}(E)$, then in condition (b) we must replace $\Omega \setminus E^\circ$ by $\Omega \setminus E$. (See Theorem 3.15 in [25].)

1.8   Carleman approximation

The background to this section is the following elegant generalization, due to Carleman [13], of the classical Weierstrass approximation theorem.

1.8 Theorem *Given any continuous functions* $f : \mathbb{R} \to \mathbb{C}$ *and* $\varepsilon : \mathbb{R} \to (0, 1]$, *there is an entire (holomorphic) function* g *such that*

$$|g(x) - f(x)| < \varepsilon(x) \quad (x \in E).$$

Motivated by this result we call (Ω, E), where $\Omega \subseteq \mathbb{R}^n$ is open and $E \subseteq \Omega$ is closed relative to Ω, a *Carleman pair* for harmonic functions if: for each $u \in C(E) \cap \mathcal{H}(E^\circ)$ and each continuous function $\varepsilon : E \to (0, 1]$, there exists a harmonic function v on Ω such that

$$|v(x) - u(x)| < \varepsilon(x) \quad (x \in E).$$

Problem *Characterize the Carleman pairs for harmonic functions.*

The solution to this problem is given in the following result, which is due to Goldstein and the author [29]. (See also Bagby and Gauthier [7] when $n = 2$. The corresponding problem for holomorphic approximation was solved by Nersesyan [35].)

1.9 Theorem *Let* $\Omega \subseteq \mathbb{R}^n$ *be a connected open set and suppose that* E *is a relatively closed proper subset of* Ω. *The following are equivalent:*

(a) (Ω, E) *is a Carleman pair for harmonic functions;*

(b)   (i) (Ω, E) *satisfies condition* (b) *of Theorem 1.6, and*

(ii) *given any compact set $K \subset \Omega$, the components of E° which intersect K are all contained in another compact set $L \subset \Omega$ (that is, the components of E° do not form "long islands").*

We note that condition (b)(ii) above excludes, when $\Omega = \mathbb{R}^2$, the set $E = \overline{S}$, where

$$S = \bigcup_{k=1}^{\infty} \left\{ (x_1, x_2) : \frac{1}{2k+1} \le x_1 \le \frac{1}{2k} \text{ and } 0 \le x_2 \le k \right\}.$$

One approach to proving the implication (b) $\Longrightarrow$ (a) in Theorem 1.9, taken from [24], may be summarized as follows. For any subset A of $\mathbb{R}^n$, let $\mathcal{U}_+(A)$ denote the collection of all functions which are positive and superharmonic on some open set which contains A (this extends our use of the notation $\mathcal{U}_+(\Omega)$ earlier). If Ω is a connected Greenian open set and $0 \in \Omega$ then, using a refined fusion lemma of the type described in Section 7 of [3], it can be shown that a further equivalent condition in Theorem 1.6 is:

(a') *for any $u \in C(E) \cap \mathcal{H}(E^\circ)$ and $\varepsilon > 0$ there exists $v \in \mathcal{H}(\Omega)$ such that*

$$|v(x) - u(x)| < \varepsilon \min\{1, G(0, x)\} \quad (x \in E),$$

where $G(0, \cdot)$ denotes the Green function for Ω with pole at 0.

Condition (a'), in turn, is easily seen to be equivalent to:

(a'') *for any $u \in C(E) \cap \mathcal{H}(E^\circ)$ and any $s \in \mathcal{U}_+(\Omega)$, there exists $v \in \mathcal{H}(\Omega)$ such that*

$$|v(x) - u(x)| < s(x) \quad (x \in E). \tag{8}$$

The next step is to obtain (8) for any $s \in \mathcal{U}_+(E)$: this is done by an inductive argument based on the observation that, if K is any compact subset of Ω and we choose $\delta > 0$ such that

$$\delta < \min\{s(x) : x \in K \cap E\},$$

then we can obtain a function $\overline{s}$ in $\mathcal{U}_+(K \cup E)$ by defining $\overline{s} = \min\{s, \delta\}$ on a suitable neighbourhood of E and $\overline{s} = \delta$ on a suitable neighbourhood of K. The argument is then completed by using condition (b) of Theorem 1.9 to construct, for any given continuous function $\varepsilon : E \to (0, 1]$, a function $s \in \mathcal{U}_+(E)$ such that $s < \varepsilon$ on E.

Finally, we note that Theorems 1.6 and 1.9 have been extended to the context of Riemannian manifolds by Bagby and Gauthier [8], and Bagby, Gauthier and Woodworth [9], respectively.

2 The Dirichlet problem with non-compact boundary

2.1 The Dirichlet problem

The purpose of this lecture is to demonstrate the power of harmonic approximation techniques by using them to solve the Dirichlet problem in the case of open sets where the boundary is non-compact. We begin by recalling some basic theory about the Dirichlet problem. (For a detailed account see Chapter 6 of [4].)

Let Ω be an open set in $\mathbb{R}^n$. The Dirichlet problem, in its simplest form, is as follows: given a continuous function f on $\partial\Omega$, find a harmonic function h_f on Ω which has boundary values f. A simple case is where Ω is the unit disc. Identifying $\mathbb{R}^2$ with $\mathbb{C}$ in the usual way, it is a familiar fact that the function

$$h_f(z) = \int_0^{2\pi} f(e^{it}) \frac{1 - |z|^2}{|z - e^{it}|^2} \frac{dt}{2\pi} \qquad (z \in \Omega) \tag{9}$$

solves the Dirichlet problem on Ω, and this solution is unique.

Now suppose that Ω_1 is the punctured disc $\{0 < |z| < 1\}$, let $f \in C(\partial\Omega_1)$, and *suppose further that there is a harmonic function h_f which solves the Dirichlet problem on Ω with boundary data f*. Since the singleton $\{0\}$ is a removable singularity for bounded harmonic functions, (9) must hold on Ω_1, and it follows that f must satisfy the condition

$$f(0) = h_f(0) = \int_0^{2\pi} f(e^{it}) \frac{dt}{2\pi}.$$

Thus, for most choices of f, we cannot achieve the desired boundary value at 0.

A similar situation arises when $\Omega = B \setminus \overline{A}$ in $\mathbb{R}^3$, where B is the open unit ball and A is the Lebesgue spine defined in (4): for most choices of $f \in C(\partial\Omega)$, we can achieve the desired boundary values only on $\partial\Omega \setminus \{0\}$.

"Bad" boundary points, like 0 in the preceding two examples, are those with very little of $\mathbb{R}^n \setminus \Omega$ near them. More precisely, it can be shown that they are those at which $\mathbb{R}^n \setminus \Omega$ is thin. We thus partition $\partial\Omega$ by defining

$$\partial\Omega_{\text{reg}} = \{y \in \partial\Omega : \mathbb{R}^n \setminus \Omega \text{ is not thin at } y\}$$

and $\partial\Omega_{\text{irr}} = \partial\Omega \setminus \partial\Omega_{\text{reg}}$. (The latter set is polar.)

The situation concerning the Dirichlet problem on *bounded* open sets is as follows. For every $f \in C(\partial\Omega)$ there is a (unique) harmonic function h_f on Ω such that

$$\left.\begin{array}{ll} h_f(x) \to f(y) & (x \to y \in \partial\Omega_{\text{reg}}) \\ \limsup\limits_{x \to y} |h_f(x)| < +\infty & (y \in \partial\Omega_{\text{irr}}). \end{array}\right\} \tag{10}$$

Further, if we fix $x \in \Omega$, then the mapping $f \mapsto h_f(x)$ defines a positive linear functional on $C(\partial\Omega)$. The Riesz representation theorem thus yields the existence of a Borel (unit) measure μ_x^Ω on $\partial\Omega$ such that

$$h_f(x) = \int_{\partial\Omega} f(y) \, d\mu_x^\Omega(y).$$

We call μ_x^Ω the *harmonic measure* for Ω and $x \in \Omega$. For example, if Ω is the unit disc, then $d\mu_z^\Omega(t)$ can be identified with

$$\frac{1 - |z|^2}{|z - e^{it}|^2} \frac{dt}{2\pi}$$

on $[0, 2\pi]$, in view of (9).

2.2   Unbounded sets

What if Ω is unbounded? If $f \in C(\partial\Omega \cup \{\infty\})$, where $\mathbb{R}^n \cup \{\infty\}$ denotes the one-point compactification of $\mathbb{R}^n$, then the situation differs little from the case of bounded open sets (the relationship between thinness and boundary behaviour at ∞ is anomalous in higher dimensions). However, if we are concerned only with boundary behaviour of the solution at points of the Euclidean boundary $\partial\Omega$, then the situation requires more careful investigation.

As a simple example, let us consider the case of the half-plane $\Omega = \mathbb{R} \times (0, +\infty)$. We again identify $\mathbb{R}^2$ with $\mathbb{C}$. If $f \in C(\mathbb{R} \cup \{\infty\})$, then a solution to the Dirichlet problem for f is given by

$$h_f(z) = \int_{-\infty}^{+\infty} f(t) \frac{y}{\pi} \frac{1}{(x-t)^2 + y^2}\, dt \quad (z = x + iy;\ y > 0);$$

that is, harmonic measure for Ω and z is given by

$$d\mu_z^\Omega(t) = \frac{y}{\pi} \frac{1}{(x-t)^2 + y^2}\, dt \quad \text{on } \mathbb{R}.$$

In fact, if f is continuous on $\mathbb{R}$ and merely satisfies

$$\int_{-\infty}^{+\infty} \frac{|f(t)|}{1+t^2}\, dt < +\infty,$$

then it is straightforward to check that f is integrable with respect to harmonic measure for Ω and that h_f has boundary values f on $\mathbb{R}$. However, much more is true, as the following striking result of R. Nevanlinna [36] shows.

2.1 Theorem *For any $f \in C(\mathbb{R})$ there is a harmonic function h_f on $\mathbb{R} \times (0, +\infty)$ with boundary values f.*

This leads to the obvious question:

Problem *Which open sets Ω have the following property: for each $f \in C(\partial\Omega)$ there exists a harmonic function h_f on Ω such that (10) holds?*

2.3   The solution

A complete solution to the above problem is contained in the next result, due to the author [22].

2.2 Theorem *Let Ω be an open set in $\mathbb{R}^n$. The following are equivalent:*

(a) *for each $f \in C(\partial\Omega)$ there exists a harmonic function h_f on Ω such that (10) holds;*

(b) *for each compact set K in $\mathbb{R}^n$, there is another compact set L which contains all bounded components ω of $\Omega \setminus K$ such that $\overline{\omega}$ intersects K.*

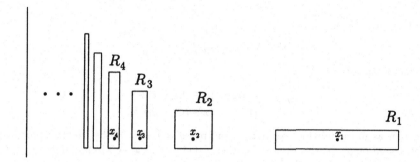

Figure 3

The easier implication to prove here is (a) $\Longrightarrow$ (b). To make the argument clear, we begin with some examples in the plane.

Firstly, consider the case of the open set $\Omega_1 = \bigcup_k R_k$, where R_k is the rectangle given by:

$$R_k = \left(\frac{1}{2k+1}, \frac{1}{2k}\right) \times (0, k) \quad (k \in \mathbb{N})$$

(see Figure 3). By considering the compact set $K = [0, 1] \times \{0\}$ we see immediately that Ω_1 fails to satisfy condition (b) of Theorem 2.2 and so the Dirichlet problem on Ω_1 is not always solvable. This is perhaps surprising at first glance, because one would expect to be able to solve the Dirichlet problem in Ω_1 by solving it in each rectangular component R_k. Such an approach would certainly lead to the desired boundary values on the boundary of each rectangle, but does not address the fact that $\{0\} \times [0, +\infty)$ is also contained in $\partial\Omega_1$. Could we fail to obtain the right values at points of this half-line? The answer is "yes". To see this, let x_k denote the point in R_k with coordinates

$$\left(\frac{4k+1}{4k(2k+1)}, \frac{1}{2}\right).$$

We choose a continuous function $f_1 : \partial\Omega_1 \to [0, +\infty)$ in such a way that $f_1 = 0$ on all of $\partial\Omega_1$ apart from the top edges of the rectangular sets ∂R_k, where it is assigned large enough values to ensure that the Dirichlet solution $h^{(k)}$ for R_k with boundary values $f_1|_{\partial R_k}$ satisfies $h^{(k)}(x_k) \geq k$. If the Dirichlet problem on Ω_1 has a solution h_{f_1} for this choice of boundary function f_1, then we would have $h_{f_1} = h^{(k)}$ on R_k by the maximum principle applied to each rectangle separately, and so would obtain the contradictory conclusion that

$$x_k \to (0, 1/2) \in \partial\Omega_1 \quad \text{and} \quad f_1(0, 1/2) = 0, \quad \text{yet} \quad h_{f_1}(x_k) \to +\infty.$$

Next we modify the above example by defining

$$\Omega_2 = \left(\bigcup_k Q_k\right) \cup (\mathbb{R} \times (-\infty, 0)),$$

where

$$Q_k = \left(\frac{1}{2k+1}, \frac{1}{2k} \right) \times (-1, k).$$

We define f_2 on $\partial\Omega_2$ by writing

$$f_2(x) = \begin{cases} f_1(x) & (x \in \partial\Omega_2 \cap \partial\Omega_1) \\ 0 & (x \in \partial\Omega_2 \setminus \partial\Omega_1), \end{cases}$$

where f_1 is as in the previous paragraph, and put $g = f + 1$. If the Dirichlet problem on Ω_2 has a solution h_g for this choice of g, then $h_g(x) \to 1$ as $x \to 0$ in Ω_2. Hence $h_g \geq 0$ on the lower side of the rectangle R_k for all sufficiently large values of k. It follows from the maximum principle applied in each such rectangle R_k that $h_g \geq h^{(k)}$ on R_k, and again we obtain the contradictory conclusion that $h_g(x_k) \to +\infty$ as $k \to \infty$.

The proof of the implication (a) $\longrightarrow$ (b) in Theorem 2.2 is now easy to explain. Suppose that condition (b) fails to hold. Then there is a compact set K and a sequence (ω_k) of bounded components of $\Omega \setminus K$ such that $\overline{\omega_k}$ intersects K and $\omega_k \not\subseteq B(k)$, where $B(r)$ denotes the open ball of centre 0 and radius r. We choose r' such that $K \subset B(r')$. By a compactness argument there exists a point x_0 in $\partial B(r')$ such that every neighbourhood of x_0 meets infinitely many of the open sets ω_k. We now imitate the argument of the previous paragraph, with x_0 in place of the point $(0, 1/2)$ and ω_k in place of the rectangle R_k to see that (a) must also fail.

2.4 Proof of Theorem 2.2 (b) $\Longrightarrow$ (a)

The implication (b) $\Longrightarrow$ (a) of Theorem 2.2 depends on a deeper argument involving harmonic approximation. We will present the main ideas without giving all the details. Suppose that condition (b) holds and let $f \in C(\partial\Omega)$. Further, suppose temporarily that Ω is (unbounded and) connected. The "obvious" way to find the desired function h_f is by means of integration with respect to harmonic measure:

$$h_f(x) = \int_{\partial\Omega} f(y) \, d\mu_x^\Omega(y) \quad (x \in \Omega).$$

The difficulty is that f may not be μ_x^Ω-integrable. A way around this problem is to modify the set Ω to achieve μ_x^Ω-integrability for f. We suppose, without loss of generality, that $B(1)$ intersects both Ω and $\mathbb{R}^n \setminus \Omega$, and choose a point q in $B(1) \cap \Omega$. If $k \in \mathbb{N}$, then let g_k denote the function equal to $|f|$ on

$$\{y \in \partial\Omega : k+1 \leq \|y\| \leq k+2\}$$

and equal to 0 elsewhere in $\mathbb{R}^n$. By defining

$$S_k = \{y \in \partial B(k) \cap \Omega : \mathrm{dist}(y, \mathbb{R}^n \setminus \Omega) \geq \varepsilon_k\}$$

and choosing $\varepsilon_k > 0$ small enough, we can arrange that the conventional solution h_{g_k} to the Dirichlet problem on $\Omega \setminus S_k$ with boundary data g_k satisfies $h_{g_k}(q) < 2^{-k}$. Now let $\omega = \Omega \setminus (\bigcup_k S_k)$ (see Figure 4). Defining $f = 0$ on $\bigcup_k S_k$, we see that $f \in C(\partial\omega)$ and

$$\int_{\partial\Omega \cap \{k+1 \leq \|y\| \leq k+2\}} |f(y)| \, d\mu_q^\omega(y) \leq h_{g_k}(q),$$

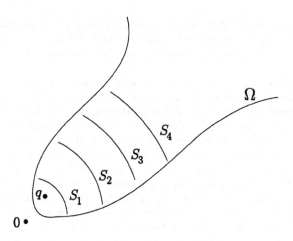

Figure 4

and so

$$\int_{\partial\omega\cap\{\|y\|\geq2\}}|f(y)|\,d\mu_q^\omega(y)=\int_{\partial\Omega\cap\{\|y\|\geq2\}}|f(y)|\,d\mu_q^\omega(y)<+\infty.$$

Thus we can define a harmonic function u on ω by writing

$$u(x)=\int_{\partial\Omega}f\,d\mu_x^\omega\quad(x\in\omega).$$

The function u has the desired boundary values at points of $\partial\Omega$. However, as a function on Ω, the problem is that u has large singularities in the form of the sets S_k.

For each k we choose an open subset ω_k of

$$\Omega\cap\left(B(k+1/2)\setminus\overline{B(k-1/2)}\right)$$

such that $S_k\subset\omega_k$ and $\overline{\omega}_k\subset\Omega$, and then (provisionally) define a relatively closed subset of Ω by writing $E=\Omega\setminus(\bigcup_k\omega_k)$. The plan is now to apply the following result of Armitage and Goldstein [5] (or see Theorem 8.2 of [3] or Corollary 3.6 of [25]).

2.3 Theorem *Suppose that Ω is connected and has a Green function $G(\cdot,\cdot)$, let E be a relatively closed subset of Ω and let $q\in\Omega$. If $\Omega^*\setminus E$ is connected and locally connected, then for each $u\in\mathcal{H}(E)$ and $\varepsilon>0$, there exists $v\in\mathcal{H}(\Omega)$ such that*

$$|(v-u)(x)|<\varepsilon\min\{1,G(q,x)\}\quad(x\in E).\tag{11}$$

Let us now apply this result to our situation (with initial reckless disregard for whether the hypotheses on $\Omega^*\setminus E$ are valid). Since $\overline{\omega}_k\subset\Omega$ for each k, we have control of the behaviour of the approximant v at points of $\partial\Omega$. Thus, if $y \in \partial\Omega_{\text{reg}}$, then as $x \to y$, we obtain $u(x) \to f(y)$ and $G(q, x) \to 0$ (a property of regular boundary points), and so $v(x) \to f(y)$ by (11). On the other hand, if $y \in \partial\Omega_{\text{irr}}$, then u is bounded near y, so the same must be true of v. Hence (10) holds.

Of course, the hypotheses on $\Omega^* \setminus E$ are certainly not valid. The most obvious problem is that each component of any of the sets ω_k is a component of $\Omega^* \setminus E$, and so this set is certainly not connected. To get around this problem we need, for each k, to join each component of ω_k to the Alexandroff point of Ω by removing a suitable connected open set U from E. (We may assume, in view of the compactness of S_k, that each set ω_k has only finitely many components.) Since we wish to retain control on the behaviour of v near $\partial\Omega$ in the above approximation step, these sets U must satisfy $\overline{U} \subset \Omega$; that is, they connect components of ω_k to ∞. With this modified definition of E the set $\Omega^* \setminus E$ is connected. We still need to ensure that this set is locally connected. Also, we must take care to ensure that the various choices of U we have just made do not cluster up near a point of $\partial\Omega$, for that would surrender control of the boundary behaviour of v at that point. The situation becomes clearer when we redefine the radii r_k at which we originally formed the spherical slits S_k. Instead of defining $r_k = k$, we proceed inductively as follows. Let $r_1 = 1$. Given r_j, let F_j denote the union of $\overline{B(r_j)} \cap \Omega$ with all the bounded components W of $\Omega \setminus \overline{B(r_j)}$ such that $\overline{W}$ intersects $\overline{B(r_j)}$. Condition (b) of Theorem 2.2 ensures that F_j is bounded so we can now choose r_{j+1} such that $F_j \subset B(r_{j+1})$. With this modification the above outline argument can be carried through satisfactorily for unbounded connected open sets Ω.

When we drop the requirement that Ω be connected, two new difficulties arise. One is that we cannot introduce spherical slits into bounded components of Ω, for there is no way of connecting these to ∞ by a connected open set which lies within Ω. This is not a serious problem since condition (b) ensures that any bounded component of Ω can intersect $\partial B(r_j)$ for at most one value of j, and so we can dispense with the portion of the slit which lies in such components. The second difficulty is that there may be infinitely many unbounded components of Ω. For example, if Ω is the union of strips given by $\Omega = \cup_k \Omega_k$, where

$$\Omega_k = \left(\frac{1}{2k+1}, \frac{1}{2k}\right) \times \mathbb{R},$$

then we have to ensure the correct boundary behaviour not only on the boundary of each individual strip but also at points of $\{0\} \times \mathbb{R}$. We can achieve this by approximating on each strip Ω_k separately to obtain

$$|(v - u)(x)| < \frac{1}{k}\min\{1, G(q_k, x)\} \quad (x \in E \cap \Omega_k),$$

where $q_k \in \Omega_k$ and G is the Green function for Ω. Thus, if $y \in \{0\} \times \mathbb{R}$ and $u(x) \to f(y)$ as $x \to y$ in Ω, then we also have $v(x) \to f(y)$ as $x \to y$ in Ω. A similar argument applies in the general case of infinitely many bounded components.

3   Approximability

3.1   The question

Following the application given in the second lecture we return to the theory of harmonic approximation. As before Ω denotes an open set in $\mathbb{R}^n$ and E is a relatively closed subset of

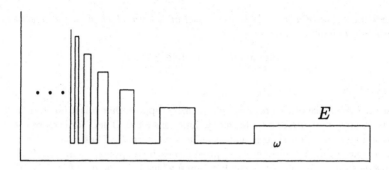

Figure 5

Ω, and we are interested in uniformly approximating functions on E by harmonic functions on Ω. Since the functions to be approximated must be in $C(E) \cap \mathcal{H}(E^\circ)$, there are two fundamental questions to be addressed:

Problem 1 *Which sets E have the property that every function in $C(E) \cap \mathcal{H}(E^\circ)$ can be uniformly approximated by functions in $\mathcal{H}(\Omega)$?*

Problem 2 *Given a particular set E, which members of $C(E) \cap \mathcal{H}(E^\circ)$ can be uniformly approximated by functions in $\mathcal{H}(\Omega)$?*

Problem 1 was answered in the first lecture. The purpose of this lecture is to give an answer to Problem 2. We note that the analogue of Problem 2 for holomorphic approximation was posed by Arakelyan [2] and solved by Stray [38] subject to some technical restrictions on Ω and E.

3.2   Accessibility

We will write the Alexandroff (one point) compactification of Ω as $\Omega^* = \Omega \cup \{\mathcal{A}\}$. Let ω be a connected open subset of Ω. We say that $\mathcal{A}$ is *accessible* from ω if there is a continuous function $p : [0, +\infty) \to \omega$ such that $p(t) \to \mathcal{A}$ as $t \to +\infty$. Next, we define $\widetilde{E}$ to be the union of E with all the connected components of $\Omega \setminus E$ from which $\mathcal{A}$ is *not* accessible. Comparing this with the set $\widehat{E}$ defined in the first lecture, we see that $\widehat{E} \subseteq \widetilde{E}$, and that this inclusion may be strict. For example, if $\Omega = \mathbb{R}^2$ and $E = \partial\omega$, where

$$\omega = \left(\bigcup_k \left(\frac{1}{2k+1}, \frac{1}{2k} \right) \times (-1, k) \right) \cup ((0, 1/2) \times (-1, 0))$$

(see Figure 5), then $\widetilde{E} = E \cup \omega$ but $\widehat{E} = E$.

Accessibility is an important concept in dealing with Problem 2 above because of the following generalized maximum principle. This result is implied by work of Fuglede on asymptotic paths of subharmonic functions [20]; an elegant and elementary proof of it has been given by Chen and Gauthier [14] (or see Theorem 3.1.10 of [4]).

3.1 Theorem *Let ω be a connected open subset of Ω from which $\mathcal{A}$ is not accessible. If s is a subharmonic function on ω and*

$$\limsup_{x\to y} s(x) \le a \quad (y \in \partial\omega \cap \Omega),$$

then $s \le a$ on ω.

To see the relevance of this result to Problem 2, suppose that u is a function on E which can be uniformly approximated by harmonic functions on Ω. Then, for each $m \in \mathbb{N}$, there is a harmonic function v_m on Ω such that $|v_m - u| < m^{-1}$ on E. Thus $|v_m - v_k| \le m^{-1} + k^{-1}$ on E, and hence on $\widetilde{E}$, by Theorem 3.1 applied to each component of $\widetilde{E} \setminus E$. It follows that (v_m) converges uniformly on $\widetilde{E}$ to some function $\widetilde{u}$ which belongs to $C(\widetilde{E}) \cap \mathcal{H}((\widetilde{E})^\circ)$. Clearly $\widetilde{u} = u$ on E, so we conclude that any such approximable function u on E has an extension to $\widetilde{E}$ which lies in $C(\widetilde{E}) \cap \mathcal{H}((\widetilde{E})^\circ)$.

3.3 The result

In order to state the solution to Problem 2 we need a little more notation. We write $C_e(E)$ for the collection of continuous functions on E which have a continuous extension to the closure of E in $\mathbb{R}^n \cup \{\infty\}$. We also write $A(E)$ for the collection of approximable functions on E; that is, $u \in A(E)$ if and only if, for each $\varepsilon > 0$, there is a harmonic function v_ε on Ω such that $|v_\varepsilon - u| < \varepsilon$ on E. If, further, the approximating functions v_ε can be chosen so that the restrictions $v_\varepsilon |_E$ belong to $C_e(E)$, then we write $u \in A_e(E)$. Clearly $A_e(E) \subseteq C_e(E)$.

The result below is due to the author [27].

3.2 Theorem *Let Ω be an open set in $\mathbb{R}^n$ and E be a relatively closed subset of Ω. The following are equivalent:*

(a) $u \in A(E)$;

(b) $u = v + h$ on E, for some $v \in A_e(\widetilde{E})$ and $h \in \mathcal{H}(\Omega)$;

(c) $u = v + h$ on E, for some $v \in C_e(\widetilde{E})$ which is finely harmonic on the fine interior of $\widetilde{E}$ and some $h \in \mathcal{H}(\Omega)$.

The notion of fine harmonicity, which is involved in condition (c) above, will be discussed shortly. We will also see that, when $n = 2$, we can omit the words "finely" and "fine" from this condition.

An easy consequence of Theorem 3.2 is that, if a function u on E is approximable by harmonic functions on Ω, then it can be arranged that the error of approximation behaves nicely on the closure of E in $\mathbb{R}^n \cup \{\infty\}$:

3.3 Corollary *Suppose that $u \in A(E)$. Then, for each $\varepsilon > 0$, there is a harmonic function w on Ω such that $|w - u| < \varepsilon$ on E and $(w - u)|_E \in C_e(E)$.*

To see how this follows from Theorem 3.2, let $u \in A(E)$. By the implication (a) $\Longrightarrow$ (b) of that result, we can write $u = v + h$ on E for some $v \in A_e(\widetilde{E})$ and $h \in \mathcal{H}(\Omega)$. Let $\varepsilon > 0$. Then there is a harmonic function v_ε on Ω such that

$$|v_\varepsilon - v| < \varepsilon \quad \text{on } \widetilde{E} \quad \text{and} \quad v_\varepsilon |_{\widetilde{E}} \in C_e(\widetilde{E}). \tag{12}$$

If we define $w = v_\varepsilon + h$, then w is harmonic on Ω and

$$w - u = v_\varepsilon + h - (v + h) = v_\varepsilon - v \quad \text{on } E,$$

so (12) yields

$$|w - u| < \varepsilon \quad \text{on } E \qquad \text{and} \qquad (w - u)\,|_E \in C_e(E),$$

as required.

3.4 Finely harmonic functions

We recall that, although harmonic functions on an open set Ω are usually introduced as solutions of Laplace's equation on Ω, they can equivalently be defined as those continuous functions u on Ω which have the spherical mean value property: whenever a closed ball of centre x is contained in Ω, the mean value of u over the boundary of the ball is equal to $u(x)$. This spherical mean value property is, of course, a special case of a more general property of harmonic functions, namely that if U is a bounded open set such that $\overline{U} \subset \Omega$, then

$$u(x) = \int u(y)\,d\mu_x^U(y) \quad (x \in U).$$

We recall that the harmonic measure μ_x^U can be characterized as the Riesz measure associated with the potential $\widehat{R}_{G(\cdot,x)}^{\Omega \backslash U}$ on Ω, where $G(\cdot,\cdot)$ is the Green function for Ω (see Section 9.1 of [4]).

The above potential differs from $G(\cdot, x)$ whenever U is a fine neighbourhood of x. Thus, if we use μ_x^U to denote the Riesz measure associated with this potential in this more general case, we can make the following definition. Let ω be a finely open set in $\mathbb{R}^n$. A function $u : \omega \to \mathbb{R}$ is said to be *finely harmonic* on ω if u is finely continuous on ω and every point x in ω has a fundamental system of fine neighbourhoods U with fine closure contained in ω such that

$$u(x) = \int u(y)\,d\mu_x^U(y). \tag{13}$$

Further, (13) holds for every bounded fine neighbourhood U of x such that the fine closure of U is contained in ω and u is bounded on U. Harmonic functions on (Euclidean) open sets are finely harmonic, and the converse is true provided we consider only locally bounded functions. (We refer to [19] and [21] for accounts of finely harmonic functions.)

We will only consider continuous functions u. The important point for us is that condition (13) is preserved by uniform convergence. Thus, in view of Theorem 3.1 and the argument which followed its statement, any function u in $A(E)$ has an extension $\widetilde{u} \in C(\widetilde{E})$ which is finely harmonic on the fine interior of $\widetilde{E}$.

Theorem 3.2 has a simpler form when $n = 2$, for the following reason. Suppose that x is in the fine interior of $\widetilde{E}$. Then (by the special property of thinness in the plane mentioned in the first lecture) there are arbitrarily small circles centred at x which are contained in $\widetilde{E}$, and by the definition of $\widetilde{E}$ it follows that $x \in (\widetilde{E})^\circ$. Thus the fine interior of $\widetilde{E}$ coincides with $(\widetilde{E})^\circ$, and fine harmonicity on this open set is equivalent to harmonicity (only continuous functions are under consideration), so we can drop the words "fine" and "finely" in condition (c) of Theorem 3.2 when $n = 2$.

3.5   A key estimate

A key tool for the proof of Theorem 3.2 is contained in the following result, which relies on delicate estimates of harmonic measure.

3.4 Lemma *Let K be a compact subset of Ω, let $a > 3$, and let w be a harmonic function on Ω such that $|w| < \varepsilon$ on E. Then there is a harmonic function s on Ω such that*

$$|w - s| < a\varepsilon \quad on \ K \cup \widetilde{E}$$

and

$$s \mid_{\widetilde{E}} \in C_e(\widetilde{E}).$$

Thus the function s above is a modification of w (that is, it approximates w on $K \cup \widetilde{E}$) which belongs to $C_e(\widetilde{E})$. We refer to [27] for a proof of this result, and merely indicate below how it yields the implication (a) $\Longrightarrow$ (b) in Theorem 3.2.

Suppose that condition (a) of Theorem 3.2 holds; that is, $u \in A(E)$. Then there is a harmonic function u_0 on Ω such that $|u - u_0| < 1$ on E. Clearly $u - u_0 \in A(E)$, so there also exists a harmonic function u_1 on Ω such that

$$|u_1 - (u - u_0)| < \frac{1}{2} \quad on \ E.$$

We proceed inductively: given $u_0, u_1, \ldots, u_{m-1}$, we choose a harmonic function u_m on Ω such that

$$|u_m - (u - u_0 - u_1 - \cdots - u_{m-1})| < \frac{1}{2^m} \quad on \ E. \tag{14}$$

Thus $|u_m| < 3/2^m$ on E and hence (by Theorem 3.1) on $\widetilde{E}$.

Now let (K_m) be an exhaustion of Ω by compact sets. For each m in $\mathbb{N}$ we apply Lemma 3.4 (with $w = u_m$, $K = K_m$, and $a = 10/3$) to obtain a harmonic function s_m on Ω such that

$$|u_m - s_m| < \frac{10}{2^m} \quad on \ K_m \cup \widetilde{E} \tag{15}$$

and

$$s_m \mid_{\widetilde{E}} \in C_e(\widetilde{E}).$$

Let

$$h = u_0 + \sum_{m=1}^{\infty} (u_m - s_m) \quad and \quad v = \sum_{m=1}^{\infty} s_m.$$

Then h converges locally uniformly on Ω to a harmonic function, in view of (15), and $v \in A_e(\widetilde{E})$ since

$$|s_m| \leq |s_m - u_m| + |u_m| < \frac{13}{2^m} \quad on \ \widetilde{E}.$$

Finally, from (14), we see that on E we have

$$u = \sum_{m=0}^{\infty} u_m$$

$$= \sum_{m=1}^{\infty} s_m + \left\{ u_0 + \sum_{m=1}^{\infty} (u_m - s_m) \right\}$$

$$= v + h,$$

as required.

3.6 A further tool

The following result is due to Debiard and Gaveau [15].

3.5 Theorem *Let K be a compact set in $\mathbb{R}^n$ and let $f : K \to \mathbb{R}$. Then*

$$f \text{ can be uniformly approximated on } K \text{ by functions in } \mathcal{H}(K)$$

if and only if

$$f \in C(K) \text{ and } f \text{ is finely harmonic on the fine interior of } K.$$

In view of the preservation of (13) under uniform convergence, the "only if" part of the above result is trivial. We refer to [15] for a proof of the converse. What is needed to complete the proof of Theorem 3.2 is a careful generalization of the "if" part of Theorem 3.5 to yield an approximation result on non-compact sets which preserves continuity at the boundary of Ω in $\mathbb{R}^n \cup \{\infty\}$.

3.6 Theorem *Suppose that $u \in C_e(E)$ is finely harmonic on the fine interior of E. Then, for each $\varepsilon > 0$, there exists an open neighbourhood U of the closure of E in $\mathbb{R}^n \cup \{\infty\}$, and a function $w \in C(\mathbb{R}^n \cup \{\infty\})$ which is harmonic on $U \cap \Omega$, such that $|w - u| < \varepsilon$ on E.*

In the case where E is bounded, it is tempting to try to prove Theorem 3.6 by applying Theorem 3.5 to $\overline{E}$, but this overlooks the fact that u need not be finely harmonic on the fine interior of $\overline{E}$. Using the techniques described in [3] (or [25]) we can approximate the above function w by harmonic functions with isolated singularities, and then use pole pushing to obtain the following corollary.

3.7 Corollary *Suppose, in addition, that $\widetilde{E} = E$. Then, for each $\varepsilon > 0$, there exists a harmonic function w on Ω such that $|w - u| < \varepsilon$ on E and $w|_E \in C_e(E)$.*

It is clear from Corollary 3.7 that (c) implies (a) in Theorem 3.2, and we discussed the implication (a) $\implies$ (b) in the previous section. Finally, the implication (b) $\implies$ (c) follows, since continuity and the condition (13) are preserved by uniform convergence.

4   Boundary behaviour of harmonic functions

4.1   Infinity sets at the boundary

In this final lecture we present two further applications of harmonic approximation. For the first of these it will be convenient to work in $\mathbb{R}^{n+1}$ ($n \geq 1$). Let $\mathcal{D}$ denote the upper half-space $\{(x,t) : x \in \mathbb{R}^n, \ t > 0\}$ and let λ denote Lebesgue measure on $\mathbb{R}^n$.

Problem   *For which sets $E \subseteq \mathbb{R}^n$ is there a harmonic function u on $\mathcal{D}$ such that*

$$u(x,t) \to +\infty \quad (t \to 0+; \ x \in E)? \tag{16}$$

When $n = 1$, the situation has been well understood for some time:

4.1 Theorem   *Let $n = 1$ and $E \subseteq \mathbb{R}$. The following are equivalent:*

(a) *there is a harmonic function u on $\mathcal{D}$ such that* (16) *holds;*

(b) *there is a holomorphic function f on $\mathcal{D}$, not identically valued 0, such that*

$$f(x+it) \to 0 \quad (t \to 0+; \ x \in E);$$

(c) *there is a superharmonic function u on $\mathcal{D}$ such that* (16) *holds;*

(d) *for each interval U, either the set $E \cap U$ is of first (Baire) category, or $\lambda(E \cap V) = 0$ for some interval $V \subseteq U$.*

Theorem 4.1 was originally proved in the context of the unit disc, in the following stages. Lusin and Privalov [34] showed that (b) implies (d), and provided some partial results in the converse direction. Given any harmonic function u on $\mathcal{D}$, we need only consider the function $f = e^{-g}$, where g is the holomorphic completion of u, to see that (a) implies (b), and so [34] includes information about the boundary behaviour of harmonic functions. Arsove [6] proved the implication (c) $\implies$ (d), which generalizes the work of Lusin and Privalov since the function $u = -\log|f|$ is superharmonic if f is holomorphic and not identically valued 0. Finally, Berman [10] completed the proof by showing that (d) implies (a).

Turning now to higher dimensions, it is natural to ask if we can simply replace "interval" by "non-empty open set in $\mathbb{R}^n$" in condition (d) and retain the equivalence of conditions (a), (c) and (d). The answer is negative. Rippon [37] has shown that, when $n = 2$, there is a superharmonic function u on $\mathcal{D}$ such that (16) holds and yet the set $\mathbb{R}^n \setminus E$ is both of λ-measure 0 and of first category. Despite the apparently hopeless prospect held out by this example, a characterization of sets $E \subseteq \mathbb{R}^n$ for which (16) holds for some superharmonic function u on $\mathcal{D}$ may be found in [26] and [18].

This leaves open the corresponding question for harmonic functions: for, as will become clear, conditions (a) and (c) cease to be equivalent in higher dimensions. Actually, the implication (a) $\implies$ (d) is true in all dimensions, as the following standard argument shows. Suppose that (a) holds and let U be a non-empty open set in $\mathbb{R}^n$. Further, let

$$E_k = \{x \in E \cap U : u(x,t) \geq 0 \text{ whenever } 0 < t \leq 1/k\} \quad (k \in \mathbb{N}).$$

Then the sequence (E_k) is increasing, and $\bigcup_k E_k = E \cap U$. If $E \cap U$ is *not* of first category, then there exists $k_0 \in \mathbb{N}$ such that $\overline{E}_{k_0}$ has non-empty interior V. By continuity, $u \geq 0$ on $V \times (0, k_0^{-1})$. We can now appeal to (a generalized form of) Fatou's theorem to see that u has a finite non-tangential limit λ−almost everywhere on the set $V \times \{0\}$ (identifying $\mathbb{R}^n$ with $\mathbb{R}^n \times \{0\}$ in the usual way). Hence $\lambda(E \cap V) = 0$, as required. (Strictly speaking, we want V to be a subset of U, but we can arrange this simply by replacing V by the non-empty open set $U \cap V$.)

Although the argument of the preceding paragraph is essentially the same as one uses in the plane, the similarity ends there: for the converse implication fails. What we need in place of condition (d) is a hybrid condition involving both the Euclidean and fine topologies. A complete solution to the problem in higher dimensions is contained in the following recent result of Hansen and the author [30].

4.2 Theorem *Let $n \geq 1$ and $E \subseteq \mathbb{R}^n$. The following are equivalent:*

(a) *there is a harmonic function u on $\mathfrak{D}$ such that (16) holds;*

(e) *there is a sequence of sets (E_k) such that $E_k \uparrow E$ and $\lambda(E \cap V_k) = 0$ for each k, where V_k denotes the fine interior of $\overline{E}_k$.*

The superharmonic functions on $\mathbb{R}$ are precisely the concave functions and so, in particular, they are all continuous. It follows that, when $n = 1$, the fine and Euclidean topologies coincide, and it not difficult to check directly in this case that condition (e) of Theorem 4.2 is equivalent to condition (d) of Theorem 4.1. We also remark that sets E which satisfy condition (e) of Theorem 4.2 have no finely interior points. This is an easy consequence of the fact that $\mathbb{R}^n$, endowed with the fine topology, is a Baire space.

4.2   Harmonic approximation and Theorem 4.2

Harmonic approximation results were crucial in formulating condition (e) of Theorem 4.2 as well as showing its sufficiency. One tool is the following generalization of Theorem 3.5, obtained by Gauthier and Ladouceur [31] using a fusion argument.

4.3 Theorem *Let D be an open set in $\mathbb{R}^n$ and F be a relatively closed subset of D. Further, let u be a continuous function on F which is finely harmonic on the fine interior of F. Then, for each $\varepsilon > 0$, there exists $v \in \mathcal{H}(F)$ such that $|v - u| < \varepsilon$ on F.*

Another tool is the following link between the fine topologies on $\mathbb{R}^n$ and $\mathbb{R}^{n+1}$, for which a proof may be found in Theorem 7.8.6 of [4]. It explains why the fine topology on $\mathbb{R}^n$ arises when studying the boundary behaviour of harmonic or superharmonic functions in $\mathfrak{D}$.

4.4 Lemma *Let $A \subseteq \mathbb{R}^n$ and $x \in \mathbb{R}^n$, where $n \geq 2$, and let $t \in \mathbb{R}$. Then $A \times \mathbb{R}$ is thin at (x,t) if and only if A is thin at x.*

The proof of the implication (e) $\Longrightarrow$ (a) in Theorem 4.2 is, in outline, as follows. Suppose that condition (e) holds and let $W = \bigcup_k (\overline{E}_k)^\circ$. Since $(\overline{E}_k)^\circ \subseteq V_k$, we see that

$$\lambda(E \cap W) \leq \sum_k \lambda(E \cap (\overline{E}_k)^\circ) \leq \sum_k \lambda(E \cap V_k) = 0.$$

Let

$$\mathfrak{W} = \{(x,t) \in \mathbb{R}^{n+1} : 0 < t \leq \operatorname{dist}(x, \mathbb{R}^n \setminus W)\}.$$

Also, let $F_k = \overline{E}_k \setminus W$, let U_k denote the fine interior of F_k (whence $U_k \subseteq V_k$), and let

$$\mathfrak{F} = \bigcup_k (F_k \times (0, k^{-1}]).$$

The first step, which is non-trivial, is to construct a non-negative continuous function u_1 on $\mathfrak{F}$ which is finely harmonic on the fine interior of $\mathfrak{F}$ and satisfies

$$u_1(x,t) \to +\infty \quad (t \to 0+; \; x \in (\bigcup_k F_k) \setminus (\bigcup_k U_k); \; x \notin P), \tag{17}$$

where $P \subseteq \bigcup_k F_k$ is polar in $\mathbb{R}^n$. Next, by Theorem 4.3 (with $\varepsilon = 1$, say), we may assume that u_1 is harmonic on a neighbourhood of $\mathfrak{F}$. We now define $u_1 = 0$ on a neighbourhood of $\mathfrak{W}$. Thus $u_1 \geq 0$ on $\mathfrak{F} \cup \mathfrak{W}$. The construction of $\mathfrak{F}$ and $\mathfrak{W}$ ensure that the set $\mathfrak{D}^* \setminus (\mathfrak{F} \cup \mathfrak{W})$ is connected and locally connected. Thus we can approximate u_1 uniformly on $\mathfrak{F} \cup \mathfrak{W}$ (see Section 1.6) by a harmonic function u_2 on $\mathfrak{D}$ such that $u_2 \geq -1$ on $\mathfrak{F} \cup \mathfrak{W}$ and (see (17))

$$u_2(x,t) \to +\infty \quad (t \to 0+; \; x \in (\bigcup_k F_k) \setminus (\bigcup_k U_k); \; x \notin P).$$

We observe that there is a positive harmonic function h on $\mathfrak{D}$ such that

$$h(y) \to +\infty \quad (y \to (x,0); \; y \in \mathfrak{D})$$

when x is in $\bigcup_k (U_k \cap E)$ or $E \cap W$ or the polar set P, since all of these sets have λ-measure zero. Finally, we define $u = u_2 + h$ and observe that

$$u(x,t) \to +\infty \quad (t \to 0+)$$

when x is in any of the sets

$$E \cap W, \qquad (\bigcup_k F_k) \setminus ([\bigcup_k U_k] \cup P), \qquad \bigcup_k (U_k \cap E), \qquad P.$$

Since E is contained in the union of these four sets, condition (a) holds as required.

4.3 Boundary cluster sets for harmonic functions

Let B denote the open unit ball of $\mathbb{R}^n$ and let $\mathcal{F}$ be some family of harmonic functions on B. A general question is as follows: which subsets E of B have the property that $\sup_E h = \sup_B h$ for all h in $\mathcal{F}$? The answer will depend, of course, on the choice of family $\mathcal{F}$. A simple case would be where $\mathcal{F} = \mathcal{H}(B)$. It is not hard to show, using Theorem 3.1 and a simple pole-pushing argument, that $\sup_E h = \sup_B h$ for all h in $\mathcal{H}(B)$ if and only if $\mathcal{A}$ (the Alexandroff point for B) is not accessible from $B \setminus \overline{E}$.

A recent result of this type concerns the class $\mathcal{D}$ of harmonic functions h on B with finite Dirichlet integral; that is,

$$\int_B \|\nabla h\|^2 \, d\lambda < +\infty.$$

Let $\mathcal{C}(\cdot)$ denote Newtonian (or logarithmic, if $n = 2$) capacity. Also, if $E \subseteq B$, let E_{NT} denote the set of points in ∂B which can be approached non-tangentially by a sequence of points in E. The following result is taken from [28]. Its holomorphic analogue is due to Stray [39], and a more general result has since been given by Böe [11].

4.5 Theorem *Let $E \subseteq B$. Then $\sup_E h = \sup_B h$ for all h in $\mathcal{D}$ if and only if $C(E_{NT}) = C(\partial B)$.*

An interesting point here is as follows. Classical boundary limit theorems such as Fatou's theorem involve the notion of non-tangential limits. On the other hand, potential theoretic limit theorems often involve the notion of fine limits. The proof of the above theorem given in [28] suggests a linkage between these two notions which we now explain.

Let $f : B \to [-\infty, +\infty]$, let $z \in \partial B$ and $l \in [-\infty, +\infty]$. We say that l is a *fine cluster value of f at z* if, for every neighbourhood N of l in $[-\infty, +\infty]$, the set $f^{-1}(N)$ is non-thin at z. The *fine cluster set of f at z* is then defined to be the set $C_F(f, z)$ of all fine cluster values of f at z. Next, the *non-tangential cluster set of f at z* is defined to be the set $C_{NT}(f, z)$ of values l in $[-\infty, +\infty]$ such that $f(x_k) \to l$ for some sequence (x_k) of points in B which approaches z non-tangentially. It is natural to ask what the relationship is between $C_{NT}(f, z)$ and $C_F(f, z)$. The proof of Theorem 4.5 can be developed (see [28] for details) to yield the following answer.

4.6 Theorem *Let $f : B \to [-\infty, +\infty]$. Then there is a Euclidean-G_δ subset A of ∂B such that $C(A) = C(\partial B)$ and*

$$C_F(f, z) \subseteq C_{NT}(f, z) \quad \text{for all } z \text{ in } A.$$

What is somewhat surprising here, and this is where harmonic approximation once again has a role to play, is that Theorem 4.6 is sharp even for harmonic functions.

4.7 Theorem *Let $A \subseteq \partial B$ be a Euclidean-G_δ set such that $C(A) = C(\partial B)$. Then there is a harmonic function h on B such that*

$$C_F(h, z) = [-\infty, +\infty] \quad \text{and} \quad C_{NT}(h, z) = \{0\}$$

whenever $z \in \partial B \setminus A$.

To see why this is so, we remark first that the capacity hypothesis on the subset A of ∂B can be shown to be equivalent to saying that A is finely dense in ∂B. Thus there is an increasing sequence (F_k) of compact sets such that $\partial B \setminus A = \bigcup_k F_k$ and $\partial B \setminus F_k$ is finely dense in ∂B for all k. We define $g_k : \partial B \to [0, 1)$ by

$$g_k(z) = \frac{\{\text{dist}(z, F_k)\}^2}{5} \quad (z \in \partial B; \ k \in \mathbb{N}).$$

Then the sets

$$E_k = \left\{ rz : z \in \partial B \setminus F_k \text{ and } 1 - r = \frac{g_k(z)}{3k - 1} \right\} \quad (k \in \mathbb{N}),$$

$$D_1 = B \cap \{rz : z \in \partial B \text{ and } 1 - r \geq g_1(z)\},$$

and

$$D_{k+1} = B \cap \left\{ rz : z \in \partial B \text{ and } \frac{g_k(z)}{3k} \geq 1 - r \geq \frac{g_{k+1}(z)}{3k + 1} \right\} \quad (k \in \mathbb{N})$$

are all closed relative to B and are pairwise disjoint. (See Figure 6, which is only indicative since the sets F_k are actually nowhere dense.) The set

$$E = \left(\bigcup_k E_k\right) \cup \left(\bigcup_k D_k\right)$$

is also a relatively closed subset of B. We define $u : E \to \mathbb{R}$ by

$$u(x) = \begin{cases} 0 & \text{if } x \in \bigcup_k D_k \\ q_k & \text{if } x \in E_k; \ k \geq 1, \end{cases}$$

where $\{q_k : k \in \mathbb{N}\} = \mathbb{Q}$. Clearly $B^* \setminus E$ is connected and locally connected, so we can apply (8) to see that there is a harmonic h function on B such that

$$|h(x) - u(x)| < 1 - \|x\| \quad (x \in E).$$

Any sequence in B which approaches a point of $\partial B \setminus A$ non-tangentially has all but finitely many points in some D_k, so $C_{\text{NT}}(h, z) = \{0\}$ whenever $z \in \partial B \setminus A$. On the other hand, if $z \in \partial B \setminus A$, then E_k is non-thin at z for all sufficiently large k (because it is a slightly distorted form of the set $\partial B \setminus F_k$), so $C_F(h, z)$ contains all but finitely many rationals and hence $C_F(h, z) = [-\infty, +\infty]$. Thus Theorem 4.7 holds.

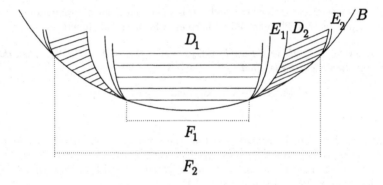

Figure 6

References

[1] N. U. Arakelyan, Uniform and tangential approximations by analytic functions, *Izv. Akad. Nauk Armjan. SSR Ser. Mat.* **3** (1968), 273-286 (Russian); English translation, *Amer. Math. Soc. Transl.* (2) **122** (1984), 85-97.

[2] N. U. Arakelyan, Approximation complexe et propriétés des fonctions analytiques, *Actes, Congrès intern. Math.* (1970), Tome 2, 595-600.

[3] D. H. Armitage, Uniform and tangential harmonic approximation, *These proceedings*.

[4] D. H. Armitage and S. J. Gardiner, *Classical Potential Theory*, Springer, London, 2001.

[5] D. H. Armitage and M. Goldstein, Tangential harmonic approximation on relatively closed sets, *Proc. London Math. Soc.* (2) **68** (1994), 112–126.

[6] M. G. Arsove, The Lusin-Privalov theorem for subharmonic functions, *Proc. London Math. Soc.* (3) **14** (1964), 260–270.

[7] T. Bagby and P. M. Gauthier, Approximation by harmonic functions on closed subsets of Riemann surfaces, *J. Anal. Math.* **51** (1988), 259–284.

[8] T. Bagby and P. M. Gauthier, Harmonic approximation on closed subsets of Riemannian manifolds, in: *Complex Potential Theory* (P. M. Gauthier, ed.), NATO ASI Ser. C Math. Phys. Sci. **439**, Kluwer, Dordrecht, 1994; 75–87.

[9] T. Bagby, P. M. Gauthier and J. Woodworth, Tangential harmonic approximation on Riemannian manifolds, in: *Harmonic Analysis and Number Theory* (S. W. Drury and M. Ram Murty, eds.), CMS Conf. Proc. **21**, Amer. Math. Soc., Providence, RI, 1997; 58–72.

[10] R. D. Berman, A converse to the Lusin-Privalov radial uniqueness theorem, *Proc. Amer. Math. Soc.* **87** (1983), 103–106.

[11] B. Böe, Sets of determination for smooth harmonic functions, *preprint*.

[12] M. Brelot, Sur l'approximation et la convergence dans la théorie des fonctions harmoniques ou holomorphes, *Bull. Soc. Math. France* **73** (1945), 55–70.

[13] T. Carleman, Sur un théorème de Weierstrass, *Ark. Mat. Astronom. Fys.* **20B** (1927), 1–5.

[14] Chen Huaihui and P. M. Gauthier, A maximum principle for subharmonic and plurisubharmonic functions, *Canad. Math. Bull.* **35** (1992), 34–39.

[15] A. Debiard and B. Gaveau, Potentiel fin et algèbres de fonctions analytiques I, *J. Funct. Anal.* **16** (1974), 289–304.

[16] J. Deny, Sur l'approximation des fonctions harmoniques, *Bull. Soc. Math. France* **73** (1945), 71–73.

[17] J. Deny, Systèmes totaux de fonctions harmoniques, *Ann. Inst. Fourier (Grenoble)* **1** (1949), 103–113.

[18] M. R. Essén and S. J. Gardiner, Limits along parallel lines and the classical fine topology, *J. London Math. Soc.* (2) **59** (1999), 881–894.

[19] B. Fuglede, *Finely Harmonic Functions*, Lecture Notes in Math. **289**, Springer, Berlin, 1972.

[20] B. Fuglede, Asymptotic paths for subharmonic functions, *Math. Ann.* **213** (1975), 261–274.

[21] B. Fuglede, Fine potential theory, *Mitt. Math. Ges. DDR* **2–3** (1986), 3–21.

[22] S. J. Gardiner, The Dirichlet problem with non-compact boundary, *Math. Z.* **213** (1993), 163–170.

[23] S. J. Gardiner, Superharmonic extension and harmonic approximation, *Ann. Inst. Fourier (Grenoble)*, **44** (1994), 65–91.

[24] S. J. Gardiner, Tangential harmonic approximation on relatively closed sets, *Illinois J. Math.* **39** (1995), 143–157.

[25] S. J. Gardiner, *Harmonic Approximation*, London Math. Soc. Lecture Note Ser. **221**, Cambridge Univ. Press, Cambridge, 1995.

[26] S. J. Gardiner, The Lusin-Privalov theorem for subharmonic functions, *Proc. Amer. Math. Soc.* **124** (1996), 3721–3727.

[27] S. J. Gardiner, Decomposition of approximable harmonic functions, *Math. Ann.* **308** (1997), 175–185.

[28] S. J. Gardiner, Non-tangential limits, fine limits and the Dirichlet integral, to appear in *Proc. Amer. Math. Soc.*

[29] S. J. Gardiner and M. Goldstein, Carleman approximation by harmonic functions, *Amer. J. Math.* **117** (1995), 245–255.

[30] S. J. Gardiner and W. Hansen, Boundary sets where harmonic functions may become infinite, preprint.

[31] P. M. Gauthier and S. Ladouceur, Uniform approximation and fine potential theory, *J. Approx. Theory* **72** (1993), 138–140.

[32] M. V. Keldyš, On the solvability and stability of the Dirichlet problem, *Uspekhi Mat. Nauk* **8** (1941), 171–231 (Russian); English translation *Amer. Math. Soc. Transl.* **51** (1966), 1–73.

[33] M. Labrèche, *De l'approximation harmonique uniforme*, Doctoral thesis, Université de Montréal, 1982.

[34] N. N. Lusin and I. I. Privalov, Sur l'unicité et la multiplicité des fonctions analytiques, *Ann. Sci. École Norm. Sup.* (3) **42** (1925), 143–191.

[35] A. A. Nersesyan, Carleman sets, *Izv. Akad. Nauk. Armjan. SSR Ser. Mat.* **6** (1971), 465–471 (Russian); English translation *Amer. Math. Soc. Transl.* (2) **122** (1984), 99–104.

[36] R. Nevanlinna, Über eine Erweiterung des Poissonschen Integrals, *Ann. Acad. Sci. Fenn. Ser. A* **24** (4) (1925), 1–15.

[37] P. J. Rippon, The boundary cluster sets of subharmonic functions, *J. London Math. Soc.* (2) **17** (1978), 469–479.

[38] A. Stray, Decomposition of approximable functions, *Ann. Math.* **120** (1984), 225–235.

[39] A. Stray, Simultaneous approximation in the Dirichlet space, to appear in *Math. Scand.*

Jensen measures

Thomas J. RANSFORD*

*Département de mathématiques et de statistique
Université Laval
Québec, Qué., G1K 7P4
Canada*

Abstract

In these notes we describe an approach to potential theory via duality. The key idea is the notion of Jensen measure, an abstraction of the sub-mean-value property of subharmonic functions.

1 Introduction

Duality plays a fundamental role in analysis. To take just one example pertinent to this conference, it provides a simple and elegant method for proving approximation theorems. The purpose of these notes is to describe a duality approach to potential theory and subharmonic functions. The basic ideas go back at least to the 1960's, but there has recently been a renewed burst of activity, both in the theory and the applications.

We shall study objects dual to subharmonic functions that are called Jensen measures.

1.1 Definition Let Ω be an open subset of $\mathbf{R}^d$, and let $x \in \Omega$. A *Jensen measure for x on* Ω is a Borel probability measure μ, supported on a compact subset of Ω, such that every subharmonic function u on Ω satisfies

$$u(x) \leq \int u \, d\mu.$$

For example, if B is a closed ball in Ω with centre x, then normalized Lebesgue measure on B is a Jensen measure for x, as is normalized surface measure on ∂B. This follows directly from the sub-mean-value property of subharmonic functions. Another simple example is $\mu = \delta_x$, the Dirac measure at x. Indeed, as we shall see, Jensen measures abound. The problem is not to produce them, but rather the reverse, namely to try to control them and, if possible, classify them.

Why are Jensen measures useful? As an illustration, let us consider the problem of calculating so-called subharmonic envelopes.

---

*Research supported by a grant from NSERC.

N. Arakelian and P.M. Gauthier (eds.), Approximation, Complex Analysis, and Potential Theory, 221–237.
© 2001 *Kluwer Academic Publishers. Printed in the Netherlands.*

1.2 Definition Given a function $\varphi: \Omega \to [-\infty, \infty]$, its *subharmonic envelope* is

$$S\varphi(x) := \sup\{u(x) : u \in \mathrm{SH}(\Omega), u \leq \varphi\} \qquad (x \in \Omega),$$

where $\mathrm{SH}(\Omega)$ denotes the family of subharmonic functions on Ω.

This quantity, which arises all over the place in potential theory, is sufficiently venerable to go also by various French names, such as the *réduite* or the *balayage* of φ. It is intimately connected to the Perron method for solving the Dirichlet problem, itself one of the reasons that subharmonic functions were first introduced. How can we calculate it? This is where Jensen measures come in.

1.3 Definition Given a function $\varphi: \Omega \to [-\infty, \infty]$, its *Jensen envelope* is

$$J\varphi(x) := \inf\left\{\int \varphi \, d\mu : \mu \in J_x(\Omega)\right\} \qquad (x \in \Omega),$$

where $J_x(\Omega)$ denotes the family of Jensen measures for x on Ω.

Note that, in order for this definition to make sense, φ has to be sufficiently 'nice' for all the integrals $\int \varphi \, d\mu$ to exist. Assuming this to be the case, the following result is then nearly obvious.

1.4 Proposition $S\varphi(x) \leq J\varphi(x) \qquad (x \in \Omega)$.

Proof Let $u \in \mathrm{SH}(\Omega)$ with $u \leq \varphi$, and let $\mu \in J_x(\Omega)$. Then

$$u(x) \leq \int u \, d\mu \leq \int \varphi \, d\mu.$$

The result follows upon taking the sup over u and the inf over μ. $\qquad\qquad\square$

What is remarkable is that, under very general conditions on φ, the inequality in Proposition 1.4 actually becomes an equality. Results of this kind are often called duality theorems, for reasons that will soon become apparent. One of the purposes of these lectures is to show how to prove them, and how to exploit them.

2   An abstract duality theorem

We shall begin by proving a duality theorem in an abstract setting. This has the advantage that it will allow us to treat simultaneously subharmonic and plurisubharmonic functions, as well as being applicable to many other situations. Equally importantly, it also brings out clearly the duality aspect.

Fix the following notation. Let X be a compact metric space, and let R be a family of continuous functions from X into $[-\infty, \infty)$ such that:

- $u, v \in R \Rightarrow u + v \in R$;

- $u \in R$, $\lambda > 0 \Rightarrow \lambda u \in R$;

- R separates points of X and contains the constants.

Given $x \in X$, an *R-measure for x* is a Borel probability measure μ on X such that

$$u(x) \le \int u \, d\mu \qquad (u \in R).$$

We write I_x for the set of all R-measures for x. Given $\varphi : X \to [-\infty, \infty]$, we define

$$R\varphi(x) = \sup \{u(x) : u \in R, \ u \le \varphi\} \qquad (x \in X),$$

$$I\varphi(x) = \inf \left\{ \int \varphi \, d\mu : \mu \in I_x \right\} \qquad (x \in X).$$

The following duality theorem is due to Edwards [E].

2.1 Theorem *Let $\varphi \colon X \to (-\infty, \infty]$ be a lower semicontinuous function. Then*

$$R\varphi(x) = I\varphi(x) \qquad (x \in X).$$

Proof As in Proposition 1.4, it is clear that $R\varphi(x) \le I\varphi(x)$ for all $x \in X$. The problem is to prove the reverse inequality.

Suppose first that $\varphi \in C(X)$, i.e. that φ is a continuous real-valued function on X. Suppose, for a contradiction, that $R\varphi(x) < I\varphi(x)$ for some $x \in X$. Adding a constant to φ, if necessary, we may assume without loss of generality that in fact $R\varphi(x) < 0 < I\varphi(x)$.

Define

$$C = \{f \in C(X) : \exists \, u \in R, \ u(x) = 0, \ u < f\}.$$

Then C is an open convex cone in $C(X)$, and $\varphi \notin C$ since $R\varphi(x) < 0$. Hence, by the Hahn-Banach theorem and the Riesz representation theorem, there exists a signed Borel measure μ on X such that

$$\int \varphi \, d\mu \le 0 < \int f \, d\mu \qquad (f \in C). \tag{1}$$

In particular, since C contains all the positive, continuous functions on X, it follows that μ must be a positive measure. Multiplying by a constant, we can suppose that μ is a probability measure.

Let us now show that μ is in fact an R-measure for x. Let $u \in R$. For $\epsilon > 0$, define $f = u - u(x) + \epsilon$. Then $f \in C$, so $\int f \, d\mu \ge 0$, whence $u(x) \le \int u \, d\mu + \epsilon$, and thus, letting $\epsilon \to 0$, we finally get $u(x) \le \int u \, d\mu$, as desired. (This argument breaks down if $u(x) = -\infty$, but in that case the final inequality is obvious anyway.)

The inequality (1) now implies that $I\varphi(x) \le 0$, which gives the required contradiction. This proves that $I\varphi(x) \le R\varphi(x)$ in the case that φ is continuous.

Now suppose that φ is merely lower semicontinuous. Let (φ_n) be a sequence in $C(X)$ which increases pointwise to φ. By what we have already proved, for each n, there exists $\mu_n \in I_x$ such that

$$\int \varphi_n \, d\mu_n < R\varphi_n(x) + \frac{1}{n}.$$

If $m \le n$, then

$$\int \varphi_m \, d\mu_n \le \int \varphi_n \, d\mu_n \le R\varphi_n(x) + \frac{1}{n} \le R\varphi(x) + \frac{1}{n}.$$

A subsequence of the (μ_n) converges weak* to another $\mu \in I_x$. Letting $n \to \infty$ through this subsequence, and then $m \to \infty$, we obtain

$$\int \varphi \, d\mu \leq R\varphi(x).$$

Hence, once again, we have $I\varphi(x) \leq R\varphi(x)$.   □

The reader is invited to experiment to see what Theorem 2.1 tells us in the special case when $\varphi = -1_F$, where F is a closed subset of X.

One might reasonably ask if Theorem 2.1 remains true if φ is *upper* semicontinuous. The answer in general is no. Indeed it is false whenever R contains a decreasing sequence (u_n) whose limit is discontinuous (this will be true in most of the cases of interest). For if we set $\varphi = \lim_n u_n$, then φ is upper semicontinuous and satisfies $I\varphi = \varphi$. However $\varphi \neq R\varphi$, because $R\varphi$, being the supremum of a family of continuous functions, is lower semicontinuous, whereas φ is not.

It is possible to take Theorem 2.1 much further, for example to prove a generalization of Choquet's theorem. For more details, see for example Chapter 1 of Gamelin's book [Gm]. The same book describes a development of a very general version of potential theory on uniform algebras, due to Gamelin and Sibony [GS], based in part on Theorem 2.1.

However, in these notes we shall restrict ourselves to applications in classical potential theory and pluripotential theory. Our next task is to make the connection between the abstract and concrete theories.

3   Duality theorems for (pluri)subharmonic functions

Let Ω be an open subset of $\mathbf{R}^d$. Recall that, given $\varphi \colon \Omega \to [-\infty, \infty]$, we write

$$S\varphi(x) = \sup\{u(x) : u \in \mathrm{SH}(\Omega),\ u \leq \varphi\} \qquad (x \in \Omega),$$

$$J\varphi(x) = \inf\left\{ \int \varphi \, d\mu : \mu \in J_x(\Omega) \right\} \qquad (x \in \Omega),$$

where $\mathrm{SH}(\Omega)$ denotes the family of subharmonic functions on Ω, and $J_x(\Omega)$ the family of Jensen measures for x on Ω.

The following duality theorem has been proved and re-proved many times. The first reference that I have been able to track down is in a paper of Koosis [Ko1, p. 77], where it was proved for the case $\Omega = \mathbf{R}^2$ (by a different method from that below).

3.1 Theorem *Let* $\varphi \colon \Omega \to \mathbf{R}$ *be a continuous function. Then*

$$S\varphi(x) = J\varphi(x) \qquad (x \in \Omega).$$

Here is the basic idea. Let X be a compact subset of Ω, and define

$$R = \{u|_X : u \in \mathrm{SH}(\Omega) \cap C(\Omega)\},$$

so that Theorem 2.1 applies. However, in order to gain useful information from this theorem, we need to relate the R-measures on X to the Jensen measures on Ω. The following lemma solves the problem.

3.2 Lemma *With the above notation*

$$I_x = \{\mu \in J_x(\Omega) : \operatorname{supp}\mu \subset X\} \qquad (x \in X).$$

Proof Let $x \in X$ and let $\mu \in J_x(\Omega)$ with $\operatorname{supp}\mu \subset X$. Given $v \in R$, we have $v = u|_X$ for some $u \in \operatorname{SH}(\Omega) \cap C(\Omega)$, so $v(x) = u(x) \leq \int_X u\,d\mu = \int v\,d\mu$. Hence $\mu \in I_x$.

Conversely, suppose that $\mu \in I_x$. Let $u \in \operatorname{SH}(\Omega)$. There exist $u_n \in \operatorname{SH}(\Omega) \cap C(\Omega)$ such that $u_n \downarrow u$ on X. (This uses a non-trivial approximation theorem [Gr, Theorem 6.1]). Put $v_n = u_n|_X$. Then $v_n \in R$, so $v_n(x) \leq \int v_n\,d\mu$. Letting $n \to \infty$, we deduce that $u(x) \leq \int u\,d\mu$. Hence $\mu \in J_x(\Omega)$. □

Proof of Theorem 3.1 Let (X_n) be a compact exhaustion of Ω. Define

$$R_n = \{u|_{X_n} : u \in \operatorname{SH}(\Omega) \cap C(\Omega)\},$$

and denote by $R_n\varphi$ and $I_n\varphi$ the corresponding envelopes of $\varphi|_{X_n}$. By Theorem 2.1 we have $R_n\varphi(x) = I_n\varphi(x)$ for all $x \in X_n$. So all we need to do is to prove that

$$\limsup_{n\to\infty} R_n\varphi(x) \leq S\varphi(x) \quad \text{and} \quad \liminf_{n\to\infty} I_n\varphi(x) \geq J\varphi(x) \qquad (x \in \Omega).$$

Given $x \in \Omega$, we eventually have $x \in X_n$, so there exists $u_n \in \operatorname{SH}(\Omega)$ such that $u_n \leq \varphi$ on X_n and $u_n(x) > R_n\varphi(x) - 1/n$. Put

$$u(y) = \limsup_{n\to\infty} u_n(y) \quad \text{and} \quad u^*(y) = \limsup_{z\to y} u(z) \qquad (y \in \Omega).$$

Then $u^* \in \operatorname{SH}(\Omega)$ and $u^* \leq \varphi$. It follows that $\limsup_{n\to\infty} R_n\varphi(x) \leq u^*(x) \leq S\varphi(x)$.

For the other limit, we use Lemma 3.2. By that lemma, if $x \in X_n$ then

$$I_n\varphi(x) = \inf\left\{\int \varphi\,d\mu : \mu \in J_x(\Omega),\ \operatorname{supp}\mu \subset X_n\right\}.$$

From this it is clear that $\lim_{n\to\infty} I_n\varphi(x) = J\varphi(x)$. □

How much of this goes through for plurisubharmonic functions? Nearly everything! Let Ω now be an open subset of $\mathbf{C}^d$. Denote by $\operatorname{PSH}(\Omega)$ the family of plurisubharmonic functions on Ω, and by $\operatorname{PJ}_x(\Omega)$ the family of pluri-Jensen measures for x on Ω (the obvious analogue of Jensen measures). For $\varphi \colon \to [-\infty, \infty]$, define

$$\operatorname{PS}\varphi(x) = \sup\{u(x) : u \in \operatorname{PSH}(\Omega),\ u \leq \varphi\} \qquad (x \in \Omega),$$

$$\operatorname{PJ}\varphi(x) = \inf\left\{\int \varphi\,d\mu : \mu \in \operatorname{PJ}_x(\Omega)\right\} \qquad (x \in \Omega).$$

3.3 Theorem *Assume that Ω is pseudoconvex. Let $\varphi \colon \Omega \to \mathbf{R}$ be continuous. Then*

$$\operatorname{PS}\varphi(x) = \operatorname{PJ}\varphi(x) \qquad (x \in \Omega).$$

Proof This is virtually identical to the proof of Theorem 3.1. There is only one tricky point. In the course of the proof of Lemma 3.2, we needed to know that, given $u \in \operatorname{SH}(\Omega)$, there exist $u_n \in \operatorname{SH}(\Omega) \cap C(\Omega)$ such that $u_n \downarrow u$ on X. The analogous result for plurisubharmonic functions is actually false in general, but it *is* true if Ω is pseudoconvex (see [FN, Theorem 5.5]). This is why we had to make that extra assumption. □

4   An application to entire functions

It is well known that an entire function with a given set of zeros must grow at at least a
certain rate. How fast? We shall see that, using Jensen measures, it is possible to give a
surprisingly precise answer to this question.

Let g be an entire function with $g \not\equiv 0$. Let $M \colon \mathbf{C} \to \mathbf{R}$ be a continuous function.
When does there exist an entire function $f \not\equiv 0$, whose zero set includes that of g (including
multiplicities), and such that $\log |f(z)| \le M(z)$ for all $z \in \mathbf{C}$?

If such an f exists, then it can be chosen so that $(f/g)(0) \neq 0$ (just replace f by $Cf(z)/z^k$
for appropriate constants C, k). The function $u = \log |f/g|$ is then subharmonic on $\mathbf{C}$ with
$u(0) > -\infty$, so for every Jensen measure μ for 0 we have

$$-\infty < u(0) \le \int u \, d\mu \le \int (M - \log |g|) \, d\mu.$$

In particular

$$\inf_{\mu \in J_0(\mathbf{C})} \int (M - \log |g|) \, d\mu > -\infty. \tag{2}$$

Remarkably, this necessary condition for the existence of f turns out to be very nearly
sufficient. This is the content of the following theorem of Khabibullin [Kh, §2, p. 1069].

4.1 Theorem *If (2) holds then, for each $\delta > 0$, there exists an entire function $f \not\equiv 0$, whose
zero set includes that of g, and which satisfies*

$$\log |f(z)| \le \max_{|\zeta - z| \le 2\delta} M(\zeta) + 3 \log(1 + |z|^2) \qquad (z \in \mathbf{C}).$$

We remark that, at the cost of some extra work, it is possible to go even further.
Khabibullin showed that in fact the term $3 \log(1 + |z|^2)$ can be eliminated. Also, it is possi-
ble to weaken somewhat the hypothesis of continuity on M. We shall follow the simplified
treatment given in Koosis's book [Ko2, Chapitre III].

The proof makes use of the following result of Lelong and Gruman [LG, Theorem 7.1]. To
give the proof, which is based on Hörmander's L^2-method for solving the $\bar{\partial}$-equation, would
take us too far afield, so we content ourselves with the statement of the theorem and refer
the reader to [LG, p. 167] or to [Ko2, p. 91] for the details of the proof. In what follows, m
denotes Lebesgue measure.

4.2 Theorem *Let u be a subharmonic function on $\mathbf{C}$ with $u \not\equiv -\infty$. Then there exists an
entire function $h \not\equiv 0$ such that*

$$\int_{\mathbf{C}} \frac{|h(z)|^2 e^{-u(z)}}{(1 + |z|^2)^3} \, dm(z) < \infty.$$

Proof of Theorem 4.1 Let ρ denote Lebesgue measure on the disk $\{|z| \le \delta\}$, normalized
to be a probability measure. Note that $\rho \in J_0(\mathbf{C})$. Define $\varphi \colon \mathbf{C} \to \mathbf{R}$ by

$$\varphi = (M - \log |g|) * \rho,$$

where $*$ denotes convolution. Then φ is continuous, and for each $\mu \in J_0(\mathbf{C})$ we have

$$\int \varphi \, d\mu = \int (M - \log|g|) \, d(\rho * \mu).$$

One easily checks that $\mu \in J_0(\mathbf{C}) \Rightarrow \rho * \mu \in J_0(\mathbf{C})$. It therefore follows from (2) that

$$\inf_{\mu \in J_0(\mathbf{C})} \int \varphi \, d\mu > -\infty.$$

In other words, $J\varphi(0) > -\infty$. Applying Theorem 3.1, we deduce that $S\varphi(0) > -\infty$. Thus there exists at least one subharmonic function u on $\mathbf{C}$ such that $u \leq \varphi$ and $u \not\equiv -\infty$. By Theorem 4.1, there exists an entire function $h \not\equiv 0$ such that

$$\int_{\mathbf{C}} \frac{|h(z)|^2 e^{-u(z)}}{(1 + |z|^2)^3} \, dm(z) < \infty.$$

Set $f = Cgh^2$, where C is a positive constant. We claim that, if C is sufficiently small, then this does the trick. Clearly f is entire, $f \not\equiv 0$ and the zero set of f includes that of g. It remains to prove the bound on $\log|f|$. For this, note that since $|f|$ is subharmonic on $\mathbf{C}$,

$$|f(z)| \leq \frac{1}{\pi\delta^2} \int_{|\zeta - z| \leq \delta} |f(\zeta)| \, dm(\zeta) \leq \frac{C}{\pi\delta^2} \int_{|\zeta - z| \leq \delta} |g(\zeta)||h(\zeta)|^2 \, dm(\zeta).$$

Now the fact that $u \leq \varphi$ translates into the inequality $(\log|g|) * \rho \leq M * \rho - u$. Furthermore, since $\log|g|$ is subharmonic, we have $\log|g| \leq (\log|g|) * \rho$. Hence $|g| \leq e^{M*\rho} e^{-u}$. Substituting this into the inequality for $|f(z)|$, we obtain

$$|f(z)| \leq \frac{C}{\pi\delta^2} \int_{|\zeta - z| \leq \delta} e^{M*\rho(\zeta)} e^{-u(\zeta)} |h(\zeta)|^2 \, dm(\zeta)$$

$$\leq \frac{C}{\pi\delta^2} \int_{\mathbf{C}} \frac{|h(\zeta)|^2 e^{-u(\zeta)}}{(1 + |\zeta|^2)^3} \, dm(\zeta) \left\{ \max_{|\zeta - z| \leq \delta} e^{M*\rho(\zeta)} (1 + |\zeta|^2)^3 \right\}$$

$$\leq CC' \max_{|\zeta - z| \leq 2\delta} e^{M(\zeta)} (1 + |z|^2)^3,$$

where C' is a constant independent of z. It thus suffices to take $C \leq 1/C'$. $\qquad\square$

Theorem 4.1, amazingly precise though it is, does beg the question: how to check (2)? Indeed, just which measures μ *are* Jensen measures? This is the next topic.

5 Harmonic measures

Let Ω be an open subset of $\mathbf{R}^d$, and let $x \in \Omega$. Recall that $J_x(\Omega)$ denotes the set of Jensen measures for x on Ω. Our aim in this section is to try to describe the elements of $J_x(\Omega)$ in terms of something more familiar.

We mentioned in the introduction that, if B is a closed ball in Ω with centre x, then the normalized surface measure on ∂B is a Jensen measure for x. More generally, given any domain $D \subset\subset \Omega$ containing x, if ω is the harmonic measure for D at x, then every subharmonic function u on Ω satisfies

$$u(x) \leq \int_{\partial D} u \, d\omega,$$

and so ω too is a Jensen measure for x. At this point it is convenient to introduce a notation.

5.1 Definition $H_x(\Omega)$ is the set of harmonic measures ω for D at x, as D runs through the domains such that $x \in D \subset\subset \Omega$.

The discussion above shows that $H_x(\Omega) \subset J_x(\Omega)$. Evidently, $J_x(\Omega)$ contains other measures as well, but the following result shows that, at least for the purposes of computing Jensen hulls, harmonic measures suffice.

5.2 Theorem *Let* $\varphi \colon \Omega \to \mathbf{R}$ *be a continuous function. Then*

$$J\varphi(x) = \inf\left\{ \int \varphi \, d\omega : \omega \in H_x(\Omega) \right\} \qquad (x \in \Omega).$$

Proof This is a special case of [CR2, Theorem 1.3] (see also Theorem 8.6 below). □

Thus, for example, in the duality theorem 3.1, and its applications, it is possible to replace Jensen measures by harmonic measures. This goes some way, at least, towards answering the question posed at the end of the previous section.

Let us explore further the consequences of this result. If μ is normalized Lebesgue measure on a (solid) ball around x, then μ is a Jensen measure for x, but obviously not a harmonic measure. Notice, however, that it can still be built up from harmonic measures (e.g. Lebesgue measures on spheres around x) by taking convex combinations of them and then proceeding to a limit. The next theorem shows that the same is possible for every Jensen measure.

Here and in what follows, we shall consider $J_x(\Omega)$ as a subset of the dual of $C(\Omega)$, endowed with the weak*-topology. We write $\overline{\mathrm{conv}}$ to denote the closed convex hull.

5.3 Theorem $J_x(\Omega) = \overline{\mathrm{conv}}\,(H_x(\Omega))$.

Proof It is easy to check that $J_x(\Omega)$ is convex and weak*-closed in $C(\Omega)^*$. Therefore it certainly contains $\overline{\mathrm{conv}}\,(H_x(\Omega))$. To prove equality, suppose the contrary, say $\mu \in J_x(\Omega)$ but $\mu \notin \overline{\mathrm{conv}}\,(H_x(\Omega))$. Then, by the Hahn-Banach theorem, there exists $\varphi \in C(\Omega)$ with

$$\int \varphi \, d\mu < \inf_{\omega \in H_x(\Omega)} \int \varphi \, d\omega.$$

This contradicts Theorem 5.2. □

Given that $J_x(\Omega)$ is convex, what can we say about $\mathrm{ext}(J_x(\Omega))$, its set of extreme points? The next result partially identifies them. As before, closures are taken with respect to the weak*-topology on the dual of $C(\Omega)$.

5.4 Theorem $H_x(\Omega) \subset \mathrm{ext}(J_x(\Omega)) \subset \overline{H_x(\Omega)}$.

Proof The right-hand inclusion is more or less a consequence of Theorem 5.3. The left-hand inclusion follows from the fact that, if $\omega \in H_x(\Omega)$, $\mu \in J_x(\Omega)$ and $\operatorname{supp}\mu \subset \operatorname{supp}\omega$, then $\int u\, d\omega \leq \int u\, d\mu$ for all $u \in \mathrm{SH}(\Omega)$. For more details see [CR2, Theorems 6.2 and 6.6]. □

Perhaps disappointingly, if $d \geq 2$, then both the inclusions in the preceding theorem turn out to be strict. In the case of the left-hand inclusion, this is obvious, since $\delta_x \in \operatorname{ext}(J_x(\Omega))$ but $\delta_x \notin H_x(\Omega)$. But the following result shows that there are also some more surprising examples.

5.5 Theorem *For each $d \geq 2$, there exists $\mu \in \operatorname{ext}(J_0(\mathbf{R}^d))$ such that $\operatorname{supp}\mu$ is the whole of the unit ball.*

Proof See [CR2, Theorem 6.5]. □

We conclude this section with an open problem.

5.6 Question Is $J_x(\Omega) = \overline{H_x(\Omega)}$?

In view of Theorem 5.3, this amounts to asking whether $\overline{H_x(\Omega)}$ is convex. In concrete terms, given two harmonic measures $\omega, \omega' \in H_x(\Omega)$, can we always find a sequence (ω_n) in $H_x(\Omega)$ such that $\omega_n \to (\omega + \omega')/2$? Observe that, if this were true, then by Theorem 5.4 $J_x(\Omega)$ would have the curious property of being equal to the closure of its extreme points.

6 Analytic-disk measures

Although Theorem 4.1 was stated and proved for entire functions on $\mathbf{C}$, it extends without difficulty to entire functions on $\mathbf{C}^d$ for $d \geq 2$. The condition (2) now becomes

$$\inf_{\mu \in \mathrm{PJ}_0(\mathbf{C})} \int (M - \log|g|)\, d\mu > -\infty,$$

where we have replaced Jensen measures by pluri-Jensen measures. This raises the problem of characterizing pluri-Jensen measures in terms of something more familiar, just as we did for Jensen measures in the previous section.

Harmonic measures are less appropriate in this context, because the analogue of the Dirichlet problem, the Monge-Ampère equation, is now non-linear. Instead, it seems more productive to exploit the link with complex analysis, and to consider so-called analytic-disk measures. These measures were introduced by Poletsky in the late 1980's, with the aim of developing a theory of plurisubharmonic functions on compacta.

Let Ω be an open subset of $\mathbf{C}^d$ and let $x \in \Omega$. Recall that $\mathrm{PJ}_x(\Omega)$ denotes the set of pluri-Jensen measures for x on Ω. We also write Δ for the open unit disc in $\mathbf{C}$, and $\mathbf{T}$ for the unit circle.

6.1 Definition An *analytic disk in Ω with centre x* is a continuous function $f \colon \overline{\Delta} \to \Omega$ such that f analytic on Δ and $f(0) = x$. A Borel measure ν on Ω is called an *analytic-disk measure for x* if it is of the form

$$\nu(E) = \lambda\left(f^{-1}(E)\right) \qquad (\text{Borel } E \subset \Omega),$$

where f is an analytic disk in Ω with centre x, and λ is normalized Lebesgue measure on the unit circle. We write $A_x(\Omega)$ for the set of analytic-disk measures for x on Ω.

It is easy to check that analytic-disk measures are pluri-Jensen measures. Indeed, let ν be as in Definition 6.1 and let u be a plurisubharmonic function on Ω. Then $u \circ f$ is upper semicontinuous on $\overline{\Delta}$ and subharmonic on Δ. Hence, for $0 < r < 1$,

$$u(x) = u \circ f(0) \leq \frac{1}{2\pi} \int_0^{2\pi} u \circ f(re^{i\theta}) \, d\theta.$$

Letting $r \to 1$ and using Fatou's lemma, we deduce that

$$u(x) \leq \frac{1}{2\pi} \int_0^{2\pi} u \circ f(e^{i\theta}) \, d\theta = \int u \, d\nu,$$

as claimed.

The following result is the plurisubharmonic analogue of Theorem 5.2. It is implicit in the work of Poletsky [P1, P2] and of Bu-Schachermayer [BS], and some further developments can be found in the recent articles [P3, P4].

6.2 Theorem *Let $\varphi \colon \Omega \to \mathbf{R}$ be a continuous function. Then*

$$\mathrm{PJ}\,\varphi(x) = \inf\left\{\int \varphi \, d\nu : \nu \in A_x(\Omega)\right\} \qquad (x \in \Omega).$$

Proof See for example [BS, Proposition III.3]. $\qquad\qquad\qquad\qquad\qquad\qquad\qquad\qquad$ □

Just as before, we can apply a Hahn-Banach argument to deduce that $\mathrm{PJ}_x(\Omega)$ is the weak*-closed convex hull of $A_x(\Omega)$. However, this time we can go one step further and give an affirmative answer to the analogue of Question 5.6. The following result is due to Bu and Schachermayer [BS, Theorem (B)].

6.3 Theorem $\mathrm{PJ}_x(\Omega) = \overline{A_x(\Omega)}$.

Proof All that is missing is to show that $\overline{A_x(\Omega)}$ is convex. This is done in [BS, Lemma III.2].
$\qquad\qquad\qquad\qquad\qquad\qquad\qquad\qquad\qquad\qquad\qquad\qquad\qquad\qquad\qquad\qquad\qquad\qquad$ □

On the other hand, the analogue of Theorem 5.4 is false: $A_x(\Omega) \not\subset \mathrm{ext}(\mathrm{PJ}_x(\Omega))$ (see [CR2, Proposition 7.3]). Just which pluri-Jensen measures are extreme remains unknown. The next section presents one approach to this problem.

7 Extreme measures and approximation

We shall work, once again, in the abstract setting of Section 2. Thus, throughout this section, X denotes a compact metric space, and R is a family of continuous functions from X into $[-\infty, \infty)$ such that:

- $u, v \in R \implies u + v \in R$;

- $u \in R, \lambda > 0 \implies \lambda u \in R$;

- R separates points of X and contains the constants.

We recall that, given $x \in X$, an *R-measure for x* is a Borel probability measure μ on X such that

$$u(x) \leq \int u \, d\mu \quad (u \in R).$$

As before, we write I_x for the set of all R-measures for x. This is a weak*-compact, convex subset of $C(X)^*$. We denote by $\text{ext}(I_x)$ its set of extreme points. The main result of this section is that the measures in $\text{ext}(I_x)$ are characterized by an approximation property.

7.1 Theorem *Let $x \in X$ and let $\mu \in I_x$. Then $\mu \in \text{ext}(I_x)$ if and only if, given $f \in C(X)$ and $\epsilon > 0$, there exist $u, v \in R$ such that*

$$\int u \, d\mu < u(x) + \epsilon, \quad \int v \, d\mu < v(x) + \epsilon \quad and \quad \int |f - (u - v)| \, d\mu < \epsilon. \tag{3}$$

Note that if $\mu \in I_x$ and if $w \in R$ with $w(x) > -\infty$, then automatically $w \in L^1(\mu)$. Thus the difference $u - v$ in the statement above makes sense. As this theorem has not been published elsewhere, we give the details of the proof here.

Proof We begin with the 'if'. Suppose that $\mu = (\mu_1 + \mu_2)/2$, where $\mu_1, \mu_2 \in I_x$. Let $f \in C(X)$, let $\epsilon > 0$, and take $u, v \in R$ satisfying (3). Using the elementary identity

$$|a - b| + (a + b) = 2 \max(a, b) \quad (a, b \in \mathbf{R}), \tag{4}$$

we have

$$\int |f - (u - v)| \, d\mu + \int (f + u + v) \, d\mu = 2 \int \max(f + v, u) \, d\mu$$
$$\geq \int (f + v) \, d\mu_1 + \int u \, d\mu_2$$
$$\geq \int f \, d\mu_1 + v(x) + u(x).$$

Using (3), we obtain

$$\int f \, d\mu \geq \int f \, d\mu_1 - 3\epsilon.$$

As ϵ and f are arbitrary, it follows that $\mu = \mu_1$. This proves that $\mu \in \text{ext}(I_x)$.

Now we turn to the 'only if'. Suppose that (3) fails. Then there exists $f \in C(X)$ and $\epsilon > 0$ such that

$$\int u \, d\mu + \int v \, d\mu + \int |f - (u - v)| \, d\mu \geq u(x) + v(x) + \epsilon \quad (u, v \in R). \tag{5}$$

Subtracting a constant from f, if necessary, we can further suppose that $\int f \, d\mu = 0$. We shall prove that we can decompose μ as $(\mu_1 + \mu_2)/2$, where $\mu_1, \mu_2 \in I_x$ and $\int f \, d\mu_1 = \epsilon$, so that $\mu \neq \mu_1$.

First, note that f is non-constant. Otherwise, the condition $\int f \, d\mu = 0$ would force $f = 0$, which is inconsistent with (5) (just take $u = v = 0$). Therefore $1, f$ are linearly independent functions, and we can define a linear functional $\varphi : \text{span}\{1, f\} \to \mathbf{R}$ by

$$\varphi(s1 + tf) = \frac{1}{2}s + \frac{1}{2}te \quad (s, t \in \mathbf{R}).$$

We next create a sublinear functional p, as follows. Set

$$C = \{g \in C(X) : \exists\, u \in R,\ u(x) = 0,\ u \leq g\}.$$

Note, in particular, that every non-negative continuous function on X belongs to C. For $h \in C(X)$, define

$$p(h) = \inf \left\{ \int g_1\, d\mu : h = g_1 - g_2,\ g_1, g_2 \in C \right\}.$$

Each $h \in C(X)$ admits at least one decomposition $h = g_1 - g_2$ with $g_1, g_2 \in C$, namely $g_1 = h^+$ and $g_2 = h^-$, so $p(h)$ is well-defined. It is easy to check that p is indeed a sublinear functional, i.e.

$$p(h_1 + h_2) \leq p(h_1) + p(h_2) \qquad (h_1, h_2 \in C(X)),$$
$$p(th) = tp(h) \qquad (h \in C(X),\ t \geq 0).$$

Next we claim that $\varphi(h) \leq p(h)$ for all $h \in \mathrm{span}\{1, f\}$. By homogeneity, this amounts to showing that

$$p(s1 + f) \geq \frac{1}{2}s + \frac{1}{2}\epsilon, \quad p(s1 - f) \geq \frac{1}{2}s - \frac{1}{2}\epsilon \quad \text{and} \quad p(s) \geq \frac{1}{2}s \qquad (s \in R).$$

Let us verify the first of these. Suppose that $s1 + f = g_1 - g_2$, where $g_1, g_2 \in C$. Then there exist $u_j \in R$ with $u_j(x) = 0$ and $u_j \leq g_j$ $(j = 1, 2)$. Thus $g_1 \geq \max(u_1, s1 + f + u_2)$. Using the identity (4) once more, it follows that

$$\int g_1\, d\mu \geq \frac{1}{2}\int |s1 + f + u_2 - u_1|\, d\mu + \frac{1}{2}\int (s1 + f + u_2 + u_1)\, d\mu.$$

Applying (5) with $u = u_1$ and $v = u_2 + s1$, we deduce that

$$\int g_1\, d\mu \geq \frac{1}{2}s + \frac{1}{2}\epsilon.$$

Finally, taking the infimum over all decompositions $g_1 - g_2$, we obtain $p(s1 + f) \geq s/2 + \epsilon/2$, as desired. The other two inequalities are proved in a similar way.

We are now in a position to apply the Hahn-Banach theorem to extend φ to a linear functional on the whole of $C(X)$, still satisfying $\varphi(h) \leq p(h)$ for all $h \in C(X)$. Let us examine the consequences of this inequality.

First of all, notice that φ is a positive linear functional. Indeed, if $h \geq 0$, then $-h = g_1 - g_2$ where $g_1 = 0 \in C$ and $g_2 = h \in C$, so

$$\varphi(-h) \leq p(-h) \leq \int g_1\, d\mu = 0,$$

in other words $\varphi(h) \geq 0$. Since $\varphi(1) = 1/2$, it follows that $\varphi = \mu_1/2$ where μ_1 is a probability measure on X. Note also that

$$\int f\, d\mu_1 = 2\varphi(f) = 2\frac{\epsilon}{2} = \epsilon.$$

Next we show that $\mu - \varphi$ is also a positive linear functional. Indeed, if $h \geq 0$, then $h = g_1 - g_2$ where $g_1 = h \in C$ and $g_2 = 0 \in C$, so

$$\varphi(h) \leq p(h) \leq \int g_1 \, d\mu = \int h \, d\mu,$$

in other words $\int h \, d\mu - \varphi(h) \geq 0$. Since $\int 1 \, d\mu - \varphi(1) = 1 - 1/2 = 1/2$, it follows that $\mu - \varphi = \mu_2/2$, where μ_2 is another probability measure on X. Thus $\mu = \mu_1/2 + \mu_2/2$.

It remains only to show that $\mu_1, \mu_2 \in I_x$. Take $u \in R$ with $u(x) = 0$. We need to show that $\int u \, d\mu_j \geq 0$. Let $n \geq 1$ and set $h_n = \max(u, -n)$. Then $-h_n = g_1 - g_2$, where $g_1 = 0 \in C$ and $g_2 = h_n \in C$, so

$$-\frac{1}{2} \int h_n \, d\mu_1 = \varphi(-h_n) \leq p(-h_n) \leq \int g_1 \, d\mu = 0.$$

Likewise, $h_n = g_1 - g_2$ where $g_1 = h_n \in C$ and $g_2 = 0 \in C$, so

$$\frac{1}{2} \int h_n \, d\mu_2 = \int h_n \, d\mu - \varphi(h_n) \geq \int h_n \, d\mu - p(h_n) \geq \int h_n \, d\mu - \int g_1 \, d\mu = 0.$$

Hence $\int h_n \, d\mu_j \geq 0$ ($j = 1, 2$). Letting $n \to \infty$, we deduce that $\int u \, d\mu_j \geq 0$ ($j = 1, 2$), as desired. This completes the proof. $\qquad\square$

It is now an easy matter to convert this abstract result into ones about Jensen measures and pluri-Jensen measures.

7.2 Corollary *Let Ω be an open subset of $\mathbf{R}^d$, let $x \in \Omega$ and let $\mu \in J_x(\Omega)$. Then $\mu \in \text{ext}(J_x(\Omega))$ if and only if, given $f \in C(\Omega)$ and $\epsilon > 0$, there exist $u, v \in \text{SH}(\Omega)$ such that*

$$\int u \, d\mu < u(x) + \epsilon, \quad \int v \, d\mu < v(x) + \epsilon \quad \text{and} \quad \int |f - (u - v)| \, d\mu < \epsilon.$$

Proof The 'if' part of the proof is exactly analogous to the 'if' in Theorem 7.1. For the 'only if', combine Theorem 7.1 with Lemma 3.2. $\qquad\square$

7.3 Corollary *Let Ω be a pseudoconvex open subset of $\mathbf{C}^d$, let $x \in \Omega$ and let $\mu \in \text{PJ}_x(\Omega)$. Then $\mu \in \text{ext}(\text{PJ}_x(\Omega))$ if and only if, given $f \in C(\Omega)$ and $\epsilon > 0$, there exist $u, v \in \text{PSH}(\Omega)$ such that*

$$\int u \, d\mu < u(x) + \epsilon, \quad \int v \, d\mu < v(x) + \epsilon \quad \text{and} \quad \int |f - (u - v)| \, d\mu < \epsilon.$$

Proof Similar. $\qquad\square$

8 Subharmonicity without upper semicontinuity

In this final section, we take a look at how Jensen measures can be used to provide a measure-theoretic approach to potential theory. To understand why there are circumstances in which this might be desirable, let us go back to the definition of subharmonic function.

Let Ω be open in $\mathbf{R}^d$, where we now assume that $d \geq 2$, and let $u: \Omega \to \mathbf{R}$ be a function. In essence, u is subharmonic iff $\Delta u \geq 0$. If $u \in C^2(\Omega)$, then this is literally true. However, C^2 is too restrictive for many applications. For example, in the Perron method of solving the Dirichlet problem, it is important that $\max(u, v)$ should be subharmonic whenever u, v are, and this would be false were we to limit ourselves to C^2-functions. It is also vital in many applications that the limit of a decreasing sequence (u_n) of subharmonic functions again be subharmonic. This too would fail if we restricted ourselves to C^2-functions, or even to continuous ones.

It is therefore customary to work with upper semicontinuous functions. The defining property $\Delta u \geq 0$ then has to be interpreted in a distributional sense. Actually, a more usual approach is to select an alternative property, equivalent to $\Delta u \geq 0$ but which can be stated without recourse to distributions. A common choice is the sub-mean-value property, namely, that for each $x \in \Omega$ and each closed ball B around x such that $B \subset \Omega$,

$$u(x) \leq \frac{1}{m(B)} \int_B u \, dm.$$

Here m denotes Lebesgue measure on $\mathbf{R}^d$. We could equally well average over spheres rather than balls: it turns out that this yields exactly the same class of functions. The resulting framework is very satisfactory in many respects.

However, there are times when upper semicontinuity is neither natural nor appropriate. To take a simple example, what if we are interested in *increasing* sequences of subharmonic functions? More generally, what if we take the supremum of a family of subharmonic functions, such as in the definition 1.2 of subharmonic envelope?

The natural thing to try is to relax upper semicontinuity yet further. An early result along these lines is the following theorem of Szpilrajn [S, Théorème 1].

8.1 Theorem *Let $u: \Omega \to [-\infty, \infty)$ be Borel and locally bounded above, and let u^* be its upper semicontinuous regularization. The following are equivalent:*

- $u(x) \leq 1/m(B) \int_B u \, dm$ *for each $x \in \Omega$ and each closed ball B in Ω around x;*

- u^* *is subharmonic on Ω and $u^* = u$ outside a set of Lebesgue measure zero.*

Recall that the *upper semicontinuous regularization* of u is defined by

$$u^*(x) = \limsup_{y \to x} u(y) \qquad (x \in \Omega).$$

It is the smallest upper semicontinuous function that majorizes u. It can happen that $u^* > u$ everywhere. The main thrust of Szpilrajn's theorem is that this is far from being the case when u satisfies the sub-mean-value property.

However, there are two unsatisfactory aspects to this theorem. Firstly, it depends on the form of the sub-mean-value property chosen. We would get a different result if we averaged over spheres instead of balls. Secondly, Lebesgue null sets, though negligible for measure theory, can actually be quite large from the point of potential theory. It would be preferable if the exceptional set, rather than being of measure zero, were of *capacity* zero, a much stronger conclusion. The solution is to use Jensen measures!

8.2 Theorem *Let* $u\colon \Omega \to [-\infty, \infty)$ *be Borel and locally bounded above, and let* u^* *be its upper semicontinuous regularization. The following are equivalent:*

- $u(x) \leq \int u\, d\mu$ *for each* $x \in \Omega$ *and each Jensen measure* μ *for* x;

- u^* *is subharmonic on* Ω *and* $u^* = u$ *outside a set of capacity zero.*

Proof See [CR1, Theorem 1.1]. □

Let us return to the question posed earlier and see what this result tells us about suprema of families of subharmonic functions. The following result is originally due to Cartan [C, Théorème 7].

8.3 Corollary *Let* (u_α) *be a family of subharmonic functions on* Ω *which is locally uniformly bounded above. Set* $u = \sup_\alpha u_\alpha$. *Then* u^* *is subharmonic and* $u^* = u$ *outside a set of capacity zero.*

Proof By Choquet's topological lemma [Kl, Lemma 2.3.4], we can reduce to the case where the family (u_α) is countable. Then u is clearly Borel and locally bounded above. Also, given $x \in \Omega$ and $\mu \in J_x(\Omega)$,

$$u(x) = \sup_\alpha u_\alpha(x) \leq \sup_\alpha \int u_\alpha\, d\mu \leq \int u\, d\mu.$$

Now apply the preceding result. □

This circle of ideas leads to the following duality theorem for measurable functions.

8.4 Theorem *Assume either that* $d \geq 3$, *or that* $d = 2$ *and* $\mathbf{R}^2 \setminus \Omega$ *is non-polar. Let* $\varphi\colon \Omega \to [-\infty, \infty)$ *be Borel and locally bounded above. Then*

$$S\varphi(x) = J\varphi(x) \qquad (x \in \Omega).$$

Proof (We give just a sketch: for more details see [CR1, Corollary 1.7].) Set $u = J\varphi$. One can show that $u(x) \leq \int u\, d\mu$ for each $x \in \Omega$ and each $\mu \in J_x(\Omega)$. By Theorem 8.2, u^* is subharmonic and $u^* = u$ outside a set of capacity zero. (Technical remark: u is perhaps not Borel, but it is still 'sufficiently measurable' for this argument to work.) By a converse to Cartan's theorem, $u = \sup_\alpha u_\alpha$ for some family of subharmonic functions (u_α). In particular $Su = u$. Since $u \leq \varphi$, this implies that $S\varphi \geq u$, i.e. $S\varphi \geq J\varphi$. The reverse inequality was proved in Proposition 1.4. □

Curiously, the result is false if $\mathbf{R}^2 \setminus \Omega$ is polar. For example, suppose that $\Omega = \mathbf{R}^2$, and take $\varphi(0) = -1$ and $\varphi = 0$ elsewhere. Then $S\varphi \equiv -1$, because, by Liouville's theorem, every subharmonic function bounded above on $\mathbf{R}^2$ is constant. On the other hand, if $x \neq 0$, then $\mu(\{0\}) = 0$ for all $\mu \in J_x(\mathbf{R}^2)$, so $J\varphi(x) = 0$. Hence $S\varphi(x) \neq J\varphi(x)$ for all $x \neq 0$.

This example indicates that, despite appearances, Theorems 3.1 and 8.4 are of a quite different nature. Whereas the former is in essence a duality theorem, the latter also relies heavily on potential theory. The difference is even more marked in the pluripotential case.

8.5 Theorem *Let Ω be a bounded pseudoconvex subset of $\mathbf{C}^d$. If $\varphi\colon \Omega \to [-\infty,\infty)$ is Borel and locally bounded above, then*

$$PS\varphi(x) = PJ\,\varphi(x) \qquad (x \in \Omega \setminus E),$$

where E is a pluripolar set. Further, there exists φ such that E is non-empty.

Proof See [CR1, Corollary 6.5]. (Note, however, that if φ is upper semicontinuous, then E is always empty; see [P1, Theorem 1] or [LS, p.3]). $\qquad\qquad\qquad\qquad\square$

In the spirit of Section 5, one might ask whether Theorem 8.2 remains true if one replaces Jensen measures by harmonic measures. The answer is yes, thanks to the following extension of Theorem 5.2.

8.6 Theorem *Let Ω be an open subset of $\mathbf{R}^d$, and let $\varphi\colon \Omega \to [-\infty,\infty)$ be Borel and locally bounded above. Then*

$$J\varphi(x) = \inf\left\{\int \varphi\,d\omega : \omega \in H_x(\Omega) \cup \{\delta_x\}\right\} \qquad (x \in \Omega).$$

Proof See [CR2, Theorem 1.3]. $\qquad\qquad\qquad\qquad\qquad\qquad\qquad\qquad\qquad\square$

Notice that we have added in the Dirac measure δ_x. Without it, the theorem would have been false (consider e.g. $\varphi = -1_{\{x\}}$). However, this is a small price to pay for the added generality, and in many applications it makes no difference. For example, in Theorem 8.2, the inequality $u(x) \le \int u\,d\mu$ is automatically satisfied if $\mu = \delta_x$. Thus the condition in that theorem becomes $u(x) \le \int u\,d\omega$ for all $\omega \in H_x(\Omega)$ and all $x \in \Omega$. In essence, this just says that on each subdomain $D \subset\subset \Omega$, the function u is majorized by its Poisson modification. Modulo some technical details, this leads to the following result.

8.7 Theorem *Let Ω be open in $\mathbf{R}^d$, and let $u\colon \Omega \to [-\infty,\infty)$ be Borel and locally bounded above. The following are equivalent:*

- *for each domain $D \subset\subset \Omega$ and each function h continuous on $\overline{D}$ and harmonic on D, if $u \le h$ on ∂D, then also $u \le h$ on D;*

- *u^* is subharmonic on Ω and $u^* = u$ outside a set of capacity zero.*

Proof See [CR2, Theorem 1.4]. $\qquad\qquad\qquad\qquad\qquad\qquad\qquad\qquad\qquad\square$

This result is an extension of the well-known characterization of subharmonic functions which gives them their name. Jensen measures do not appear in the statement, yet they played a vital role in the proof.

References

[BS]   S. Bu and W. Schachermayer, Approximation of Jensen measures by image measures under holomorphic functions and applications, *Trans. Amer. Math. Soc.* **331** (1992), 585–608.

[C]   H. Cartan, Théorie du potentiel newtonien: énergie, capacité, suites de potentiels, *Bull. Soc. Math. France* **73** (1945), 74–106.

[CR1]   B. J. Cole and T. J. Ransford, Subharmonicity without upper semicontinuity, *J. Funct. Anal.* **147** (1997), 420–442.

[CR2]   B. J. Cole and T. J. Ransford, Jensen measures and harmonic measures, *J. Reine Angew. Math.* (to appear).

[E]   D. A. Edwards, Choquet boundary theory for certain spaces of lower semicontinuous functions, in: *Function Algebras* (F. Birtel, ed.), Scott, Foresman and Co., Chicago, IL, 1966, 300–309.

[FN]   J. E. Fornaess and R. Narasimhan, The Levi problem on complex spaces with singularities, *Math. Ann.* **248** (1980), 47–72.

[Gm]   T. W. Gamelin, *Uniform Algebras and Jensen Measures*, Cambridge University Press, 1978.

[GS]   T. W. Gamelin and N. Sibony, Subharmonicity for uniform algebras, *J. Funct. Anal.* **35** (1980), 64–108.

[Gr]   S. Gardiner, *Harmonic Approximation*, Cambridge University Press, 1995.

[Kh]   B. N. Khabibullin, Sets of uniqueness in spaces of entire functions of a single variable, *Math. USSR Izv.* **39** (1992), 1063–1084.

[Kl]   M. Klimek, *Pluripotential Theory*, Oxford University Press, 1991.

[Ko1]   P. Koosis, La plus petite majorante surharmonique et son rapport avec l'existence des fonctions entières de type exponentiel jouant le rôle de multiplicateurs, *Ann. Inst. Fourier (Grenoble)* **33** (1983), 67–107.

[Ko2]   P. Koosis, *Leçons sur le théorème de Beurling et Malliavin*, Les Publications CRM, Montréal, 1996.

[LS]   F. Lárusson and R. Sigurdsson, Plurisubharmonic functions and analytic discs on manifolds, *J. Reine Angew. Math.* **501** (1998), 1–39.

[LG]   P. Lelong and L. Gruman, *Entire Functions of Several Complex Variables*, Springer, Berlin, 1986.

[P1]   E. Poletsky, Plurisubharmonic functions as solutions of variational problems, *Proc. Sympos. Pure Math.* **52** Part 1, Amer. Math. Soc., Providence, RI, 1991, 163–171.

[P2]   E. Poletsky, Holomorphic currents, *Indiana Univ. Math. J.* **42** (1993), 85–144.

[P3]   E. Poletsky, Disk envelopes of functions I, in: *Complex Analysis in Contemporary Mathematics* (E. M. Chirka, ed.), Fasis, Moscow, 1997.

[P4]   E. Poletsky, Disk envelopes of functions II, *J. Funct Anal.* **163** (1999), 111–132.

[S]   E. Szpilrajn, Remarques sur les fonctions sousharmoniques, *Ann. of Math.* **34** (1933), 588–594.

Simultaneous approximation in function spaces

Arne STRAY

Department of Mathematics
University of Bergen
5008 Bergen
Norway

Abstract

Some general problems in joint approximation, originally raised by Lee Rubel, are discussed. The emphasis is on spaces of analytic functions in the unit disc D. Besides giving a survey of known results for the Hardy spaces and the space $H(D)$ consisting of all analytic functions in D, we give some new results for the Dirichlet space $\mathcal{D}$ consisting of all analytic functions in D having finite Dirichlet integral.

1 Introduction

Consider topological function spaces E_1 defined on a set D_1, and E_2 defined on a set D_2. Let E consist of all functions f on $D_1 \cup D_2$ such that $f|_{D_i} \in E_i$ for $i = 1, 2$. Suppose there is a subset P of E such that $P|_{D_i}$ is dense in E_i for $i = 1, 2$. Given $g \in E$, is there a net $\{p_\alpha\} \subset P$ such that $p_\alpha|_{D_i} \to g|_{D_i}$ for $i = 1, 2$? In the following we consider the more special situation where E_1 is a space A consisting of analytic or harmonic functions on the unit disc $D = \{z : |z| < 1\}$, and $E_2 = A|_X \cap B = \{f|_X : f \in A\} \cap B$, where B is some topological funtion space on X, and X is a relatively closed subset of D.

Consider the case where B consists of all uniformly continuous functions on X. We call X a *Mergelyan set* for $[A, P]$ if for any $f \in A$ with $f|_X \in B$, there are $p_n \in P$, $n = 1, 2, \ldots$, such that $p_n \to f$ in A and $p_n \to f$ uniformly on X.

Another important case occurs if B denotes the bounded continuous functions on X with the topology of bounded pointwise convergence. In this case X is called a *Farrell set* for $[A, P]$ if the above-mentioned approximation problem is solvable.

These two definitions are motivated by Mergelyan's theorem and The Farrell-Rubel-Shields theorem concerning uniform and pointwise bounded approximation by polynomials, respectively [G, pp. 48 and 151].

The set P of approximating functions will in almost all cases we study consist of polynomials in the complex variable z or harmonic polynomials in x and y if A is a space of harmonic functions. In these cases we will speak about Farrell and Mergelyan sets for A without mentioning P.

The main problem is to find geometrical or topological characterizations of Farrell and Mergelyan sets. It turns out that such a characterization is going to depend heavily on the function space itself. We end this introductory part by giving a less specific condition that

239

N. Arakelian and P.M. Gauthier (eds.), Approximation, Complex Analysis, and Potential Theory, 239–261.
© 2001 *Kluwer Academic Publishers. Printed in the Netherlands.*

may characterize Farrell and Mergelyan sets for a rather broad class of function spaces. We consider a convex balanced neighbourhood K of the origin in A, and define the functions

$$\Gamma_K(z) = \sup\{|f(z)| : f \in K, |f| \le 1 \text{ on } X\}$$

and

$$\gamma_K(z) = \sup\{|f(z)| : f \in K \cap P, |f| \le 1 \text{ on } X\}$$

if $z \in D$.

1.1 Conjecture *X is a Farrell set for A if and only if $\Gamma_K(z) = \gamma_K(z)$ for all K and all $z \in D$.*

1.2 Remark We suspect the conjecture has an affirmative answer in the Hardy spaces $H^p(D), 1 < p \le \infty$. We get a stronger conjecture by replacing the word "all" by "some". We are not so sure that this stronger version holds in the Hardy spaces.

We shall call X a *set of determination* for A if

$$\sup\{|f(z)| : z \in X\} = \sup\{|f(z)| : z \in D\}$$

for any f in A. Note that if the point evaluations $f \to f(z)$, $z \in D$, are continuous on A, then X is a set of determination for A if and only if X is a Farrell set for A with the additional property $\overline{X} \supset \partial\overline{D}$.

Evidently if X is a set of determination for A, and $A \supset A_1$, then X is also a set of determination for A_1. Such a monotonic property does not hold for the wider classes of Farrell or Mergelyan sets.

For all function spaces in the unit disc where the classes of Farrel and Mergelyan sets have been described, it turns out that these classes are identical. One might wonder if this is a completely general phenomenon. This turns out not to be the case. In a paper by A. A. Danielyan [Da], where the simultaneous approximation problem for $[H^\infty(G), \mathcal{P}]$ is studied for more general plane open set G, and $\mathcal{P}$ denotes the polynomials, an example is given of a set X being a Farrell set but not a Mergelyan set for $[H^\infty(G), \mathcal{P}]$.

For more on simultaneous approximation by polynomials on general plane sets, see the paper [FPG] by A. G. O'Farrell and F. Perez-Gonzalez.

Nevertheless, it seems reasonable to conjecture that the classes of Farrell and Mergelyan sets coincide for function spaces in the unit disc D. So far this has been proved to be true for all the Hardy spaces, the Dirichlet space, and the space $H(D)$ consisting of all analytic functions in D.

2 The Brown-Shields-Zeller story

Before we describe the Farrell and Mergelyan sets for the Hardy spaces, it is natural to discuss the work by L. Brown, A. L. Shields and K. Zeller published in [BroSZ]. A central theme there is a description of the sets of determination for the algebra $H^\infty(D)$ consisting of all bounded analytic functions in D.

Before formulating one of their main results, we have to define the *nontangential closure* X_{nt} of our relatively closed subset X of D. It consists of all $z \in \partial D$ such that there is a sequence w_n in X where $w_n \to z$ and $|w_n - z| \leq C(1 - |w_n|)$ for some constant C independent of n. We also define the *tangential closure* X_t of X as $\overline{X} \cap T \setminus X_{\mathrm{nt}}$, where $T = \partial D$ denotes the unit circle. Brown, Shields, and Zeller proved among other things:

2.1 Theorem *Let $X = \{\alpha_n\}$ denote a sequence in D clustering only on ∂D. The following statements are equivalent:*

(i) *X is a set of determination for $H^\infty(D)$.*

(ii) *The linear measure of $\partial D \setminus X_{\mathrm{nt}}$ is zero.*

(iii) *There are constants c_n with $\sum |c_n| < \infty$ and some $c_n \neq 0$, such that $\sum c_n e^{\alpha_n z} \equiv 0$ for $z \in D$.*

Let us give a brief proof of (i) $\Longleftrightarrow$ (ii). If $z = r e^{i\theta} \in D$ and $t > 1$, we define

$$I_z^t = \{ e^{i\phi} : |\phi - \theta| < t(1 - |z|) \}$$

and also

$$S_m^t(X) = \bigcup I_z^t, \quad |z| > 1 - m^{-1}, \quad z \in X, \quad m = 1, 2, \ldots .$$

Observe first that (ii) $\Longrightarrow$ (i) is an immediate consequence of Fatou's theorem [Du, p. 14]. Since

$$\bigcap_{m=1}^{\infty} S_m^t(X) \subset X_{\mathrm{nt}},$$

there is a set $K \subset \partial D$ with $|K| > 0$, with $K \cap S_m^t = \emptyset$ for m large enough if (ii) fails. Let u denote the harmonic extension to D of the characteristic function χ_K of K, and let $f = e^{u + i\tilde{u}}$, where $\tilde{u}$ denotes the harmonic conjugate of u vanishing at the origin. Clearly $\|f\|_X = \sup\{|f(z)|, \ z \in X\} < \|f\|_D$ and we have proved that (i) $\Longrightarrow$ (ii).

Note the following bonus of the proof: for any $t > 1$, the sets X_{nt} and $\bigcap_{m=1}^{\infty} S_m^t(X)$ differ only by a set of zero linear measure. (This can also be seen directly using Lebesgue's theorem on differentiation of the integral.)

For the proof of (ii) $\Longleftrightarrow$ (iii), we refer to [BroSZ]. We remark that the proof by Brown, Shields and Zeller gives that (iii) may be replaced by

(iv) *For any entire function $f = \sum d_n z^n$ with $d_n \neq 0$ for $n \geq 0$, there are constants c_n, not all of them zero, such that $\sum c_n f(\alpha_n z) \equiv 0$ for $z \in D$ and $\sum |c_n| < \infty$.*

2.2 Remark In [Bo1] Bonsall studied sets of determination for bounded harmonic functions in D, and proved results related to Theorem 2.1 for such functions. Later, S. Gardiner [Ga] proved similar results for harmonic functions in several dimensions. Sets of determination for differences of positive harmonic functions in D were characterized by W. K. Hayman and T. J. Lyons [HayL], and later Gardiner [Ga] extended their results to higher dimensions.

In [Bo1] Bonsall raised the problem of describing the sequences α_n in D such that $|\alpha_n| \to 1$ and

$$\sum_{N=1}^{\infty} c_n P_{\alpha_n} \equiv 0 \quad \text{for some sequence } c_n \text{ with } \quad 0 < \sum_{n=1}^{\infty} |c_n| < \infty \tag{2.1}$$

holds in $L^1(d\theta)$, where $d\theta$ denotes Lebesgue measure on the unit circle T, and $P_z(\theta) = (1 - |z^2|)/|e^{i\theta} - z|^2$ denotes the Poisson kernel representing z. By duality arguments and the results in [Bo1], Bonsall showed that (2.1) follows if condition (ii) in Theorem 2.1 holds for $X = \{\alpha_n\}$.

In [BoW] the problem was solved, and the proof of the reverse implication depended on the Brown-Shields-Zeller paper in an essential way. Bonsall and Walsh proved:

(2.1) *holds if and only if* $X = \{\alpha_n\}$ *satisfies condition* (ii) *in Theorem* 2.1.

There are two ways to see that condition (ii) in Theorem 2.1 follows from (2.1) which are different from the one given in [BoW] and may be worth seeing.

The first one is extremely short: consider e^{zw} as a function of z with w as a parameter. If we integrate e^{zw} against both sides in (2.1), we get

$$\sum c_n e^{\alpha_n w} \equiv 0,$$

and (ii) follows by Theorem 2.1.

The second proof depends less on the Brown-Shields-Zeller theorem: assume that (ii) in Theorem 2.1 fails. By the main result in [Str4], there is $f \in H^\infty$ such that f is close to 1 at z_1, while f is close to 0 at z_n for $n > 1$. If $c_1 \neq 0$, as of course we may assume, we have a contradiction to (2.1).

2.3 Problem Prove that (2.1) implies the analogue of condition (ii) with the unit disc D replaced by the open unit ball in $\mathbb{R}^n$ when n exceeds 2.

The problem is interesting because it seems to require new ideas. All proofs of the corresponding result in two dimensions depend on complex analytic methods. That (2.1) is a consequence of condition (ii) is true in the open unit ball in $\mathbb{R}^n$ for all n. This is contained in Gardiner's paper [Ga].

3   Hardy spaces and BMO

If $0 < p < \infty$, we recall that the Hardy space $H^p(D)$ consists of all analytic functions f in D such that

$$||f||_p = \lim_{r \to 1^-} \int_T |f(re^{i\theta})|^p \, d\theta = \int_T |f(e^{i\theta})|^p \, d\theta < \infty.$$

If $p \geq 1$, $H^p(D)$ is a Banach space, while if $p < 1$, $H^p(D)$ is an F-space with the translation-invariant metric

$$d(f, g) = ||f - g||_p^p.$$

A good reference giving more details about $H^p(D)$ is Duren's book [Du]. The failure of the Hahn-Banach theorem for $H^p(D)$ when $0 < p < 1$, is described in Chapter 7, where some results about F-spaces also are given.

The next theorem summarises various results about Farrel and Mergelyan sets for $H^p(D)$.

3.1 Theorem *Let* X *denote a relatively closed subset of the unit disc* D. *The following statements are equivalent:*

(i) X is a Farrell set for $H^p(D)$, $0 < p \le \infty$.

(ii) X is a Mergelyan set for $H^p(D)$, $0 < p \le \infty$.

(iii) The linear measure of $\overline{X} \cap \partial D \setminus X_{\mathrm{nt}}$ is zero.

(iv) If $f \in H^p(D)$ is bounded on X and g is a uniformly continuous function on X, there are polynomials p_n such that $p_n \to f$ in $H^p(D)$ and $\|p_n - g\|_X \to \|f - g\|$ as $n \to \infty$.

3.2 Remark If $p < \infty$, the topology on $H^p(D)$ is assumed to be induced from the metric described above. If $p = \infty$, we may use the weak*-topology or the topology of pointwise bounded convergence.

Theorem 3.1 was proved over several years. The equivalence of (i) and (iii) when $p = \infty$, was first done by A. M. Davie using functional analysis, and then by A. Stray by constructive methods, see [Str2]. That (ii) $\Longrightarrow$ (iii) was proved by J. Detraz [De]. The converse, (iii) $\Longrightarrow$ (ii), is most easily obtained using Davie's argument for (iii) $\Longrightarrow$ (i).

Since Davie's proof with some modifications turned out to be useful in various different cases we shall consider, we present his argument here:

So we assume $f \in H^\infty$, $\|f\|_\infty = 1$ and $\|f\|_X < \eta$. Let

$$N = \{g \in C(\overline{D}) : \|g\|_X \le \eta, \ \|g\|_{\overline{D}} \le 1\}.$$

We have to show that g is in the closure of $N \cap P$ in the topology of uniform convergence on compact subsets on D.

By the Riesz representation theorem, it is sufficient to prove that $|\mu(g)| \le 1$ whenever μ is a regular complex measure with compact support in D such that $|\mu(h)| \le 1$ for all $h \in N \cap P$. ($N \cap P$ is a convex set in the space $C(D)$ with the topology of uniform convergence on compact subsets of D, and the dual space consists of the regular complex measures with compact support in D.)

N is the unit ball in $C(\overline{D})$ with respect to some norm which is equivalent to sup norm on $\overline{D}$ since $\eta > 0$. Hence we can extend the functional $p \to \mu(p)$ to $C(\overline{D})$, and represent it by a measure ν on $\overline{D}$ such that $\nu(f) \le 1$ for all $f \in N$.

But the last fact implies

$$|\nu|(\overline{D} \setminus \overline{X}) + \eta|\nu|(\overline{X}) \le 1,$$

where $|\nu|$ denotes the total variation of ν. Since $\mu - \nu$ is orthogonal to the polynomials, the Rudin-Carleson theorem [G] gives that $\nu(K) = 0$ if $K \subset \partial D$ has zero linear measure.

But then our original function g is a well-defined element of $L^\infty(\nu)$, and the above inequality gives $|\nu(g)| \le 1$ since $|g| \le \eta$ almost everywhere w.r.t. ν on $\overline{X}$ in the light of Fatou's theorem [Du, p. 14], and condition (iii) in Theorem 3.1. Let $g_n(z) = g((1 - n^{-1})z)$, $z \in D$, $n = 1, 2, \ldots$. Then we get by dominated convergence $\mu(g) = \lim \mu(g_n) = \lim \nu(g_n) = \nu(g)$, and the proof is complete.

The motivation for introducing the formally stronger condition (iv) came from a paper on rational approximation by A. M. Davie, T. W. Gamelin and J. Garnett, where certain distance estimates are related to approximation problems [DGG]. The equivalence of (iii) and (iv) was given for $p \ge 1$ in [RS] by using methods from functional analysis found in [DGG]. Later, a constructive argument was used in [PGS] giving the same equivalence for all $p > 0$. For $0 < p < 1$, it is well known that duality methods break down in the study of $H^p(D)$, see [Du, Chapter 7]. One of the fascinating aspects of the simultaneous approximation problem in the unit disc is the interplay between functional analysis and more constructive arguments. Modifications of Davie's argument as well as the constructive methods have found applications for many function spaces, where the geometric characterization is quite different from the one given in Theorem 3.1.

The problem of characterizing Farrell and Mergelyan sets for a close relative of $H^p(D)$, namely the space BMOA, consisting of analytic functions of bounded mean oscillation, was raised in [RS]. Recently, the Farrell sets were characterized for both BMOA and VMOA (functions of vanishing mean oscillation). The main new idea came in a paper by A. Nicolau and J. Orotbig. The methods of Nicolau and Orotbig are constructive and related to the John-Nirenberg theorem. Nicolau and Orotbig obtained the following:

3.3 Theorem *Let A denote a measurable subset of the unit circle, $z_0 \in \overline{A}$ and suppose A is of zero density at z_0. Then there is a nonnegative function g in VMO such that the radial limit of g at z_0 is zero, while $g(z) \to \infty$ as $z \to z_0$ inside A.*

Moreover, there is a nonnegative function f in VMO such that $f \leq 1$ on A, while $\lim_{r \to 1^-} f(rz_0) = \infty$.

This real-variable result is the main ingredient in proving that the Farrell sets for VMOA (with the norm-topology) are characterized by the geometric condition (iii) in Theorem 3.3. Full details of the work by Nicolau and Orotbig are published in [NO]. Farrell and Mergelyan sets for BMOA (with the w^*-topology) were also characterized in [PGTS] by methods based on functional analysis, but for VMOA the results were incomplete. It sems still to be an open problem whether a Mergelyan set X for VMOA must satify condition (iii) in Theorem 3.1.

Let us consider an alternative way to characterize Farrell sets mentioned briefly in the introduction. Suppose $0 < r \leq \infty$ and let $K = \{f \in H^r : \|f\|_r \leq c\}$ for some $c > 1$. Define

$$\Gamma_K^r(z) = \sup\{|f(z)| : f \in K, \ |f| \leq 1 \text{ on } X\}$$

and

$$\gamma_K^r(z) = \sup\{|f(z)| : f \in K \cap P, \ |f| \leq 1 \text{ on } X\},$$

where P denotes the polynomials. It is now easy to see that

$$\Gamma_K^r(z) = \gamma_K^r(z), \ z \in D \iff X \text{ is a Farrell set for } H^r.$$

Only the implication "$\Longrightarrow$" requires a proof. If X fails to be Farrell, we find as in the proof of Theorem 2.1 a set $L \subset \overline{X} \cap \partial D \setminus X_{nt}$ with $|L| > 0$ and define $f = e^{U + i\tilde{U} - 1}$, where U is the harmonic extension of the characteristic function of L to D, and $\tilde{U}$ its conjugate. Then $1 = |f| > \|f\|_X$ almost everywhere on L with respect to linear measure, while $|p| \leq \|p\|_X$ on L for any polynomial p by continuity. The function $h = cf^a$, where a is chosen such that $|h| \leq 1$ on X, now shows that

$$\Gamma_K^r(z) > \gamma_K^r(z)$$

at least if $r = \infty$, provided z is close to L so that $U(z)$ is close to 1. For $1 < r < \infty$ we get the same conclusion with a little bit more work. We omit the details.

3.4 Problem Is there a proof that $\Gamma_K^r(z) \equiv \gamma_K^r(z)$ if and only if F is Farrell *not* depending on the geometric characterization of Farrell sets?

Before leaving this section, let us depart from polynomial approximation for a while and consider Farrell and Mergelyan sets for $[H^p(D), H^\infty(D)]$, where $p < \infty$. By considering the inner-outer factorization it is trivial that *any* closed subset X of D is a Farrell set for $[H^p(D), H^\infty(D)]$, where $H^p(D)$ has the usual metric topology. In fact we have :

3.5 Theorem *The Farrell and Mergelyan sets for* $[H^p(D), H^\infty(D)]$ *are identical and consist of all closed subsets* X *of* D.

Proof Let us first show that any closed subset X of D is a Farrell set. We recall (see [Du]) that any $f \in H^p(D)$ can be factorized as $f = IF$, where I is inner and F is outer. What is important here is the explicit expression for F:

$$F(z) = \exp\left(\frac{1}{2\pi} \int_0^{2\pi} \frac{e^{i\theta} + z}{e^{i\theta} - z} \log |f(e^{i\theta})| \, d\theta\right).$$

Let us define $F_n \in H^\infty(D)$ by

$$F_n = \exp\left(\frac{1}{2\pi} \int_0^{2\pi} \frac{e^{i\theta} + z}{e^{i\theta} - z} u_n(e^{i\theta}) \, d\theta\right),$$

where

$$u_n(e^{i\theta}) = \min\{\log |f(e^{i\theta})|, n\}.$$

Then $f_n = IF_n \in H^\infty(D)$ and $|f_n| \le |f|$ pointwise in D. Moreover, it is not hard to see that $f_n \to f$ in $H^p(D)$. Since $\log |f| \in L^1(d\theta)$, $u_n \to \log |f|$ in $L^1(d\theta)$, and this implies that $f_n \to f$ in measure. By monotone convergence, $\int_0^{2\pi} |f_n|^p \, d\theta \to \int_0^{2\pi} |f|^p \, d\theta$ and this implies that $f_n \to f$ in $H^p(D)$ for $0 < p < \infty$.

We now prove that X also is a Mergelyan set for $[H^p(D), H^\infty(D)]$: if $f \in H^p(D)$ and $f|_X$ is uniformly continuous, we shall find $h \in H^\infty(D)$ (even with the property that $h|_X$ is uniformly continuous) such that h is approximating f in $H^p(D)$ as well as uniformly on X. Let

$$\widehat{X}^\infty = \{z \in D : |g(z)| \le \sup_{w \in X} |g(w)|, g \in H^\infty(D)\}$$

denote the "H^∞-hull" of X. It follows from the maximum principle (see the proof of Lemma 1 in [Str4]) that $f|_{\widehat{X}^\infty} \in C_{\mathrm{ua}}(\widehat{X}^\infty)$, where $C_{\mathrm{ua}}(\widehat{X}^\infty)$ denotes the uniformly continuous functions on $\widehat{X}^\infty$ being analytic at interior points. We first show that the algebra

$$B = \{f \in H^\infty : f|_{\widehat{X}^\infty} \in C_{\mathrm{ua}}(\widehat{X}^\infty)\}$$

is uniformly dense in $C_{\mathrm{ua}}(\widehat{X}^\infty)$. As in [Str4] it suffices to approximate $(z - z_0)^{-1}$ uniformly on $\widehat{X}^\infty$ if $z_0 \in D \setminus \widehat{X}^\infty$, by functions from B. Let h in H^∞ satisfy $1 = h(z_0) > \|h\|_{\widehat{X}^\infty}$. Using the construction following Lemma 1 in [Str4], we may even assume $h|_{\widehat{X}^\infty} \in C_{\mathrm{ua}}(\widehat{X}^\infty)$. It follows as in [Str4] that $(z - z_0)^{-1}$ is uniformly approximable on $\widehat{X}^\infty$ by functions in B and, using known results from rational approximation (see [Str4, pp. 67-68]), we get that B is dense in $C_{\mathrm{ua}}(\widehat{X}^\infty)$.

So, given f as above and $\epsilon > 0$, we first pick $h \in B$ such that $||f - h||_{\widehat{X}\infty} < \epsilon$. Then we pick $h_0 \in B$ such that h_0 approximates $f - h$ as well as we please in H^p while $||h_0||_{X\infty} < 2\epsilon$. This is possible by Lemma 1 and the subsequent construction in [Str4], and thus $h + h_0$ is the required approximant to f from $H^\infty \cap C_{\mathrm{ua}}(\widehat{X}^\infty)$.   □

3.6 Problem Describe the Farrell and Mergelyan sets for $[A, A \cap H^\infty]$ whenever $A \cap H^\infty$ is dense in A.

In particular this is interesting when A is the Dirichlet space or the Bergman class in D.

4 $H(D)$

The problem of describing the Mergelyan sets for $H(D)$, the space of all analytic functions in D equipped with the topology of uniform convergence on compact subsets of D, was raised around 1970 by L. A. Rubel. Originally, the consept of a Farrell set was reserved for the Hardy space $H^\infty(D)$, but it soon became natural to study Farrell sets in a more general context including $H(D)$. The first results about Mergelyan sets for $H(D)$ appeared in [RST], and the year after, a complete characterization was found [Str1].

Before formulating the main result, let us make an easy observation about sets of determination for $H(D)$:

> A closed set $X \subset D$ is a set of determination for $H(D)$ if and only if no point in $D \setminus X$ can be connected to ∂D by an arc in $D \setminus X$.

Proof If X fails to be a set of determination, there is $f \in H(D)$ such that $1 = f(z) > ||f||_X$ for some $z \in D \setminus X$. The maximum principle now easily gives the existence of an arc γ, where $\gamma(0) = z$ and $|\gamma(t)| \to 1$ as $t \to \infty$.

On the other hand, if z and γ exist, an easy pole-pushing argument gives that X is not a set of determination.   □

So the class of all sets of determination for $H(D)$ is certainly smaller than the corresponding class for the Hardy spaces, as the tangential closure X_t is empty when X is a set of determination for $H(D)$.

When we pass to Mergelyan or Farrell sets, there are no such simple inclusions any more. Rubel conjectured that Mergelyan sets should be described in terms of the polynomial hull. This turned out to be correct. If A is any space of functions on D containing analytic or harmonic polynomials, let us define the b-*hull*

$$\widehat{X}^b = \{z \in D : |f(z)| \le ||f||_X, \ f \in A, \ f|_X \text{ is bounded}\},$$

and the u-*hull*

$$\widehat{X}^u = \{z \in D : |f(z)| \le ||f||_X, \ f \in A, \ f|_X \text{ is uniformly continuous}\},$$

and finally the usual *polynomial hull*

$$\widehat{X} = \{z \in C : |p(z)| \le ||p||_X, \ p \text{ polynomial}\}.$$

Observe that $\widehat{X}^{\mathrm{b}}$ and $\widehat{X}^{\mathrm{u}}$ are subsets of D, while $\widehat{X}$ is normally not. Using the maximum principle and a simple pole-pushing argument one proves:

$$\widehat{X}^{\mathrm{u}} = \widehat{X}^{\mathrm{b}} = X \cup \left(\bigcup V_i\right),$$

where $\{V_i\}$ is the collection of all components V of $D \backslash X$ such that no $z \in V$ can be connected to ∂D by an arc in V.

Now to the main result of this section:

4.1 Theorem *The following statements are equivalent:*

(i) X *is a Mergelyan set for* $H(D)$.

(ii) X *is a Farrell set for* $H(D)$.

(iii) *If* $K \subset D$ *is compact, then* $\widehat{X \cup K}^{\mathrm{b}} = \widehat{X \cup K}$.

Let us see that Farrell and Mergelyan sets for $H(D)$ and the Hardy spaces are completely different classes.

Let X consist of two different line segments in D meeting at some point $z \in \partial D$. Then X is Farrell for $H^p(D)$, $0 < p < \infty$, while X is not Farrell for $H(D)$. On the other hand, let X consist of the points z_n in a sequence $z_n \subset D$, clustering only at a proper subset $S \subset \partial D$ of positive linear measure. Then we have $\widehat{X \cup K}^{\mathrm{b}} = \widehat{X \cup K}$ for any compact subset K of D. But if $\sum_n (1 - |z_n|) < \infty$, then X_{nt} has zero linear measure (prove this!), and hence X is not Farrell for $H^p(D)$, $0 < p \leq \infty$.

Since the characterizations of Farrell and Mergelyan sets as well as the corresponding classes are completely different for $H(D)$ and $H^p(D)$, it may be of some interest to observe that the proof of (iii) $\Longrightarrow$ (i) in Theorem 4.1 required a generalization of the methods used in the proof of Theorem 3.1 to more general planar domains than the unit disc D. See [Str1] for details.

While the results for $H(D)$ are rather complete, the following may be of interest to study:

4.2 Problem Describe the Farrell sets for $h(D)$, when $h(D)$ are the harmonic functions in D equipped with the topology of uniform convergence on compact subsets of D, and D is the unit disc or the ball in $\mathbb{R}^n$, $n \geq 3$.

4.3 Remark In Problem 4.2, the approximating functions must of course be real harmonic polynomials. The Mergelyan sets for $h(D)$ have already be dealt with by S. Gardiner in [Ga2], and Gardiner's characterization is similar to the equivalence of (i) and (ii) in Theorem 4.1. It seems, however, that Gardiner's method is not easily adapted to pointwise bounded approximation.

We close this section by considering another generalization of Theorem 4.1. Let $\mathcal{A}$ denote a closed algebra of analytic functions on the unit disc D such that

$$A(D) \subset \mathcal{A} \subset H^\infty(D),$$

where $A(D)$ denotes the disc algebra consisting of all continuous functions on $\overline{D}$ being analytic in D. We assume that if $f \in \mathcal{A}$ and $w \in D$ and $f(w) = 0$, then $f = (z - w)g$, where $g \in \mathcal{A}$.

If X is a relatively closed subset of D, we define the $\mathcal{A}$-hull of X to be

$$\widehat{X}^{\mathcal{A}} = \{z \in D : |f(z)| \le \|f\|_X, \ f \in \mathcal{A}\}.$$

We now state:

4.4 Theorem *The following statements are equivalent:*

(i) X *is a Mergelyan set for* $[H(D), \mathcal{A}]$.

(ii) X *is a Farrell set for* $[H(D), \mathcal{A}]$.

(iii) *If* $K \subset D$ *is compact, then* $\widehat{X \cup K}^{\text{b}} = \widehat{X \cup K}^{\mathcal{A}}$.

Proof Suppose (i) holds. Let $z \in D \setminus \widehat{X \cup K}^{\text{b}}$. Then there is $f \in H(D)$ with a uniformly continuous restriction to $X \cup K$ such that $|f(z)| > \|f\|_{X \cup K}$. This follows since $\widehat{X \cup K}^{\text{b}} = \widehat{X \cup K}^{\text{u}}$. By (i) there is a sequence a_n in $\mathcal{A}$ converging uniformly to f on $X \cup K$ and on compact subsets of D. It follows that $z \in D \setminus \widehat{X \cup K}^{\mathcal{A}}$. But then (iii) follows since $\widehat{X \cup K}^{\text{b}} \subset \widehat{X \cup K}^{\mathcal{A}}$. That (ii) $\Longrightarrow$ (iii) is proved in the same way.

Now assume (iii) holds. Let $Y = \widehat{X \cup K}^{\mathcal{A}}$. If V is a component of $D \setminus Y$ and $w \in V$, there is $a \in \mathcal{A}$ such that $1 = a(w) > \|a\|_{X \cup K}$. Then $1 - a = (z - w)g$ with $g \in \mathcal{A}$, and hence

$$(z - w)^{-1} = g \sum_{n \ge 0} a^n$$

is a uniform limit on $X \cup K$ of functions from $\mathcal{A}$. It follows that the function algebra $R(\overline{Y})$ is contained in the same closure, where $R(\overline{Y})$ denotes the functions on $\overline{Y}$ that are uniformly approximable on $\overline{Y}$ by rational functions with the poles off $\overline{Y}$. (Strictly speaking, we must consider $R(\overline{Y})|_Y$ since functions in $\mathcal{A}$ are defined inside D.)

But $R(\overline{Y})|_Y$ is uniformly dense in the uniformly continuous functions on Y analytic at interior points. This is seen from Vitushkin's theory on rational approximation (see p. 68 in [Str4]) and hence (i) follows from (iii). That (iii) implies (ii) is proved in the same way as one proves (iii) $\Longrightarrow$ (ii) in Theorem 4.1 (see [Str1]). The key point is that in our situation $R(\overline{Y})$ is a Dirichlet algebra on $\partial \overline{Y}$, so the proof in [Str1] can be copied almost word for word. $\square$

5   The Dirichlet space and related Besov spaces

Let $\mathcal{D}$ denote all analytic functions f in the unit disc D such that the Dirichlet integral

$$D(f) = \iint |f'|^2 \, dx \, dy < \infty$$

is finite. It is well known that $\mathcal{D}$ is a subset of the Hardy space $H^2(D)$, but the norms are of course different. The functions in $\mathcal{D}$ satisfy a Fatou-type theorem about nontangential limits, but in a more precise form than for $H^p(D)$-spaces, in the sense that the exceptional sets now have zero logarithmic capacity (see [Be] for details). The boundary values on ∂D

are finely continuous, and for this reason one should expect weaker conditions on Farrell sets for $\mathcal{D}$ than for $H^p(D)$. We equip $\mathcal{D}$ with the norm

$$\|f\| = [|f(0)|^2 + D(f)]^{1/2},$$

making it a Hilbert space with the set $\mathcal{P}$ of polynomials as a dense subset.

We shall identify the unit circle T with the interval $[0, 2\pi)$ on the real line. If $E \subset T$ is measurable, we define

$$\operatorname{cap}(E) = C_{1/2,2}(E),$$

where $C_{\alpha,p}$ denotes the capacity associated with the Bessel potential space $L^{p,\alpha}(R)$. For details about this capacity and $L^{p,\alpha}(R)$, we refer to the book [AH] by D. Adams and L. I. Hedberg.

We remark that $\operatorname{cap}(I)$ is comparable to $(\log(1/|I|))^{-1}$ if I is an interval and $|I|$ denotes its length, and that $\operatorname{cap}(E)$ is equivalent to logarithmic capacity.

A set E is called *thin* at z_0 if

$$\sum_{n \geq 1} \operatorname{cap}(E \cap \{z : |z - z_0| < 2^{-n}\}) < \infty.$$

Otherwise E is called *thick* at z_0.

A property is said to hold *quasi-everywhere* (q.e.) if it holds outside a set of zero capacity. The Beurling-Tsuji theorem says:

Each $f \in \mathcal{D}$ has a nontangential limit $f(e^{i\theta})$ q.e. on T.

Now take $f \in \mathcal{D}$ and consider the restriction $f|_T$. Suppose the nontangential limit $f(e^{i\theta})$ exists. Then f is called *finely continuous* at $e^{i\theta}$ if $\{e^{i\phi} : |f(e^{i\theta}) - f(e^{i\phi})| \geq \epsilon\}$ is thin at $e^{i\theta}$ for all $\epsilon > 0$. It is an important fact that for any $f \in \mathcal{D}$, its restriction $f|_T$ to T is finely continuous q.e. on T. (We refer to [AH, Chapter 6] for a complete discussion of thinness and fine continuity.)

The Dirichlet integral $D(f)$ can also be expressed according to Douglas's formula as a boundary integral:

$$D(f) = \frac{1}{\pi} \int_T \int_T \left(\frac{f(e^{i\theta}) - f(e^{i\phi})}{e^{i\theta} - e^{i\phi}} \right)^2 d\theta \, d\phi,$$

and if $f = u + iv$, with u being the real part of $f \in \mathcal{D}$, we have

$$D(u) = \iint_D \left[\left(\frac{\partial u}{\partial x} \right)^2 + \left(\frac{\partial u}{\partial y} \right)^2 \right] dx \, dy = \frac{1}{2\pi} \int_T \int_T \left(\frac{u(e^{i\theta}) - u(e^{i\phi})}{e^{i\theta} - e^{i\phi}} \right)^2 d\theta \, d\phi.$$

Consider now $u \in \operatorname{Re} \mathcal{D}$, and let u_M be defined by $u_M = u$ if $|u| \leq M$, $u_M = M$ if $u > M$, and finally $u_M = -M$ if $u < -M$ (u_M is called a *truncation* of U).

From the last formula it follows immediately that $u_M \in \mathcal{D}$ and that $D(u_M) \leq D(u)$. The same inequality clearly also holds if we only truncate u from above or below.

If X is a closed subset of D, we recall the definition of $S_m^t(X)$, X_{nt} and X_t from Section 2. We now formulate a characterization of Farrell sets for $\mathcal{D}$:

5.1 Theorem *The following statements are equivalent for a closed subset X of D:*

(i) X *is a Farrell set for* $\mathcal{D}$.

(ii) *For* $m = 1, 2, \ldots$, $S_m^t(X) \cup X_{nt}$ *is thick at almost all* $z \in \overline{X} \cap T$. *The exceptional set has zero capacity.*

(ii) *For* $m = 1, 2, \ldots$, $S_m^t(X)$ *is thick at almost all* $z \in \overline{X} \cap T$. *The exceptional set has zero capacity.*

(iv) X *is a Mergelyan set for* $\mathcal{D}$.

5.2 Remark The formulation of Theorem 5.1 contains a real number $t > 1$. It is part of the proof that statements (ii) and (iii) are independent of t in the sense that the validity of them for one t is equivalent to the validity for all $t > 1$. Also: the precise meaning of the word "thick" in this connection is given by Wiener's criterion [AH] as formulated above.

Most of Theorem 5.1 was proved in [Str3]. There the equivalence of (i), (ii) and (iii) was obtained as well as (i) $\implies$ (iv). We indicate the main ideas from the proofs in [Str3], and give more details about proving that (iv) $\implies$ (ii). B. Bøe (unpublished) has described Farrell sets in a similar way for a wider range of Besov spaces.

For the proof of Theorem 5.1 we need:

5.3 Proposition *Let* X *be a relatively closed suset of* D. *Then* X *is a Farrell set for* $\mathcal{D}$ *if and only if* $|f| \leq \|f\|_X$ *q.e. on* $\overline{X} \cap T$ *for all* $f \in \mathcal{D}$.

The proof of Proposition 5.3 is rather long and will not be given here. It is based on a duality argument modeled on Davie's argument which we gave in Section 3.

Proof of Theorem 5.1 To see that (i) $\implies$ (ii), we argue as follows. If (ii) fails, there is an integer m such that $S_m^t(X)$ fails to be thick at a subset K of X_t having positive capacity. It then follows that there is $u \in \operatorname{Re} \mathcal{D}$ such that $u = 1$ q.e. on a subset K' of K with $\operatorname{cap}(K') > 0$, while $u \leq r$ q.e. on $S_m^t(X)$ for some $r < 1$. By truncation if necessary, we may assume $u \leq 1$ on T, and if u is extended to D by Poisson's integral formula and $\tilde{u}$ is a harmonic conjugate it is easy to see that $f = e^{u + i\tilde{u}}$ violates Proposition 5.3.

The proof that (iii) $\implies$ (i) is more difficult. We assume (iii) is true and that (i) fails and look for a contradiction. We first use Proposition 5.3 and find $f \in \mathcal{D}$ such that $\|f\|_X < 1$, while $|f| \geq 1$ on a subset K of X_t having positive capacity. By truncation we may assume $|f| \leq 1$ in D, and by replacing f by f^N for some N if nessecary, we may assume $\|f\|_X < \epsilon$ and still have $|f| = 1$ on K. From f we build $u \in \operatorname{Re} \mathcal{D}$ of the form

$$u = \frac{1}{2}(1 + e^{i\alpha} f).$$

Then $0 \leq u \leq 1$ and $|u - 1/2| < \epsilon$ on X, while we may choose α such that $u \geq 1 - \epsilon/2$ on a subset K' of K having positive capacity. If we had $u \leq 1 - \epsilon$ on $S_m^t(X)$, we would have the desired contradiction because of (iii) and fine continuity of u. The same contradiction would be obtained if X_{nt} is thick q.e. in X_t by the Beurling-Tsuji theorem. So let us assume $\zeta \in X_t$ and that $S_m^t(X)$ is thick at ζ. We also assume $t > 2$, and get rid of this assumption at the end of the argument. Since $t > 2$, it is easy to see that there is a subset $E \subset I$, for each interval I in the union defining $S_m^t(X)$, such that $|E| \geq k|I|$, and $|u| \leq 1 - \epsilon$ on E, where k is a numerical constant. Now there is a basic inequality

$$\operatorname{cap}\left(\bigcup_\nu I_\nu\right) \leq k_0 \operatorname{cap}\left(\bigcup_\nu E_\nu\right),$$

where k_0 depends only on k, provided $\{I_\nu\}$ is a finite collection of disjoint intervals with corresponding subsets $\{E_\nu\}$ satisfying the above inequality $|E_\nu| \geq k|I_\nu|$. (For a proof, see Lemma 5 in [Str3] or Lemma 9.8.3 in [AH].)

Using the above inequality one deduces that there is a subset E^* of $S_m^t(X)$ which is thick at ζ and such that $u \leq 1 - \epsilon$ on E^*. We may assume that u is finely continuous at ζ, and thus we have a contradiction.

We finally remark that the basic inequality used above also gives that $S_m^t(X)$ and $S_m^{t_1}(X)$ are thick at the same points if $t > 1$ and $t_1 > 1$ are arbitrary. This means that the assumption $t > 2$ can be removed.

The proof that (i) $\Longrightarrow$ (iv) is rather long, and we refer to [Str3] for the details. The general idea behind the proof has turned out to be useful for other function spaces as well, and will be outlined in the final section.

It remains to prove (iv) $\Longrightarrow$ (ii), and since this is not contained in [Str3], we give more details.

Let $K \subset X_t \setminus S_m^t(X)$ be compact, and assume $S_m^t(X)$ is thin at all $z \in K$. Our goal is to prove that $\operatorname{cap}(K) = 0$. To obtain this we use a construction from [BroC], going back to Carleson [C3].

Let

$$K_\epsilon = \{z \in T : \ d(z, K) \leq \epsilon\}.$$

By the Choquet property (see [AH, p. 175]) we have

$$\operatorname{cap}(S_m^t(X) \cap K_\epsilon) \to 0$$

as $\epsilon \to 0$. Choose $K_n = K_{\epsilon_n}$ such that

$$\sum_n [\operatorname{cap}(S_m^t(X) \cap K_n)]^{1/2} < \infty.$$

Now let U_n be the conductor potential for $S_m^t(X) \cap K_n$ with respect to the kernel $k(r) = \log^+(4/r)$. The capacity defined by this kernel is equivalent to logarithmic capacity (see [C3]). So

$$U_n(z) = \int_{S_m^t(X) \cap K_n} \log^+\left(\frac{4}{|z - \zeta|}\right) d\mu_n(\zeta),$$

where μ_n is supported on $S_m^t(X) \cap K_n$, and its total mass is comparable to $\operatorname{cap}(S_m^t(X) \cap K_n)$.

The following properties of U_n are now easily verified: $U_n \in \operatorname{Re} \mathcal{D}$, $U_n \geq 0$, $U_n = 1$ q.e. on $S_m^t(X) \cap K_n$, and

$$\|U_n\| \sim (\operatorname{cap}(S_m^t(X) \cap K_n)^{1/2}.$$

Now let

$$u(z) = \sum_{n=1}^\infty \lambda_n U_n(r_n z),$$

where r_n and λ_n are increasing sequences converging to 1 and ∞, respectively. We can choose λ_n converging so slowly that $u \in \operatorname{Re}\mathcal{D}$ and u is harmonic near $T \setminus K$.

Now let

$$f_K = e^{-(u+i\tilde{u})},$$

where $\tilde{u}$ is a harmonic conjugate to u. It is easy to see that $u(z) \to \infty$ if $d(z, K) \to 0$. But then $f_K|_X$ is uniformly continuous, and $f_K(z) \to 0$ if $z \in K$ and $d(z, K) \to 0$.

Since K is a compact subset of X_t, K is nesscarily totally disconnected. We split K into $K = K_1 \cup K_2$, where K_1 and K_2 are disjoint compact sets.

Now repeat the construction above to obtain f_{K_1}, and form

$$g = (f_{K_1})^{1/N} \in \mathcal{D}$$

with N large.

Since X is a Mergelyan set for $\mathcal{D}$, we can choose a polynomial p such that

$$\|g - p\| + \|g - p\|_X < \epsilon,$$

where ϵ is given. By truncating $\operatorname{Re} p$, we can find $h \in (\operatorname{Re}\mathcal{D}) \cap C(T)$ such that $|h - 1| < 3\epsilon$ on K_1, and $|h| < 3\epsilon$ on K_2, and moreover,

$$\|h\| + \|h\|_\infty < C_\epsilon,$$

where C_ϵ only depends on ϵ. This follows from a result by C. Sundberg on truncation of BMO functions [Su]. Since K is totally disconnected, it follows that there are constants $\delta \in (0, 1)$ and $M < \infty$, such that if $w \in C(K)$ is real valued and bounded by 1 on K, there is $h \in (\operatorname{Re}\mathcal{D}) \cap C(T)$ such that

$$\|w - h\|_K < \delta,$$

and

$$\|h\| + \|h\|_\infty < M.$$

But it is well known that the last two inequalities imply that $(\operatorname{Re}\mathcal{D}) \cap C(T)_K = C(K)$. By a theorem of T. Sjødin [S], we must have $\operatorname{cap}(K) = 0$, and so Theorem 5.1 is proved. $\quad\square$

The sets of determination have a somewhat simpler characterization than the one in Theorem 5.1.

5.4 Theorem *The following statements are equivalent for a closed subset X of D:*

(i) *For all $f \in \mathcal{D}$ the supremum of $|f|$ on X equals the supremum of $|f|$ on D.*

(ii) $\operatorname{cap}(T) = \operatorname{cap}(T \setminus X_t)$.

Proof Theorem 5.1 gives that (i) in Theorem 5.4 is equivalent to

$$\operatorname{cap}(T \setminus X_m) = \operatorname{cap}(T)$$

for all $m \geq 1$, where $X_m = X_t \setminus S_m^t(X)$. So clearly (ii) $\Longrightarrow$ (i).

To see that (i) $\Longrightarrow$ (ii), we assume $\operatorname{cap}(T \setminus X_m) = \operatorname{cap}(T)$ for $m = 1, 2, \ldots$, while $\operatorname{cap}(T \setminus X_t) < \operatorname{cap}(T)$, and look for a contradiction. By regularity there is a compact subset

K of X_t such that $\mathrm{cap}(T \setminus K) < \mathrm{cap}(T)$. By a theorem of L. Ahlfors and A. Beurling (see [Ah, Be] or [C2, pp. 82–84]), there is a nontrivial analytic function defined in the complement of K having finite Dirichlet integral there. By shrinking K if necessary, we may assume that f does not admit analytic continuation across any part of K. We put $K_m = K \setminus S_m^t(X)$. Then $K \setminus K_m$ is open in K. But $K \setminus K_m$ is also dense in K because of the way we shrank K. By the Baire category theorem, $\bigcap_{m=1}^{\infty}(K \setminus K_m)$ is dense in K. But this is a contradiction, since obviously $\bigcap_{m=1}^{\infty}(K \setminus K_m) = \emptyset$. □

It may be of some interest to compare the results about Farrell and Mergelyan sets obtained for $\mathcal{D}$ and the Hardy space $H^p(D)$.

For $H^p(D)$ we have:

X is a set of determination if and only if X_{nt} has positive Lebesgue density at almost all $z \in T$.

The corresponding statement for $\mathcal{D}$ is:

X is a set of determination if and only if X_{nt} has positive Wiener density at all $z \in T$, except for a set of zero capacity.

There is a similar analogy for Farrell (Mergelyan) sets:
For $H^p(D)$ we have:

X is a Farrell set if and only if $|f| \le 1$ a.e. on $\overline{X} \cap T$ if $|f|$ is bounded by 1 on X.

The corresponding statement for $\mathcal{D}$ is:

X is a Farrell set if and only if $|f| \le 1$ q.e. on $\overline{X} \cap T$ if $|f|$ is bounded by 1 on X.

On the other hand, the geometric properties of Farrell sets may be quite different for $H^p(D)$ and $\mathcal{D}$. As an example, let $K \subset T$ have positive Lebesgue measure $|K|$, and at the same time satisfy $\mathrm{cap}(T \setminus K) = \mathrm{cap}(T)$. Such sets exist (see [Ah, Be] or [AH, Remark 2, p. 314]). If $\{J_\nu\}$ denotes the intervals in T complementary to K, we can push each J_ν slightly into D, and obtain intervals $\tilde{J}_\nu$ such that if

$$X = \bigcup_{\nu=1}^{\infty} \tilde{J}_\nu,$$

then X is a Farrell set for $\mathcal{D}$ even though $|X_t| > 0$, so that F is not a Farrell set for $H^p(D)$.

5.5 Remark B. Bøe has described sets of determination for harmonic functions in a half-space in $\mathbb{R}^n$ belonging to a wide range of Besov spaces [B]. S. Gardiner [Ga3] has treated sets of determination for harmonic functions having finite Dirichlet integral in the ball in $\mathbb{R}^n$.

6 The Nevanlinna class

The *Nevanlinna class* N consists of all analytic functions f in D such that

$$||f|| = \lim_{r \to 1} \int_0^{2\pi} \log(1 + |f(Re^{i\theta})|) \, d\theta < \infty.$$

With the metric $d(f,g) = ||f - g||$, N is complete, but unfortunately not a topological vector space. Even so, we may introduce a certain notion of convergence in N such that the set $\mathcal{P}$ of analytic polynomials is dense in N, and such that this notion of convergence implies uniform convergence on compact subsets of D.

To do this, we need a well-known factorization formula for N (see [Du, p. 25]). If μ is a positive finite measure on ∂D, singular with respect to Lebesgue measure on ∂D, the singular inner function I_μ is given by

$$I_\mu(z) = \exp\left(-\int_0^{2\pi} H(z, e^{i\theta})\, d\mu(\theta)\right),$$

where

$$H(z, e^{i\theta}) = \frac{e^{i\theta} + z}{e^{i\theta} - z}$$

is the Herglotz kernel. Now any function f in N has a canonical factorization

$$f = B\frac{I_\mu}{I_\nu}F,$$

where B is a Blaschke product, I_μ and I_ν are singular inner functions where μ and ν are mutual singular, and F is the so-called outer part of f, i.e.

$$F(z) = \exp\left(\frac{1}{2\pi}\int_0^{2\pi} \log|f(e^{i\theta})|\, d\theta\right).$$

Sometimes we will write $\mu = \mu_f$ and $\nu = \nu_f$ to indicate the dependence of the measures on f. More details than we give here, can be found in [PS].

With the above factorization in mind, we define a measure ϕ_f associated with f as follows:

$$d\phi_f = \frac{1}{2\pi}\log(1 + |f(e^{i\theta})|)\, d\theta + \nu_f.$$

6.1 Definition A sequence f_n in N is said to *converge to f in N* if $d\phi_{f_n} \to d\phi_f w^*$ in the space $\mathcal{M}(T)$ of measures on T, and $f_n(e^{i\theta}) \to f(e^{i\theta})$ in measure as $n \to \infty$.

We can now state:

6.2 Lemma *If $f \in N$, then $f_r \to f$ in N as $r \to 1^-$.*

For the proof of Lemma 6.2, we refer to [PS].
From Lemma 6.2 we immediately get

$$||f|| = \frac{1}{2\pi}\int_0^{2\pi} \log|f(e^{i\theta})|\, d\theta + \nu_f(T).$$

Hence it follows that convergence in N implies

$$||f_n|| \to ||f||.$$

Secondly, convergence in N implies uniform convergence on compact subsets of D. This follows directly from the Khinchin-Ostrowski theorem (see [HaJ, p. 278] for details about the Khinchin-Ostrowski theorem).

We recall that the *Smirnov class* N^+ consists of all analytic functions $f \in D$ of the form

$$f = BIF,$$

where B is a Blaschke product, I a singular inner function, and F is the outer function corresponding to f:

$$F(z) = \exp\left(\frac{1}{2\pi}\int_0^{2\pi}\frac{e^{i\theta}+z}{e^{i\theta}-z}\log|f(e^{i\theta})|\,d\theta\right).$$

The metric in N^+ is given by

$$d_{N^+}(f,g) = \frac{1}{2\pi}\int_0^{2\pi}\log(1+|f(e^{i\theta})-g(e^{i\theta})|)\,d\theta.$$

Our next results shows that our notion of convergence is a natural extension to N of the convergence defined by the above metric in N^+.

6.3 Lemma *Let f and f_n belong to N^+. Then $f_n \to f$ in N if and only if $d_{N^+}(f_n, f) \to 0$.*

Proof If $d_{N^+}(f_n, f) \to 0$, clearly $f_n(e^{i\theta}) \to f(e^{i\theta})$ in measure as $n \to \infty$. Moreover, since $|\log(1+|f|) - \log(1+|f_n|)| \le \log(1+|f-f_n|)$, it follows that $f_n \to f$ in N.
The converse follows from

$$\log(1+|f-f_n|) \le \log(1+|f|) + \log(1+|f_n|)$$

and dominated convergence. $\qquad\qquad\qquad\qquad\qquad\qquad\qquad\qquad\qquad\qquad\qquad$ $\square$

The problem of characterizing Farrell and Mergelyan sets for N is far from solved, but a few results are known (see in particular [PS]). If X is Farrell for N, it can be proved that $|X_t| = 0$. Thus any Farrell set for N is a Farrell set for $H^p(D)$, $0 < p \le \infty$. But the converse is not true as the following simple example shows: let

$$X = \{w : |w - 1/2| = 1/2\} \cap D.$$

Now let I_μ be the inner function corresponding to μ being the Dirac measure at $z = 1$. If $p_n \in \mathcal{P}$ and $p_n \to 1/I_\mu$ in N, while p_n is uniformly bounded on X, p_n will be uniformly bounded inside $\{w : |w - 1/2| = 1/2\}$, and hence p_n will not converge pointwise to $1/I_\mu$ in D.

From this example it is evident that something else than $|X_t| = 0$ is needed in order to describe Farrell sets for N. The right tool seems to be the concept of minimal thinness used by W. K. Hayman and T. J. Lyons in their characterization of sets of determination for the harmonic functions $h_0(D)$ which are differences of positive harmonic functions in D (see [HayL]). Minimal thinness can be expressed in various ways. If $0 < \rho < 1$, let $X_\rho = \{w : \sigma(w, X) < \rho\}$ where σ is the hyperbolic metric in D. Then

$$X \text{ is not minimal thin at } \zeta \in T \iff \iint_{X_\rho}\frac{1}{|\zeta - z|^2}\,dx\,dy(z) = \infty.$$

W. Hayman and T. Lyons proved:

6.4 Theorem *A closed subset X of D is a set of determination for $h_0(D)$ if and only if X is not minimally thin at any $\zeta \in \partial D$.*

From the Hayman-Lyons theorem it follows that a relatively closed subset Y of D is a set of determination if and only if $Y \setminus \{w : |B(w)| < t\}$ is a set of determination for $h_0(D)$, for some $t > 0$ and any Blaschke product B.

A more satisfactory statement can be given if Y is sufficiently thick:

> If X is a set of determination for $h_0(D)$, and $Y = \{z \in D : \sigma(z, X) < t\}$ for some $t > 0$, then Y is a set of determination for N.

There does not seem to be a similar simple characterization of Farrell sets for N. To indicate some of the complications preventing a simple characterization, let us modify the example preceding Theorem 6.4 and define

$$X = \{w : |w - 1/2| = 1/2\} \cap \{z \in D : \operatorname{Im} z \geq 0\}.$$

Then it can be proved that X is Farrell for N even though X is minimal thin at $z = 1$. This is proved in [PS]. Let us just show that I^{-1} can be approximated in N by polynomials p_n such that $\|p_n\|_X \to \|I^{-1}\|_X$. The argument goes as follows. Let ϕ_n denote a sequence of smooth positive functions on $\{e^{i\theta} : -\pi/2 < \theta < 0\}$ converging weak* to the Dirac measure at $z = 1$. Form

$$a_n = \exp\left(\frac{1}{2\pi} \int_0^{2\pi} \frac{e^{i\theta} + z}{e^{i\theta} - z} \phi_n(e^{i\theta})\, d\theta\right).$$

Since $1/a_n \to I$ uniformly on compact subsets of $\overline{D} \setminus \{1\}$, clearly $a_n \to I^{-1}$ in measure. Moreover, $a_n \to I^{-1}$ in N if and only if $\log(1 + |a_n|) - \log|a_n| \to \log 2$ in the w^*-topology. But

$$\log\left(\frac{1 + |a_n|}{|a_n|}\right) \to \log 2$$

by dominated convergence, so the weak*-convergence follows. We may assume that the measure $d\phi_n\, d\theta$ has mass one, and then it follows that $|a_n| \leq |I^{-1}|$ pointwise on X, so the asserted convergence is proved.

Thus it may be hard to find a purely geometric characterization of Farrell sets for N. An alternative might be the following. Let t and s be positive and define

$$K_{s,t} = \{f \in N : \|f\| \leq s, \ \|f\|_X \leq t\}.$$

6.5 Conjecture X *is Farrell for* $N \iff \sup_{f \in K_{s,t}} |f(z)| = \sup_{p \in K_{s,t} \cap \mathcal{P}} |p(z)|$, *for all* $z \in D$.

Let us finally point out another open problem: it can be proved that X Farrell for $N \implies X$ is Mergelyan for N, but the converse is not known.

7   More open problems and further research

There are many well-known function spaces where very little is known about the corresponding Farrell and Mergelyan sets.

The *Bloch space* $\mathcal{B}$ consists of all analytic functions in D such that

$$\|f\|_\mathcal{B} = \sup\{(1 - |z|^2)|f'(z)| : z \in D\} < \infty$$

with the norm

$$\|f\| = |f(0)| + \|f\|_\mathcal{B}.$$

$\mathcal{B}$ is a Banach space, and it can be shown that under the pairing

$$(f, g) = \int_D f(z)\overline{g(z)}\, dA(z),$$

$\mathcal{B}$ can be realized as the dual of the space $L_a^1(D)$ consisting of all analytic functions f in D which are integrable in D with respect to area measure $dA(z)$ (see [Z, Chapter 5] for more about $\mathcal{B}$).

7.1 Problem Describe the Farrell and Mergelyan sets for $[\mathcal{B}, \mathcal{P}]$, where $\mathcal{P}$ denotes the polynomials and $\mathcal{B}$ has the w^*-topology.

7.2 Remark That $\mathcal{P}$ is w^*-dense in $\mathcal{B}$ is easy to see combining the density of $\mathcal{P}$ in $L_a^1(D)$ (see [HKZ, p. 4]) and expanding $\int_D f(rz)\overline{g(z)}\, dA(z)$ in terms of the Taylor coefficients of f and g, where $0 < r < 1$.

The *little Bloch space* $\mathcal{B}_0$ consists of all $f \in \mathcal{B}$ such that $(1 - |z|^2)|f'(z)| \to 0$ as $n \to \infty$. With the norm induced from $\mathcal{B}$, $\mathcal{B}_0$ is itself a Banach space containing $\mathcal{P}$ as a dense subset (see [Z, p. 84]). We may therefore ask for the description of Farrell and Mergelyan sets for $[\mathcal{B}_0, \mathcal{P}]$, where $\mathcal{B}_0$ has the norm topology. In particular, one might wonder whether the sets of determination for $\mathcal{B}$ and $\mathcal{B}_0$ are the same, in analogy with what is known for BMOA and VMOA.

There is a crucial property of $\mathcal{B}$ that may be relevant for our problem. Let

$$\sigma(z, w) = \frac{|z - w|}{|1 - \overline{w}z|}, \quad z, w \in D.$$

denote the pseudohyperbolic metric in D. The Bergman or hyperbolic metric is given by

$$\beta(z, w) = \log \frac{1 + \sigma(z, w)}{1 - \sigma(z, w)}.$$

In [Z, p. 81] it is proved that $f \in \mathcal{B}$ if and only if

$$|f(z) - f(w)| \le C\beta(z, w)$$

for some constant C. So subsets of D which are uniformly dense in D in the β-metric, are certainly sets of determination for $\mathcal{B}$, but the problem is to find a nessecary and sufficient condition.

The *Bergman space* $A^2(D)$, consists of all analytic functions f in D such that

$$\|f\|_2 = \left(\int_D |f|^2\, dx\, dy\right)^{1/2} < \infty.$$

Very little seems to be known about Farrell and Mergelyan sets for $A^2(D)$. A good start might be to look at sets of determination first. Probably the work of K. Seip on interpolation and sampling is relevant for this problem. A good reference in this connection is Chapters 4 and 5 in [HKZ], where further references to Seip's work can be found. We believe that any Farrell set for $A^2(D)$ also is a Mergelyan set for $A^2(D)$ (see the scheme for approximation described below in this section).

We end this collection of open problems in simultaneous approximation by looking at yet another problem for the Hardy spaces. We consider a uniformly closed algebra B, where $A(D) \subset B \subset H^\infty$. We ask for a description of the Farrell sets for $[H^p, B]$. In the light of what is known if $B = A(D)$ and $B = H^\infty(D)$, there should be good hope for progress on this problem in various cases. In particular, if B consists of all $f \in H^\infty(D)$ having a continuous extension to a fixed measurable subset E of T, the Farrell sets X could very well be described by the condition $|X_t \cap (T \setminus E)| = 0$.

A scheme for approximation

In many cases where the polynomials $\mathcal{P}$ are dense in some linear function space A, it may be proved that a Farrell set for A is also a Mergelyan set for A, even without knowledge of a geometric characterization of Farrell sets. We outline the main ideas in such an argument (see [PGS] and [Str3] for two examples where the scheme works).

Suppose X is a Farrell set for A. Instead of proving that X is a Mergelyan set for A, we actually look for a formally stronger statement to hold:

> If g is a uniformly continuous function on X and $f \in A$, there exist polynomials p_n such that $p_n \to f$ in A and $\|p_n - g\|_X \to \|f - g\|_X$.

The scheme for approximation consists of three steps:

Step I: *A should have a "Runge" property.* Given $f \in A$ bounded on X, there are $f_n \in A$ such that:

(1) f_n is analytic across $\partial D \setminus (\overline{X})$.

(2) $f_n \to f$ in A.

(3) $\|f_n - f\|_X \to 0$.

One convenient way to establish the Runge property is to apply the so-called T_ϕ-operator which is part of Vitushkin's method in rational approximation. A good introduction to this operator can be found in [G, Chapter VIII]. In the following steps, f is assumed to be analytic across $\partial D \setminus (\overline{X})$.

Step II: *Localization.* Cover $X \cap \partial D$ by discs Δ_j such that $g \approx g_j$ (constant) in $X \cap \Delta_j$. Use that F is a Farrell set for A to find $p_n^j \in \mathcal{P}$ such that $p_n^j \to f - g_j$ in A and $\|p_n^j\|_{X \cap \Delta_j} \to \|f - g_j\|_{X \cap \Delta_j}$.

In order to prove that X is a Mergelyan set for A, the condition on p_n^j on $X \cap \Delta_j$ may be weakend to the following:

$$\|p_n^j\|_{X \cap \Delta_j} \le K\|f - g_j\|_{X \cap \bar\Delta_j},$$

where $\tilde{\Delta}_j$ has the same center as Δ_j, but twice the radius, and the constant K is independent of f, g, and Δ_j.

Step III: *"Glueing together"*. Define a "polynomial spline" q_n^j as follows. We put

$$q_n^j = p_n^j \quad \text{in} \quad D \cap \Delta_j \setminus \bigcup_{i<j} \Delta_i$$

for $j = 1, 2, \ldots$ (assume $\Delta_j = \emptyset$ if $i < 1$), and moreover,

$$q_n^j = p_n^1 \quad \text{in} \quad D \cap \Delta_j.$$

Given $\epsilon > 0$, one can choose $\{\Delta_j\}$ such that $\|q_n^j - g\|_X < \|f - g\|_X + \epsilon$.

Finally on has to replace the polynomial spline q_n^j by a true polynomial $\tilde{q}_n^j$ such that

$$\|\tilde{q}_n^j - g\|_X < \|f - g\|_X + 2\epsilon$$

and getting $\tilde{q}_n^j$ close to f in A for n large enough.

The reason why $\tilde{q}_n^j$ often can be obtained from q_n^j, is that $q_n^j \to f$ uniformly on compact subsets of D, so that the "jump"-discontinuities of q_n^j inside D get smaller as n increases.

References

[AH]   D. R. Adams and L. I. Hedberg, *Function Spaces and Potential Theory*, Grundlehren Math. Wiss. **314**, Springer-Verlag, Berlin, 1996.

[Ah]   L. V. Ahlfors, *Conformal Invariants*, McGraw-Hill, New York, 1973.

[AhB]   L. V. Ahlfors and A. Beurling, Conformal invariants and function theoretic null sets, *Acta Math.* **83** (1950), 101–129.

[B]   B. Bøe, Sets of determination for smooth harmonic functions, preprint, Univ. of Bergen.

[Be]   A. Beurling, Ensembles exceptionnels, *Acta Math.* **72** (1940), 1–13.

[Bo1]   F. F. Bonsall, Domination of the supremum of a bounded harmonic function by its supremum over a countable subset, *Proc. Edinburgh Math. Soc.* **30** (1987), 471–477.

[Bo2]   F. F. Bonsall, Some dual aspects of the Poisson kernel, *Proc. Edinburgh Math. Soc.* **33** (1988), 207–232.

[BoW]   F. F. Bonsall and D. Walsh, Vanishing l^1-sums of the Poisson kernel, and sums with positive coefficients, *Proc. Edinburgh Math. Soc.* **32** (1989), 431–447.

[BrC]   D. A. Brannan and J. G. Clunie, *Aspects of Contemporary Complex Analysis*, Academic Press, New York, 1980.

[BroC] L. Brown and W. Cohn, Some examples of cyclic vectors in the Dirichlet space, *Trans. Amer. Math. Soc.* **96** (1960), 162–183.

[BroSZ]   L. Brown, A. L. Shields, and K. Zeller, On absolutely convergent exponential sums, *Proc. Amer. Math. Soc.* **95** (1985), 42–46.

[C1]   L. Carleson, A representation formula for the Dirichlet Integral, *Math. Z.* **56** (1960), 190–196.

[C2]   L. Carleson, *Selected Problems on Exceptional Sets*, Van Nostrand, 1967.

[C3]   L. Carleson, Sets of uniqueness for functions regular in the unit circle, *Acta Math.* **87** (1952), 325–345.

[Da]   A. A. Danielyan, Certain problems arising from Rubel's problem on simultaneous approximation, *Dokl. Akad. Nauk* **341** (1) (1995), 10–12; English transl.: *Russian Acad. Sci. Dokl. Math.* **51** (1995), 164–165.

[DGG]   A. M. Davie, T. W. Gamelin, and J. Garnett, Distance estimates and pointwise bounded density, *Trans. Amer. Math. Soc.* **175** (1973), 37–68.

[De]   J. Detraz, Algèbres de fonctions analytiques dans le disque, *Ann. Sci. École Norm. Sup.* **4** (1970), 313–352.

[Du]   P. Duren, *Theory of H^p-Spaces*, Academic Press, New York, 1970.

[FPG]   A. G. O'Farrell and F. Pérez-González, Pointwise bounded approximation by polynomials, *Math. Proc. Cambridge Philos. Soc.* **112** (1992), 147–155.

[G]   T. W. Gamelin, *Uniform Algebras*, Prentice Hall, Englewood Cliffs, NJ, 1969.

[Ga]   S. J. Gardiner, Sets of determination for harmonic functions, *Trans. Amer. Math. Soc.* **338** (1993), 233–243.

[Ga2]   S. J. Gardiner, Mergelyan pairs for harmonic functions, *Proc. Amer. Math. Soc.* **126** (1998), 2699–2703.

[Ga3]   S. Gardiner, Nontangential limits, fine limits and the Dirichlet integral, to appear in *Proc. Amer. Math. Soc.*

[PGS]   F. Pérez-González and A. Stray, Farrell and Mergelyan sets for H^p-spaces, *Michigan Math.* **36** (1989), 379–386.

[PGTS]   F. Pérez-González, R. Trujillo-Gonzalez, and A. Stray, Joint Approximation in BMOA and VMOA, *J. Math. Anal. Appl.* **237** (1999), 128–138.

[PS]   F. Pérez-González and A. Stray, On Farrell sets in the Nevanlinna class, preprint.

[HKZ]   H. Hedenmalm, B. Korenblum, and K. Zhu, *Theory of Bergman Spaces*, Grad. Texts in Math. **199**, Springer-Verlag, New York, 2000.

[HaJ]   V. Havin and B. Jøricke, *The Uncertainty Principle in Harmonic Analysis*, Ergeb. Math. Grenzgeb. (3) **28**, Springer-Verlag, New York, 1994.

[HayL]   W. K. Hayman an T. J. Lyons, Bases for positive continuous functions, *J. London Math. Soc.* **42** (1990), 292–308.

[He]   L. I. Hedberg, Removable singularities and condenser capacities, *Ark. Mat.* **12** (1974), 181–201.

[NO]   A. Nicolau and J. Orotbig, Joint approximation in BMO, *J. Funct. Anal.* **173** (2000), 21–48.

[R]   L. A. Rubel, Bounded convergence of analytic functions, *Bull. Amer. Math. Soc.* **77** (1971), 13–20.

[RS]   L. A. Rubel and A. Stray, Joint approximation in the unit disc, *J. Approx. Theory* **37** (1983), 44–50.

[RST]   L. A. Rubel, A. L. Shields, and B. A. Taylor, Mergelyan sets and the modulus of continuity, *J. Approx. Theory* **15** (1975), 23–40.

[S]   T. Sjödin, Capacities of compact sets in linear subspaces of R^n, *Pacific J. Math.* **78** (1978), 261–266.

[St]   E. M. Stein, The characterization of functions arising as potentials, *Bull. Amer. Math. Soc.* **67** (1961), 102–104.

[Str1] A. Stray, Characterization of Mergelyan sets, *Proc. Amer. Math. Soc.* **44** (1974), 347–352.

[Str2] A. Stray, Pointwise bounded approximation by functions satisfying a side condition, *Pacific J. Math.* **51** (1974), 301–305.

[Str3] A. Stray, Simultaneous approximation in the Dirichlet space, to appear in *Math. Scand.*

[Str4] A. Stray, A Mergelyan type theorem for some function spaces, *Publ. Mat.* **39** (1995), 61–69.

[Su]   C. Sundberg, Truncations of BMO functions, *Indiana Math. J.* **33** (1984), 749–771.

[T]   M. Tsuji, *Potential Theory in Modern Function Theory*, Maruzen, Tokyo, 1959.

[Z]   K. Zhu, *Operator Theory in Function Spaces*, Marcel Dekker, New York and Basel, 1990.

Index